Zeolites and Porous Materials: Insight into Catalysis and Adsorption Processes

Zeolites and Porous Materials: Insight into Catalysis and Adsorption Processes

Editors

Maja Milojević-Rakić
Danica Bajuk-Bogdanović

MDPI • Basel • Beijing • Wuhan • Barcelona • Belgrade • Manchester • Tokyo • Cluj • Tianjin

Editors
Maja Milojević-Rakić
University of Belgrade
Belgrade, Serbia

Danica Bajuk-Bogdanović
University of Belgrade
Belgrade, Serbia

Editorial Office
MDPI
St. Alban-Anlage 66
4052 Basel, Switzerland

This is a reprint of articles from the Special Issue published online in the open access journal *Catalysts* (ISSN 2073-4344) (available at: https://www.mdpi.com/journal/catalysts/special_issues/zeolite_porous).

For citation purposes, cite each article independently as indicated on the article page online and as indicated below:

LastName, A.A.; LastName, B.B.; LastName, C.C. Article Title. *Journal Name* **Year**, *Volume Number*, Page Range.

ISBN 978-3-0365-8154-5 (Hbk)
ISBN 978-3-0365-8155-2 (PDF)

Contents

About the Editors

Maja Milojević-Rakić

Dr. Milojević-Rakić is an Associate Professor at the Faculty of Physical Chemistry, University of Belgrade, Serbia. She has published two university books and 35 papers in international scientific journals. She has participated in four international and five national projects. She is acting as the supervisor of four Ph.D. theses.

Danica Bajuk-Bogdanović

Dr. Danica Bajuk-Bogdanović is a Senior Research Associate at the Faculty of Physical Chemistry, University of Belgrade, Serbia. Dr. Bajuk-Bogdanović has published over 100 papers in international scientific journals. She has participated in eight international and five national projects. She is acting as the supervisor of two Ph.D. theses.

Preface to "Zeolites and Porous Materials: Insight into Catalysis and Adsorption Processes"

This Reprint "Zeolites and Porous Materials: Insight into Catalysis and Adsorption Processes" aims to provide a comprehensive overview of the current knowledge and the latest advancements in the field. It brings together a collection of contributions from leading experts and researchers who share their insights, experiences, and findings in this rapidly evolving area of science and technology.

This Reprint contains several research papers, each focusing on different aspects of zeolites and porous materials, including synthesis methods, characterization techniques, and applications in catalysis and adsorption. It begins with an Editorial introduction to the topic, highlighting the principles that govern the structure and properties of porous materials. Furthermore, it highlights the characterization techniques essential for understanding the structure, porosity, and surface properties of these materials. From spectroscopic and microscopic techniques to computational modeling, the contributors present state-of-the-art methods to elucidate the structure–property relationships involved, and unravel the intricate mechanisms underlying the catalytic and adsorption processes.

The role of advanced materials is explored in various catalytic reactions, such as hydrocarbon conversions, biomass transformations, and environmental remediation. Additionally, the Reprint sheds light on the importance of adsorption processes in water treatment, pharmaceutical formulations, and energy storage, elucidating the potential of zeolites and porous materials as efficient adsorbents.

Throughout this Special Issue, the contributors not only highlight the achievements and breakthroughs in the field but also address the challenges and future directions. They discuss the current limitations and propose innovative strategies to overcome them, opening up new avenues for research and development. Moreover, this Reprint emphasizes the importance of sustainability and green chemistry, showcasing how zeolites and porous materials can contribute to a more sustainable and environmentally friendly future.

We hope that this Special Issue will serve as a valuable resource for researchers, scientists, engineers, and students interested in zeolites, porous materials, catalysis, and adsorption processes. We believe that the diverse perspectives presented within these pages will inspire new ideas, foster collaborations, and contribute to the advancement of this exciting and rapidly evolving field.

Maja Milojević-Rakić and Danica Bajuk-Bogdanović
Editors

catalysts

Editorial

Recent Advances in Zeolites and Porous Materials Applications in Catalysis and Adsorption Processes

Maja Milojević-Rakić * and Danica Bajuk-Bogdanović *

Faculty of Physical Chemistry, University of Belgrade, 11000 Belgrade, Serbia
* Correspondence: maja@ffh.bg.ac.rs (M.M.-R.); danabb@ffh.bg.ac.rs (D.B.-B.)

Citation: Milojević-Rakić, M.; Bajuk-Bogdanović, D. Recent Advances in Zeolites and Porous Materials Applications in Catalysis and Adsorption Processes. *Catalysts* **2023**, *13*, 863. https://doi.org/10.3390/catal13050863

Received: 28 April 2023
Accepted: 6 May 2023
Published: 9 May 2023

Zeolites and porous materials are some of the most promising materials for various applications. Zeolites with highly ordered pore systems, high specific surface area and negatively charged framework enable the introduction of different functionalizing phases that can boost their performance. In addition, excellent adsorption and catalytic properties are often witnessed, with further research increasingly being transferred to the domain of biological applications.

The Special Issue "Zeolites and Porous Materials: Insight into Catalysis and Adsorption Processes" gives recent advances in developing zeolite-based and porous materials generally employed in adsorption and catalytic systems within ten research articles and three review contributions.

Zeolites and different porous materials are employed as perspective heterogeneous catalysts in several critical industrial and environmental processes. To contribute to a better understanding of factors affecting catalysis, Xian et al. studied Y zeolites with different textural properties as porous catalysts for the production of benzene, toluene, ethyl-benzene and xylene (BTEX) [1]. The hydrocracking of tetralin was investigated in a fixed-bed microreactor operated under high pressure. The main features affecting catalytic performance depended on aluminum content and the ratio between micro/mesoporosity in zeolite samples. A great extent of microporosity is beneficial for BTEX molecules synthesis, while the cracking was determined by thermodynamics and zeolites' strong acidity, affecting equilibrium temperature [1].

Hierarchical zeolites may be synthesized to achieve mesoporosity development in addition to microporosity. Hierarchical structures are proposed for essential processes of fine chemicals production, methanol conversions, carbonization and biomass transformation. Different degrees of porosity may be induced through a consecutive fluorine/alkaline treatment of ZSM-5 zeolite, as presented by Guo et al. [2]. A specific delay in mesopore formation is associated with the fluorination procedure altering spontaneous porosity formation in an alkaline solution, enabling the prolonged formation of mesopores. Favorable electronic and steric conditions near aluminum sites play a dominant role in porosity development [2].

Textural properties are often addressed when hierarchical zeolites govern adsorption and catalysis. However, a hydrothermal synthesis procedure produces zeolites in powdered form. This drawback emerges due to pressure lowering in fixed-bed reactors operating in industrial processes. Different methods may be used to address this issue. Zeolites may be prepared in various geometries and particle sizes through the extrusion process and clay and oxides addition as binders. The development of hierarchical structures allows facilitated diffusion and mass transfer inside pores. A review by Asgar Pour et al. enabled insight into the extrusion procedure with binder as a new and emerging approach in shaping hierarchical zeolites conducted without binder [3]. This field requires novel investigations and refinements that would enable the preservation of initial zeolite structure and procedure scale-up to assist industry requirements [3].

Post-synthetic procedures, such as desilication, dealumination and sequential treatments, may be applied to further increase zeolites' applicability. This is especially important for their performance in various industrial applications and petroleum processing. This specific tailoring of textural properties and active site evolution is assessed for different zeotypes, including MFI, FAU and SAPO frameworks. The modification of SSZ-32 zeolites by desilication and sequential desilication/dealumination steps treatments is proposed by Yang et al. [4]. The developed surface and acidity of modified zeolite supported on Pt were utilized for the catalytic hydroisomerization of long-chain n-alkanes. This study offers valuable insights into the fabrication of high-performance bifunctional catalysts by understanding the significance of the catalyst support in the hydroisomerization of long-chain n-alkanes. Moreover, it suggests an alternate approach for augmenting the low-temperature performance of diesel/lubricating oils in petroleum processing [4].

Petroleum processing extensively uses catalysts of large porosity, which are afterward disposed of without extensive recycling. This brings the necessity of dealing with spent ZSM-5 catalysts for methanol conversion to hydrocarbons. A route proposed by Yong et al. involves partial acid-induced dealumination of spent zeolite combined with alkali treatment to increase the specific surface of the resulting material [5]. Although treated spent adsorbent maintains the MFI framework, the dealumination procedure depends on acid use. Oxalic acid may be selected as it removes both framework and non-framework aluminum, increasing porosity and surface area. Furthermore, alkali treatment changes spent catalysts' morphological features, giving rise to its proposed adsorption performance for p-xylene and n-hexane removal. These findings support the idea of subsequent environmental application of spent catalysts/adsorbents, a novel perspective in sustainable solutions [5].

The production of pyridine is an example of a process that requires specific physicochemical characteristics and porosity of the catalysts. Pala-Rosas et al. investigated the catalyzed reaction of acrolein/ammonia amino-cyclization using a group of FAU zeolites with different Si/Al [6]. The catalytic activity showed that acrolein conversion favors higher Al content in Y zeolite in line with an acidity increase. Catalysts with substantial surface areas and developed porosity produced pyridines yield inversely dependent on total acidity. The high selectivity for 3-picoline was associated with an equal number of Brønsted and Lewis sites. On the other hand, a greater extent of Lewis acidity yields predominantly pyridine products. A detailed deactivation study supports the premise of the crucial role of zeolite large pore volume and BET surface affecting deactivation rate. These textural properties enable polyaromatics deposition on the catalyst, which lowers activity [6].

Catalyst deactivation due to coke deposition poses a significant problem in industrial processes. To investigate coke formation, ethanol dehydration was investigated as a model process by Verdeş et al. [7]. The 12-tungstophosphoric acid (HPW) and its Cs salt supported on SBA-15 support were employed as catalysts, and the additional effect of Pb doping was analyzed. Porous catalyst support enables higher conversion and selectivity and lowers coke formation. Pb doping has different effects; conversely, it supports coke formation for HPW catalysts while disabling coke formation when Cs is present. Thus, a co-doping procedure for catalysts and their dispersion over support may benefit industrial dehydration processes [7].

Porous materials are efficiently implemented in environmentally significant reactions, such as eliminating nitrogen oxides. For this purpose, a selective catalytic reduction (SCR) of nitrogen oxides with ammonia is recognized as a highly efficient method. Zhao et al. investigated the role of copper sites in the nitrogen oxide SCR with ammonia [8]. Several porous materials were tested to support copper catalytic sites: SSZ-13, ZSM-5 and Beta zeolites, with a targeted Si/Al ratio of 14 and Cu/Al set at 0.4. The amount of isolated Cu^{2+} ions in the catalysts determines Lewis acidity and is related to nitrate ions as intermediates in SCR at low temperatures. On the other hand, Cu^{+}, along with Bronsted acid sites, is associated with SCR performance at high temperatures. Cu-containing SSZ-13 zeolite performed optimally in the ammonia SCR process [8].

Another aspect of zeolites' application to atmospheric processes may be seen through vapor extraction with a perspective of clean water production. Such a specific design may be seen in a composite setup consisting of a 13X zeolite and graphene binder proposed by Gildernew et al. [9]. Tailoring of substrate structure to suit efficient sorption performance was accomplished by polymer guarding pore network during binder infiltration. Obtained composites with sustained channeling displayed a specific heat flux of around 7700 W/kg, with substantial improvement in mass transport [9].

Expensive and tedious procedures for synthesizing porous materials can fail to produce materials with substantial adsorption capacity and/or catalytic efficiency. Thus, the researcher's focus is shifted to utilizing or modifying readily available and affordable natural materials.

The porosity of catalysts is essential for the aforementioned acid/base catalysis and the redox processes sensitive to catalysts' textural properties. The oxygen reduction reaction is of utmost importance for fuel cell design. Bio-based carbons are at the forefront of this investigation due to their environmental soundness, conductivity, ease of doping and textural features, as summarized by Rupar et al. [10]. Biopolymers are selected as readily available precursors for carbon-designed ORR electrode materials. The preparation procedure for these functional carbons governs porosity development and influences their performance to a large extent [10].

Adsorption on zeolites and other porous materials is successfully applied both for pollutants removal from the environment and as support of pharmaceuticals. Organic dyes are often used as model compounds to test the availability of active sites on designated adsorbents. Different pristine silicate samples containing bentonite, kaolin, clinoptilolite and diatomite were applied for methylene blue removal [11]. Extensive dye removal was accomplished with the bentonite sample reaching a capacity of nearly 240 mg/g. To make a sustainable circle of contaminants removal from water, saturated adsorbents were carbonized in an inert atmosphere. This procedure provided a nitrogen-doped carbonaceous surface layer at the bentonite surface, producing a carbon/silicate composite. This novel composite sample was used as electroactive material for oxygen reduction under alkaline conditions. As the reduction pathway was assessed as a two-electron reaction, the main species were hydroperoxyl radicals, which can degrade methylene blue. After 60 min of reduction in the electrochemical reactor, only 30% of the dye remained. The proposed process of adsorption, carbonization of saturated adsorbent and production of electrode material for oxidative degradation of pollutants reported by Popadić et al. is a perspective route for sustainable environmental solutions focusing on porous materials [11].

Natural alumosilicates are often functionalized as their synthetic counterparts to suit different pharmaceutical applications. Modified phillipsite samples can be treated to attain monolayer and bilayer coverage with surfactants–octadecyldimethylbenzylammonium chloride and dodecylamine as proposed by Smiljanić et al. [12]. Composites obtained in such procedures are used as supports for the non-steroid anti-inflammatory drug diclofenac sodium. A homogeneous adsorption site distribution is accomplished as adsorption isotherms follow the Langmuir model, with substantial drug loading of nearly 40 mg/g. Philipsite samples functionalized with benzyl or pyridine groups are promising materials for pharmaceutical delivery systems [12].

To further highlight the diversity of zeolite features and applicability, a critical review by Mijailović et al. gives an overview of zeolite adsorption performance for pharmaceuticals [13]. This feature is addressed in two aspects-zeolites employed as efficient environmental adsorbents for pharmaceutical waste and, at the same time, promising drug delivery systems due to their favorable adsorption capacities and drugs' sustained release. Zeolites' aforementioned applications indicate the necessity of thermodynamic and kinetic investigation following the adsorption process. Detailed study of intermolecular interactions supported by computation methods may guide future investigations of adsorption-aided solutions to environmental and pharmaceutical issues [13].

To conclude, the research and review papers published in the "Zeolites and Porous Materials: Insight into Catalysis and Adsorption Processes" issue address novel trends in industrial and environmental processes enabled by adsorption and/or catalysis involving porous materials. All contributions are valuable in the growing zeolite and porous materials studies field and will hopefully inspire new advanced solutions. As guest editors, we would like to thank all authors for their contributions and the Editorial team for their support during the realization of this issue.

Author Contributions: Conceptualization, M.M.-R. and D.B.-B.; writing—original draft preparation, M.M.-R. and D.B.-B.; writing—review and editing, M.M.-R. and D.B.-B. All authors have read and agreed to the published version of the manuscript.

Funding: This research was funded by the Ministry of Science, Technological Development and Innovation of the Republic of Serbia, Grants no. 451-03-47/2023-01/200146.

Conflicts of Interest: The authors declare no conflict of interest.

References

1. Xian, C.; Mao, Y.; Long, X.; Wu, Z.; Li, X.; Cao, Z. The Effect of Y Zeolites with Different Pores on Tetralin Hydrocracking for the Production of High-Value Benzene, Toluene, Ethylbenzene and Xylene Products. *Catalysts* **2022**, *12*, 848. [CrossRef]
2. Guo, Z.; Hong, M.; Yu, Y.; Liu, G.; Zang, J.; Zhang, D.; Gong, H.; Yang, K.; Huang, S. Interventions to the Spontaneous Fabrication of Hierarchical ZSM-5 Zeolites by Fluorination-Alkaline Treatment. *Catalysts* **2022**, *12*, 954. [CrossRef]
3. Asgar Pour, Z.; Abduljawad, M.M.; Alassmy, Y.A.; Cardon, L.; Van Steenberge, P.H.M.; Sebakhy, K.O. A Comparative Review of Binder-Containing Extrusion and Alternative Shaping Techniques for Structuring of Zeolites into Different Geometrical Bodies. *Catalysts* **2023**, *13*, 656. [CrossRef]
4. Yang, X.; Zhao, W.; Liu, L.; Niu, X.; Wang, Q. Tuning the Structure and Acidity of Pt/Hierarchical SSZ-32 Catalysts to Boost the Selective Hydroisomerization of n-Hexadecane. *Catalysts* **2023**, *13*, 702. [CrossRef]
5. Yong, X.; Su, H.; Zhao, N.; Jin, Z.; Yao, M.; Ma, Y. Xylene and N-Hexane Adsorption Performance of a Waste Methanol-to-Propylene Catalyst under Acid-Base Treatment. *Catalysts* **2022**, *12*, 1028. [CrossRef]
6. Pala-Rosas, I.; Contreras, J.L.; Salmones, J.; López-Medina, R.; Angeles-Beltrán, D.; Zeifert, B.; Navarrete-Bolaños, J.; González-Hernández, N.N. Effects of the Acidic and Textural Properties of Y-Type Zeolites on the Synthesis of Pyridine and 3-Picoline from Acrolein and Ammonia. *Catalysts* **2023**, *13*, 652. [CrossRef]
7. Verdeş, O.; Popa, A.; Borcănescu, S.; Suba, M.; Sasca, V. Thermogravimetry Applied for Investigation of Coke Formation in Ethanol Conversion over Heteropoly Tungstate Catalysts. *Catalysts* **2022**, *12*, 1059. [CrossRef]
8. Zhao, W.; Shen, M.; Zhu, Y.; Ren, X.; Li, X. Insights into Synergy of Copper and Acid Sites for Selective Catalytic Reduction of NO with Ammonia over Zeolite Catalysts. *Catalysts* **2023**, *13*, 301. [CrossRef]
9. Gildernew, E.; Tareq, S.; Yang, S. Three-Dimensional Graphene with Preserved Channeling as a Binder Additive for Zeolite 13X for Enhanced Thermal Conductivity, Vapor Transport, and Vapor Adsorption Loading Kinetics. *Catalysts* **2022**, *12*, 292. [CrossRef]
10. Rupar, J.; Tekić, D.; Janošević Ležaić, A.; Upadhyay, K.K. ORR Catalysts Derived from Biopolymers. *Catalysts* **2022**, *13*, 80. [CrossRef]
11. Popadić, D.; Gavrilov, N.; Ignjatović, L.; Krajišnik, D.; Mentus, S.; Milojević-Rakić, M.; Bajuk-Bogdanović, D. How to Obtain Maximum Environmental Applicability from Natural Silicates. *Catalysts* **2022**, *12*, 519. [CrossRef]
12. Smiljanić, D.; Daković, A.; Obradović, M.; Ožegović, M.; Marković, M.; Rottinghaus, G.E.; de Gennaro, B. Influence of the Type and the Amount of Surfactant in Phillipsite on Adsorption of Diclofenac Sodium. *Catalysts* **2022**, *13*, 71. [CrossRef]
13. Mijailović, N.R.; Nedić Vasiljević, B.; Ranković, M.; Milanović, V.; Uskoković-Marković, S. Environmental and Pharmacokinetic Aspects of Zeolite/Pharmaceuticals Systems—Two Facets of Adsorption Ability. *Catalysts* **2022**, *12*, 837. [CrossRef]

catalysts

Article

The Effect of Y Zeolites with Different Pores on Tetralin Hydrocracking for the Production of High-Value Benzene, Toluene, Ethylbenzene and Xylene Products

Ce Xian [1,2], Yichao Mao [2,*], Xiangyun Long [2], Ziming Wu [1], Xiang Li [3] and Zhengkai Cao [1,*]

1 Dalian Research Institute of Petroleum and Petrochemicals, SINOPEC, Dalian 116045, China; xiance.fshy@sinopec.com (C.X.); wuziming.fshy@sinopec.com (Z.W.)
2 Research Institute of Petroleum Processing, SINOPEC, Beijing 100083, China; longxy.ripp@sinopec.com
3 College of Chemical Engineering and Materials Science, Tianjin University of Science and Technology, Tianjin 300457, China; lixiang@tust.edu.cn
* Correspondence: maoyc.ripp@sinopec.com (Y.M.); xiaocao19910926@163.com (Z.C.)

Abstract: A series of Y zeolites with different pore properties was prepared as a support for hydrocracking catalysts for the production of BTEX (benzene, toluene, ethyl-benzene, and xylene) from tetralin. Some important characterizations, including N_2 adsorption–desorption, NH_3-TPD, Py-IR, and HRTEM, were applied to obtain the properties of different catalysts. Meanwhile, the tetralin hydrocracking performances of those catalysts were investigated on a high-pressure fixed-bed microreactor. The results showed that Si/Al ratio is the core property of zeolites and that the increase in the V_{micro}/V_{meso} of zeolites could facilitate the formation of BTEX products by hydrocracking tetralin. The method of hydrocracking tetralin was proposed. It was also found that the hydrogenation–cracking path was controlled by aromatic saturation thermodynamics, and strong acidity aided the backward shift of equilibrium temperature.

Keywords: Y zeolite; porosity; tetralin; hydrocracking; BTEX

Citation: Xian, C.; Mao, Y.; Long, X.; Wu, Z.; Li, X.; Cao, Z. The Effect of Y Zeolites with Different Pores on Tetralin Hydrocracking for the Production of High-Value Benzene, Toluene, Ethylbenzene and Xylene Products. *Catalysts* **2022**, *12*, 848. https://doi.org/10.3390/catal12080848

Academic Editors: Maja Milojević-Rakić and Danica Bajuk-Bogdanović

Received: 1 July 2022
Accepted: 30 July 2022
Published: 2 August 2022

1. Introduction

With the continuous implementation of the carbon-neutrality policy and the development trend in energy transformation, the petrochemical industry has undergone great changes in terms of product structure adjustment and the nature of the products produced. From a long-term perspective, the demand for transport fuel will gradually decrease. In addition, the severity of the use of crude oil continues to worsen, which is a greater challenge for the petroleum processing and refining industry.

Light cycle oil (LCO), with a boiling point similar to that of diesel, is derived from fluid catalytic cracking, which has high aromatic, sulfur, and nitrogen contents, as well as a low cetane index [1,2]. There is a large gap between the nature of LCO and the China national VI diesel standard. If LCO is used to produce the blending component of clean diesel fuel, it will consume a large amount of hydrogen and cause high CO_2 emissions, which means it does not meet the requirements of environmental protection policies, the process is not economical, and the product has low commercial value. Therefore, many researchers use LCO, which is rich in aromatic hydrocarbons, to produce high-octane gasoline and light aromatic BTEX (benzene, toluene, ethyl-benzene, and xylene) components through selective hydrocracking technology [3,4]. This process can effectively reduce hydrogen consumption and greatly improve the commercial value of the product. To date, many institutions have developed a number of technologies using this principle, such as FD2G, RLG, and LCO-X.

According to previously reported research, most of the poly-aromatics in LCO are di-aromatic hydrocarbons. Researchers have investigated the reaction regularity of model compounds, such as naphthalene and tetralin, and have studied the reactivity of hydro cracking catalysts [5,6]. Sato et al. [7] explored the hydrocracking of tetralin over NiW/USY catalysts; the reaction mechanisms of tetralin were investigated according to the product distribution, and they highlighted the importance of strengthening the connection between the active hydrogenation site and the acid site to improve the performance of the catalyst. Active metal components are the source of hydrogenation activity. Typically, noble metals and non-noble metal mixtures are used to hydrocrack catalysts. Noble metals can provide much higher hydrogenation activity; however, they are sensitive to S and N compounds and are easily poisoned to the point of deactivation. They are not suitable for LCO feeds with high sulfur content. Upare et al. [8] compared a monometallic Mo/Beta catalyst and bimetallic CoMo/Beta catalyst in the selective hydrocracking of tetralin into monocyclic aromatic hydrocarbons; the results indicated that the bimetallic CoMo Beta catalyst not only improved the catalytic activity in hydrocracking reactions but also showed better long-term stability. This study revealed that CoMo/Beta with SiO_2/Al_2O_3 of 25 gave the highest desired mono-aromatic hydrocarbon yield of 62.6% at the 99.5% conversion of tetralin for more than 140 h of reaction time. Laredo et al. [9] used tetralin to explore the effect of an NiMo/Al_2O_3/ZSM-5 catalytic system and operating conditions on BTX formation. It was disclosed that high BTX selectivity in the liquid phase with suitable tetralin conversion could be achieved using this catalytic system. One reason for this result is that Ni-Mo can effectively reduce coke formation and minimize catalysts' deactivation. Cao et al. [10] studied the selective hydrocracking of naphthalene and tetralin into mono-aromatic hydrocarbons over NiMo-AY and CoMo-AY grading catalysts. The results showed that the CoMo catalyst could hydro-saturate more naphthalene to tetralin but exhibit a lower yield of light aromatics. The NiMo catalyst presented higher selectivity in converting naphthalene into cyclanes. A CoMo-AY/NiMo-AY catalyst grading system with low carbon deposition and high stability was shown to be capable of maintaining a high percentage of active phases, which was more efficient in the conversion of LCO to high-octane gasoline.

The carrier of hydrocracking catalysts is not only the source of acid sites, but also, reactant and product diffusion depend on it. As has been previously reported, commonly used carrier types mainly include Y zeolite [11], Beta zeolite, ZSM-5 zeolite [12], MCM-41 zeolite, γ-Al_2O_3, and amorphous silica alumina (SiO_2-Al_2O_3) [13]. Some researchers have studied the conversion of LCO over porous aromatic framework carrier catalysts [14]. Each catalyst carrier has different acid properties and pore characteristics, thus showing different reaction characteristics [15]. Researchers have used a variety of modification methods to optimize the performance of carriers [16,17]. Nakajima et al. [18] used a variety of zeolites to study the activity and selectivity of the tetralin ring-opening reactions; the influence of the acidity and textural properties of the zeolites on the activity and selectivity was shown. The results showed that a 12-ring on the BEA zeolite and strong Brønsted acid sites exhibited higher tetralin conversion; moreover, BEA zeolites increased the number of Brønsted acid sites while maintaining the selectivity. Ren et al. [19] used fluorination−alkaline treatment-modified USY zeolites as the carrier of the hydrocracking catalyst and studied the hydrogenation and ring-opening performance of naphthalene. The results showed that the modified USY zeolite increased the volume of mesopore and pore diameter and increased the number of strong acid sites. The ring-opening product yield of the naphthalene hydrocracking reaction was greatly increased, and the content of poly-aromatics was effectively reduced. Youngseok et al. [20] used a combination of H-ZSM-5, H-Beta, and mesoporous H-Y zeolite as a carrier for the hydrocracking catalysts. Their study revealed that the tetralin conversion rate was 97.4%, and the total yields of BTX and alkylbenzenes were close to 56.1%. This was due to the fact that hybrid zeolite contains a 10-ring channel, 12-ring channel, and mesopores, while the appropriate acid properties aid the production of a much higher yield of BTX-rich light aromatics. Above all, there are many ways to adjust the channel structure of a catalyst carrier. However, the topological structure and specific characteristics of different carriers vary greatly, and the influence of other properties, especially acid properties, cannot be explained when exploring the influence of catalyst pore properties.

As reported, Y zeolite, which is the main carrier of industrial hydrocracking catalysts, has a suitable pore structure, adjustable acidity, and favorable stability. In order to reduce the influence of the acidic properties of a carrier, Y zeolite was modified using appropriate methods, and two series of Y zeolites with similar acid properties but significantly different pore structures were prepared, and the hydrocracking catalyst was prepared using these Y zeolites as carriers. The influence of the pore properties of Y zeolite on the selective hydrocracking of tetralin to produce light aromatics was investigated. The influence of Y zeolite's pore characteristics on the key conversion steps of tetralin was clarified, which played a guiding role in the development of hydrocracking catalysts.

2. Results and Discussion

2.1. Characterization of Zeolites and Catalysts

Y zeolite has a different order of Si–O–Al or Si–O–Si linkages in the framework, and the molar Si/Al ratio of zeolite result in variations in the acidity of the Si(OH)Al group [21]. The acid properties of Y zeolites are mainly related to the content and distribution of aluminum atoms. Moreover, the amount of aluminum in the framework affects the unit cell constant. In this work, a series of zeolite Y was divided into two parts according to the value of the Si/Al ratio characterized by the XRF and the unit cell constant characterized by the XRD, as shown in Table 1 and Figure 1.

Table 1. Unit cell constant of zeolite Y1–Y8.

Sample	Si/Al	Unit Cell Constant/(Å)	Sample	Si/Al	Unit Cell Constant/(Å)
Y1	7.88	24.39	Y5	3.34	24.56
Y2	8.89	24.32	Y6	3.68	24.62
Y3	8.40	24.37	Y7	3.59	24.54
Y4	8.25	24.35	Y8	3.49	24.65

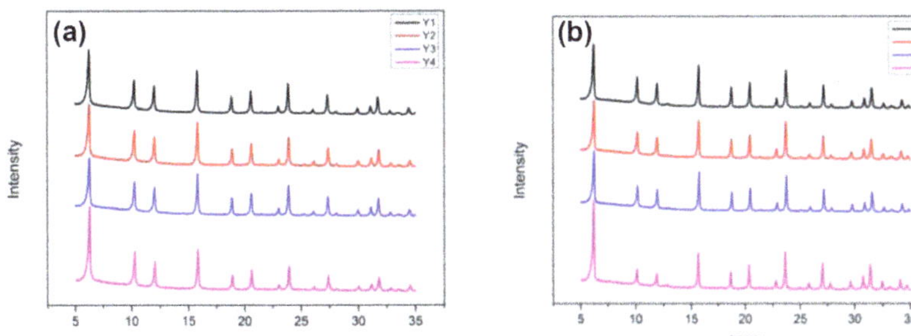

Figure 1. (**a**) XRD patterns of Y1–Y4 and (**b**) XRD patterns of Y5–Y8.

The key factor determining both the unit cell constant and acidity is the Si/Al ratio of zeolites. It can be seen from Table 1 that the Si/Al ratio and the unit cell constants of Y1–Y4 zeolites were very similar, and the Si/Al ratio and the unit cell constants of Y5–Y8 zeolites were also relatively similar. This means that the acidities of the samples may have been similar in each group. From Figure 1, it can be seen that each group of samples showed a typical Y zeolite crystal structure, indicating that the framework structure of all the samples was not destroyed. NH_3-TPD was used to estimate the acid site strength distribution using a single temperature programmed desorption method; desorption peaks at less than 200 °C, 200–350 °C and higher than 350 °C represented the acidic sites with weak strength, medium strength, and high strength [22]. The results are shown in Figure 2.

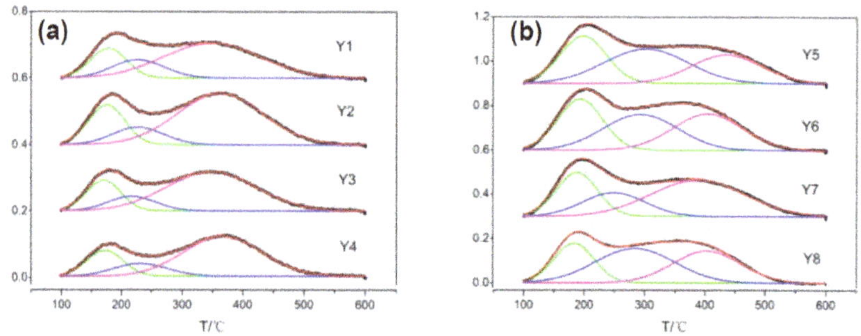

Figure 2. (**a**) NH_3-TPD profiles of Y1–Y4 and (**b**) NH_3-TPD profiles of Y5–Y8. Black line: the desorption peak of zeolite acid sites obtained by testing. Green line: the desorption peak of weak acid sites obtained by fitting. Blue line: the desorption peak of medium acid sites obtained by fitting. Pink line: the desorption peak of strong acid sites obtained by fitting. Red line: the desorption peak of zeolite acid sites obtained by fitting.

In order to further analyze the difference in the acid strength of each sample, the profiles in Figure 2 were treated according to the highest degree of fitting, and the acid amount was calculated according to the peak area and the desorption ammonia titer. The results are shown in Table 2.

Table 2. Acid strength distribution of Y1–Y8.

Sample	Total Acid Amount /(μmol·g^{-1})	Weak Strength /(μmol·g^{-1})	Medium Strength /(μmol·g^{-1})	Strong Strength /(μmol·g^{-1})
Y1	1400	280	230	890
Y2	1620	330	230	1060
Y3	1470	300	190	980
Y4	1500	300	220	980
Y5	1860	520	810	530
Y6	1950	500	780	670
Y7	1840	470	650	720
Y8	1920	450	870	600

It can be seen from Table 2 that the total acid amounts of the samples could also be divided into two groups according to the unit cell constant, and the acidities of zeolites within each group were very similar. The total acid amounts in theY5, Y6, Y7 and Y8 samples were higher than those in the Y1, Y2, Y3, and Y4 samples. However, the amounts of strong acid in Y5, Y6, Y7, and Y8 were lower than those in the Y1, Y2, Y3, and Y4 samples; a similar trend was shown in the Si/Al ratios. Therefore, it can be concluded that a high ratio of Si and Al can facilitate more strong acid sites but fewer total acid sites.

The standard deviation of most samples was less than 10%. In terms of the different strengths of acidity, Y1, Y2, Y3, and Y4 were classified as strong acid zeolites (SAZ), and Y5, Y6, Y7, and Y8 could be defined as medium acid zeolites (MAZ). In a comparison between SAZ and MAZ, it was indicated that the medium acid sites may be converted from strong ones in the preparation of zeolite dealumination to reduce the unit cell constant. The IR spectra of pyridine adsorbed onto zeolite at various temperatures is usually used to identify the Brϕnsted acid sites and Lewis acid sites with different strengths [23]; the results are shown in Table 3.

Table 3. Py-IR results of Y1–Y8.

Sample	200 °C		350 °C	
	L Acid Amount /(μmol·g^{-1})	B Acid Amount /(μmol·g^{-1})	L Acid Amount /(μmol·g^{-1})	B Acid Amount /(μmol·g^{-1})
Y1	155	282	118	239
Y2	81	533	65	482
Y3	75	428	64	385
Y4	34	323	34	320
Y5	39	480	24	412
Y6	30	641	19	531
Y7	13	550	9	486
Y8	22	589	12	537

Table 3 suggests that the zeolites could still be divided into two parts according to the Brϕnsted acid sites and Lewis acid sites; the difference in the acid amount in the SAZ series was small after Y1 was removed. The MAZ produced similar results when Y7 was removed. Compared with MAZ, the SAZ had more Lewis acid and less Brϕnsted acid, which was quite different to the NH_3-TPD results. The differences within each group were also higher than those observed in the NH_3-TPD results. This can be attributed to the different absorbed molecular sizes of ammonia and pyridine. Through external and internal diffusion, the probe molecule could reach distinct active sites for absorption; therefore, the difference was usually due to the porous property of zeolites. The isotherms of nitrogen physisorption and the pore size distribution of the Y zeolites are presented in Figures 3 and 4.

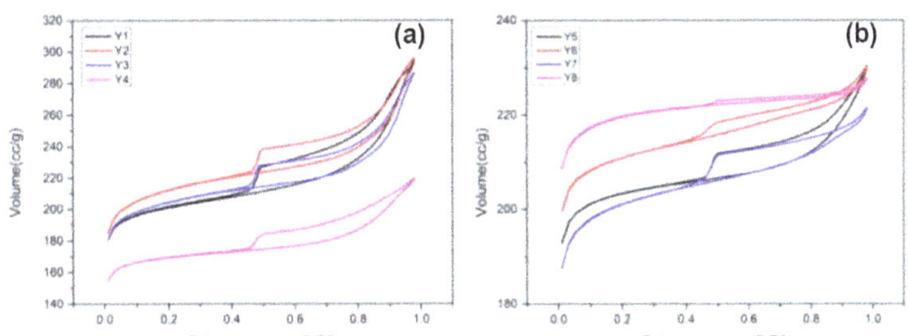

Figure 3. The isotherms of nitrogen physisorption of zeolites (**a**) Y1–Y4 and (**b**)Y5–Y8.

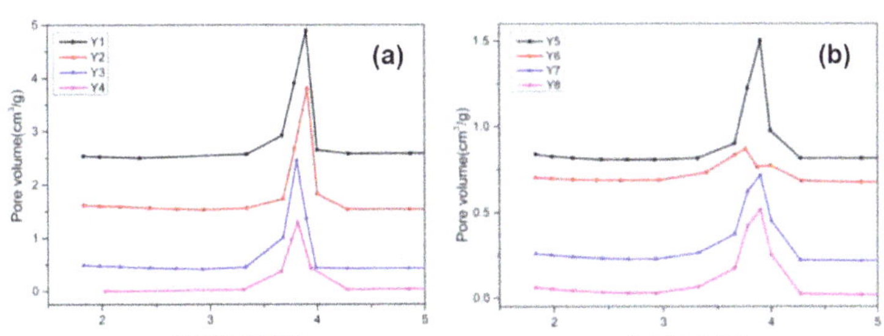

Figure 4. The pore size distribution of zeolites (**a**) Y1–Y4 and (**b**)Y5–Y8.

It can be clearly seen that the N_2 adsorption–desorption isotherms of all the zeolites were type IV with a typical H_2 type hysteresis loop, indicating that these zeolites possess microporous and mesoporous structures [24]. The porous properties are shown in Table 4.

Table 4. Porosity of Y zeolites.

Sample	S_{BET}/ ($m^2 \cdot g^{-1}$)	S_{micro}/ ($m^2 \cdot g^{-1}$)	S_{meso}/ ($m^2 \cdot g^{-1}$)	V_{total}/ ($mL \cdot g^{-1}$)	V_{micro}/ ($mL \cdot g^{-1}$)	V_{meso}/ ($mL \cdot g^{-1}$)	V_{micro}/ V_{meso}
Y1	571	537	34	0.34	0.25	0.09	2.77
Y2	681	609	72	0.45	0.28	0.17	1.64
Y3	694	640	54	0.44	0.30	0.14	2.14
Y4	721	659	62	0.46	0.31	0.15	2.06
Y5	708	678	30	0.37	0.32	0.05	6.40
Y6	678	650	28	0.34	0.30	0.04	7.50
Y7	682	662	20	0.36	0.31	0.05	6.20
Y8	704	690	14	0.34	0.31	0.03	10.33

The results from Table 4 show that, in SAZ, the S_{micro} increased while the S_{meso} changed randomly; in the MAZ series, the opposite was observed. In addition, the Y8 zeolite showed a significantly high value of V_{micro}/V_{meso}, which could also be verified from the super flat hysteresis loop in Figure 4b.

Due to the high surface tension in the capillary condensation process, transition metal compounds are usually located on the outer surfaces of catalyst microparticles during wet impregnation preparation of the catalyst and further form layered sulfides in the sulfidation process. The sizes of bulk particles from active metal sulfides are usually used to identify the hydrogenation activity of catalysts [25]; here, the catalysts in the sulfide form were investigated using the HRTEM method. Figure 5 shows microphotographs of the catalysts, and the average particle length ($\overline{L}$) and stacking number ($\overline{N}$) were calculated, as shown in Table 5.

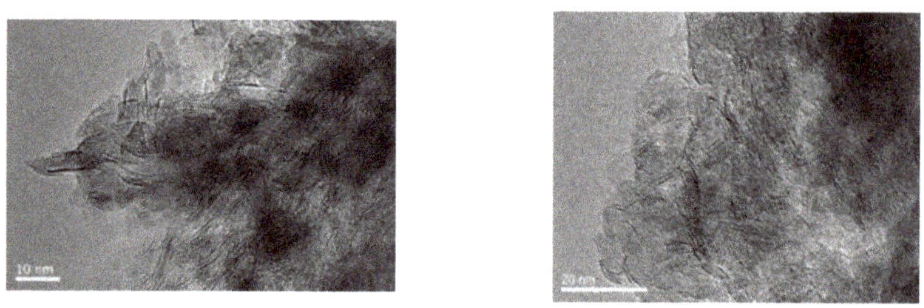

(a) (b)

Figure 5. HRTEM images of sulfide catalysts (**a**) CAT3 and (**b**) CAT7.

Table 5. The average particle length ($\overline{L}$) and stacking number ($\overline{N}$) of CAT3 and CAT7.

Simple	Average Particle Length ($\overline{L}$)/nm	Average Stacking Number ($\overline{N}$)
CAT3	5.91	1.99
CAT7	5.67	1.73

The particle length and stacking number values were close to the results found in common hydrocracking catalysts [26]. The size of bulk particles from the Y7 sample was slightly smaller than that of Y3; this can be explained by the fact that there were stronger Brϕnsted sites, which led to strong metal–support interaction, which ensured it was easy to retain the single-layer structure during the sulfidation and reaction procedure.

2.2. Hydrocracking of Tetralin

Tetralin hydrocracking over catalysts takes place through a complex reaction scheme, as shown in Figure 6, which is related to this research [7,27].

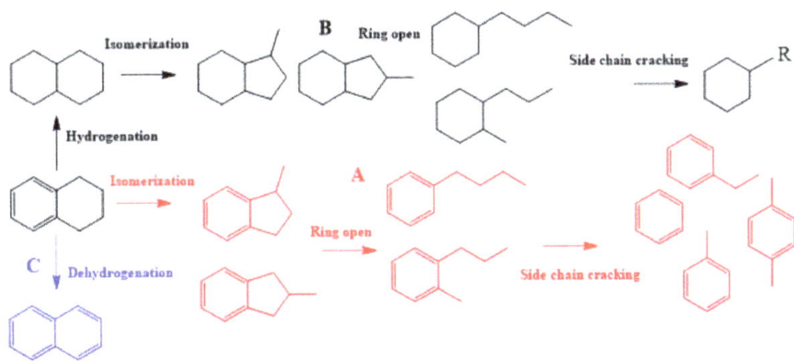

Figure 6. Reaction scheme for tetralin hydrocracking over bifunctional catalysts. A: Isomerization-cracking; B: hydrogenation-cracking; and C: dehydrogenation.

The hydrocracking of tetralin is based on the mechanism of a carbonium reaction, including hydrogenation, isomerization, β cracking, and dehydrogenation. In addition to hydrogenation, tetralin can undergo naphthenic ring isomerization or dehydrogenation; these schemes are differentiated in the first step in the reaction of tetralin. In path A and B, the naphthenic ring opens, followed by cracking reactions to form BTEX or single naphthenic compounds. To facilitate the analysis of the results of the catalytic tests, the identified compounds were assembled in different groups, as indicated in Table 6.

Table 6. Compounds assembled in different groups.

Group	Compounds
HEAVY	naphthalene, alkyl-naphthalene
ARO-ISO	methyl-indan
ARO-RO	butyl-benzene, methylpropenyl-benzene
ARO-C	benzene, toluene, xylenes, ethyl-benzene
H	decalin
HYD-RO	butyl-cyclohexane, methyl-butylcyclopentane
HYD-C	methyl-cyclopentane, ethyl-cyclohexane

In order to reasonably explain the influence of the pore properties of Y zeolite on the hydrocracking process of tetralin, the conversion rates, yields of products and selectivity are defined by the following:

$$C_{tetralin} = \frac{\omega_i - \omega_f}{\omega_i} \times 100\% \quad (1)$$

where ω_i and ω_f represent the content of tetralin in the material and the product, respectively.

$$Y_{iso-cracking} = \sum \omega_{ARO-ISO} + \omega_{ARO-RO} + \omega_{ARO-C} \quad (2)$$

$$Y_{hydro-cracking} = \sum \omega_{H} + \omega_{HYD-RO} + \omega_{HYD-C} \quad (3)$$

$$Y_{BTEX} = \sum \omega_{ARO-C} \quad (4)$$

$$Y_{heavy} = \sum \omega_{HEAVY} \quad (5)$$

$$S_{iso-cracking} = \frac{Y_{iso-cracking}}{C_{tetralin}} \times 100\% \quad (6)$$

$$S_{heavy} = \frac{\omega_{HEAVY}}{C_{tetralin}} \times 100\% \quad (7)$$

$$S_{ring-opening} = \frac{\omega_{ARO-RO} + \omega_{ARO-C}}{C_{tetralin}} \times 100\% \quad (8)$$

$$S_{sidechain-cracking} = \frac{\omega_{ARO-C}}{C_{tetralin}} \times 100\% \quad (9)$$

Tetralin hydrocracking conversion over different catalysts is shown in Figure 7a,b. It can be seen that the conversion of tetralin obviously increased with the increase in temperature, and the conversion of the SAZ group was significantly higher than that of the MAZ group at the same temperature. This was mainly due to the high Si/Al ratio in the SAZ group, which had higher strong acid strength. Sato et al. [7] also found that a higher ratio of strong acid sites contributes to the improvement in the conversion depth of tetralin. In the SAZ group, the conversion of tetralin was mainly proportional to S_{BET} and V_{total}. In the MAZ group, it can be seen that CAT5 displayed higher conversion than the others, and this was mainly due to the high S_{BET} and V_{total} of Y5. Moreover, it can be seen that the conversion curve converged in the high reaction temperature for CAT6, CAT7 and CAT8; this indicated that the reaction barrier was overcome by the diffusion improvement [28].

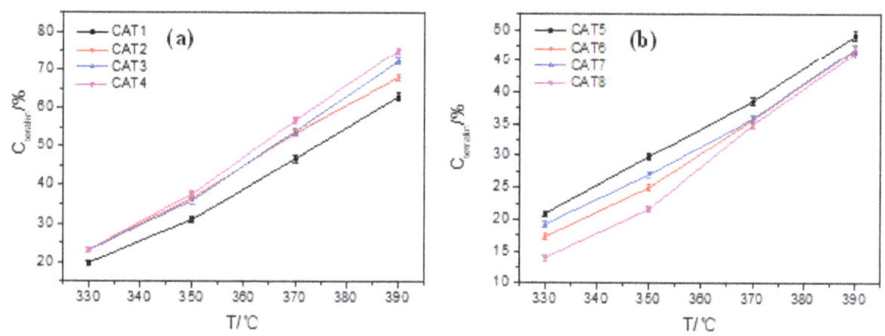

Figure 7. The relationship of conversion and temperature over catalysts (**a**) CAT1–CAT4 and (**b**) CAT5–CAT8.

Figure 8 shows the relationship between $Y_{iso-cracking}$ and temperature over different catalysts. When the reaction temperature was 330 °C and 350 °C, the $Y_{iso-cracking}$ values of the SAZ and MAZ group were similar. When the reaction temperature rose to 370 °C and 390 °C, the $Y_{iso-cracking}$ value of the SAZ group was higher than that of the MAZ group. It can be seen that a higher temperature and more strong acid sites can facilitate the occurrence of the iso-cracking reaction. Strong acid strength is more conducive to the occurrence of the iso-cracking reaction pathway, and the SAZ group with a higher Si/Al ratio performed better. From Figure 8a, it is obvious that the $Y_{iso-cracking}$ value of CAT4 was lower than that of the others, and the $Y_{iso-cracking}$ values of CAT1, CAT2, and CAT3 were similar. This may have been related to pore size distribution. It can be seen from Figure 4 that Y1, Y2, and Y3 had concentrated pore size distributions which were all concentrated from 3.5 nm to 4.0 nm. The pore size distribution of Y4 was relatively dispersed, which was not conducive to the diffusion reactants. From Figure 8a, it can be seen that the $Y_{iso-cracking}$ values of CAT6, CAT7, and CAT8 were more similar. In terms of the pore distribution, Y7 and Y8 had similar pore concentration characteristics; their $Y_{iso-cracking}$ value was consistent at 390 °C.

Figure 9 shows a comparison of the yields of BTEX in the SAZ group and MAZ group; it can be seen that the yields the two groups were similar. Generally, there are two main reasons for this. One reason is that in terms of acidity, ring-opening and sidechain-breaking reactions in the iso-cracking pathway do not require many strong acid sites, and more medium acid sites are favorable for the occurrence of these two ideal reactions. This is easily illustrated by the Si/Al ratio of the zeolites: the MAZ group had a lower Si/Al ratio and more moderate acid strength and B-acid content, which are very favorable characteristics for both reaction paths. The other reason is the effect of pore properties. From Table 4, it can be seen that the V_{total} and V_{meso} of the SAZ group were larger than those of the MAZ group, and the V_{micro} of the MAZ group was larger than that of the SAZ group. Trends in the yields of BTEX in the SAZ group and MAZ group were similar to the yields of iso-cracking. However, the MAZ group could produce semblable yields of BTEX, which means that the presence of more micropores is beneficial to the ring-opening and sidechain-breaking reactions.

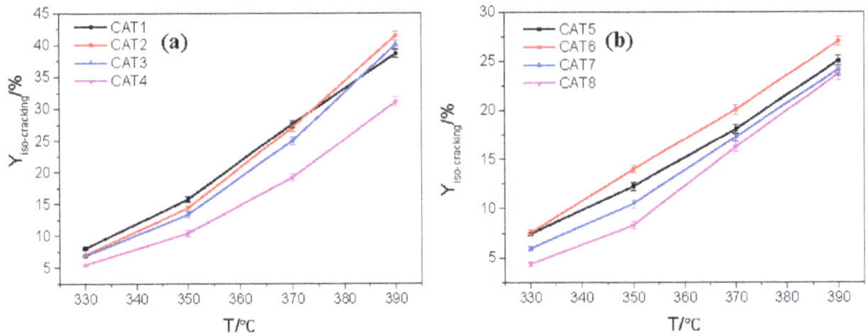

Figure 8. The relationship of yields of iso-cracking and temperature over catalysts (**a**) CAT1–CAT4 and (**b**) CAT5–CAT8.

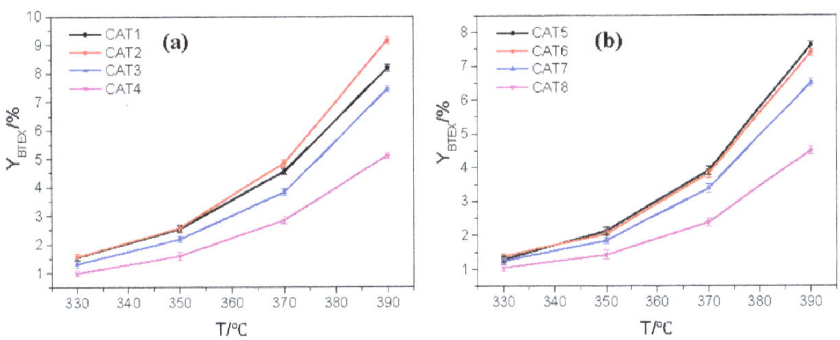

Figure 9. The relationship of yields of BTEX and temperature over catalysts (**a**) CAT1–CAT4 and (**b**) CAT5–CAT8.

The hydro-cracking reaction pathway is an important component of tetralin conversion. Figure 10 shows $Y_{hydro-cracking}$ over different catalysts. The $Y_{hydro-cracking}$ of all the catalysts increased first and then decreased. This inflection point is mainly related to the thermodynamics of aromatics saturation [29]. It can be seen that the SAZ group had the highest $Y_{hydro-cracking}$ value at 370 °C, and most catalysts in the MAZ group had the highest $Y_{hydro-cracking}$ value at 350 °C. This shows that the acid properties of zeolite directly affect the hydro-cracking pathway performance of tetralin. Strong acid sites in zeolites help to delay the aromatics saturation point; therefore, a suitable Si/Al ratio is necessary to reduce the selectivity of the hydro-cracking reaction pathway of aromatics, and this feature can be effectively used to design the development of related catalysts.

The dehydrogenation reaction is the main source of heavy aromatics. From Figure 11, it can be seen that Y_{HEAVY} is proportional to the reaction temperature. This is because the dehydrogenation reaction is endothermic, and raising the temperature favors dehydrogenation. Moreover, the Y_{HEAVY} of the SAZ group was larger than that of the MAZ group. This is mainly because the SAZ group had a higher Si/Al ratio and higher acid strength, which provided suitable conditions for the dehydrogenation reaction. Gutiérrez et al. [30] investigated the role of acid sites in LCO hydrogenation and coking deactivation; it was found that strong acid sites are considered to be the main coking sites of LCO, and there is a linear relationship between the amount of coking and the number of strong acid sites. In the SAZ group, their V_{meso} was larger than that of the MAZ group, and more mesoporous pores provided a reaction space for dehydrogenation. Y2 and Y4 had lower V_{micro}/V_{meso} than Y1 and Y3; this was consistent with the higher yield of CAT2 and CAT4.

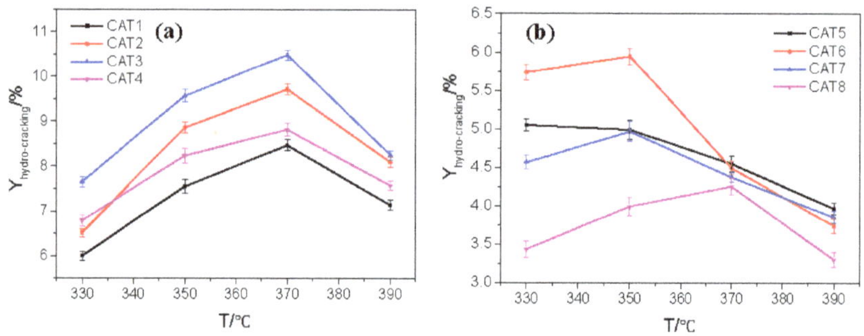

Figure 10. The relationship between yields of $Y_{hydro-cracking}$ and temperature over catalysts (**a**) CAT1–CAT4 and (**b**) CAT5–CAT8.

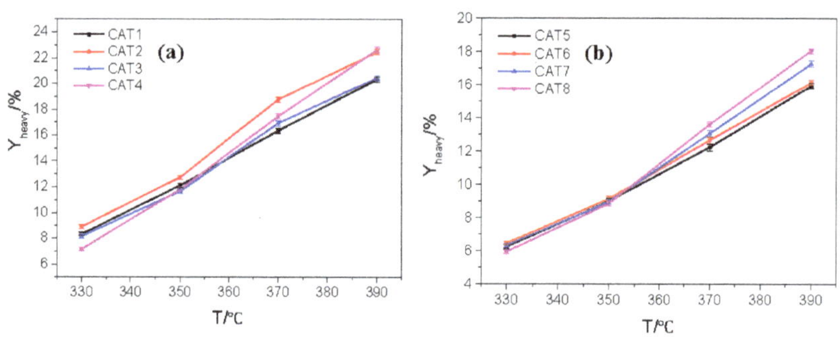

Figure 11. The relationship between yields of Y_{heavy} and temperature over catalysts (**a**) CAT1-CAT4 and (**b**) CAT5-CAT8.

By comparing the conversion rate of tetralin and the yield of product, it was found that the size of the micropores and mesopores was the main influencing factor in zeolites with similar acid properties. The presence of more micropores in zeolites is favorable to the formation of BTX products, and the presence of more mesopores is favorable to the dehydrogenation reaction. Therefore, the relationship between microporous and mesoporous molecular sieves should be adjusted to a reasonable range.

In order to investigate the effect of the microporous and mesoporous relationship on reaction selectivity, the selectivity at the same conversion rate was calculated. Due to the high activity of the SAZ series, the selectivity at a 60% conversion rate was calculated and correlated with the pore properties. The conversion of the MAZ series was low even at a high reaction temperature, so 45% conversion was used as the benchmark; the results are shown in Figures 12 and 13.

From the results shown in Figures 12 and 13, it can be seen that although acid properties display a certain influence, the selectivity of the catalyst and the ratio of V_{micro}/V_{meso} still correspond well. With the increase in V_{micro}/V_{meso}, the dehydrogenation selectivity increased and the isomerization cracking selectivity decreased. The isomerization selectivity of the MAZ molecular sieve reached a constant value faster. The selectivity of ring-opening and selectivity of sidechain-breaking decreased almost synchronously.

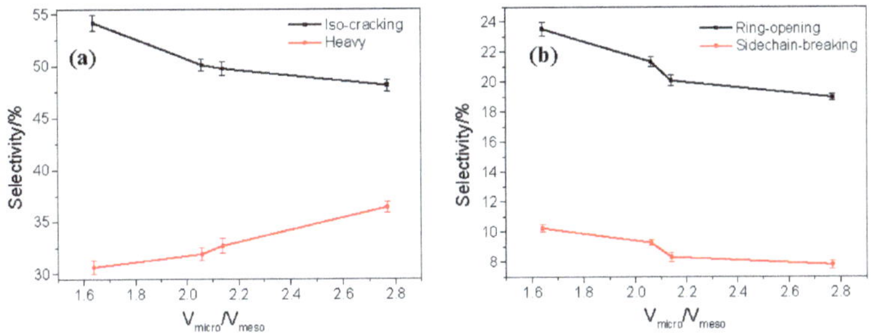

Figure 12. The path selectivity of SAZ zeolite at 60% conversion rate vs. V_{micro}/V_{meso}. (**a**) The selectivity of Iso-cracking and Heavy vs. V_{micro}/V_{meso} and (**b**) The selectivity of Ring-opening and Sidechain-breaking vs. V_{micro}/V_{meso}.

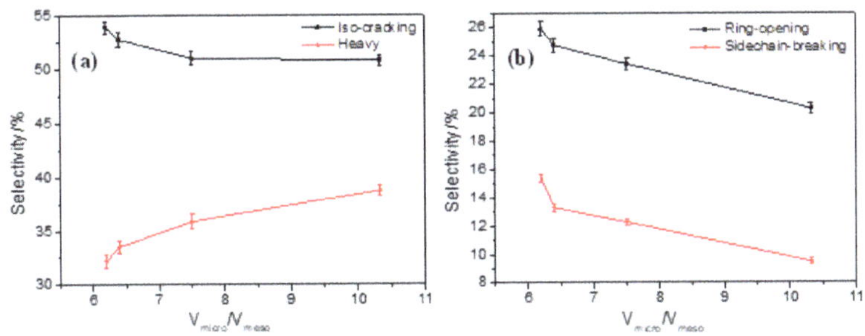

Figure 13. The path selectivity of MAZ zeolite at 45% conversion rate vs. V_{micro}/V_{meso}. (**a**) The selectivity of Iso-cracking and Heavy vs. V_{micro}/V_{meso} and (**b**) The selectivity of Ring-opening and Sidechain-breaking vs. V_{micro}/V_{meso}.

3. Materials and Methods

3.1. Materials

The following reagents were used without further purification: USY zeolite (Sinopec Catalyst Co., Ltd., Beijing, China), The eight USY zeolites that were labeled Y1, Y2, Y3, Y4, Y5, Y6, Y7, and Y8 were Pseudoboehmite (SB, SASOL), Tetralin (98%, ACROS), Cyclohexane (analytically pure, Beijing Fine Chemicals Co., Ltd., Beijing, China) and CS_2 (analytical pure, Beijing Chemical Factory Products.).

3.2. Preparation of Catalysts

The catalysts were prepared according to the method of an industrial catalyst, except the type of zeolites, the composition of the support and the preparation method were consistent. The carrier was extruded in the proportion of Y zeolite (65 wt%) and alumina (35 wt%); then, it was dried at 393 k for 180 min and calcinated at 823 K for 180 min. The metal components in the catalyst were NiMo bimetals which were loaded via impregnation; then, they were dried at 393 k for 180 min and calcinated at 723 K for 120 min. The corresponding catalysts prepared using the zeolites were labeled CAT1, CAT2, CAT3, CAT4, CAT5, CAT6, CAT7, and CAT8.

3.3. Characterization

X-ray diffraction (XRD) measurements [12] were performed on a Bruker D5005 X′ Pert diffractometer equipped with Cu Kα radiation (λ = 1.5406 Å).

N_2 adsorption–desorption measurements were performed on a Micromeritics ASAP 2400 instrument [13].

The composition of zeolites was quantified using X-ray fluorescence (XRF) conducted on a PANalytical AxiosMAX analyzer.

The acidity was determined via the IR spectroscopy of adsorbed pyridine (Py-IR) by using a Nicolet Magna-IR 560 ESP spectrophotometer (USA) with a resolution of 1 cm^{-1}. Purified pyridine gas flowed through the catalyst at room temperature for 40 min. The IR spectra were recorded and degassed at the temperatures of 250 °C and 350 °C, respectively. The amount of acid was calculated by the following formula [23]:

$$C\ (\text{B-sites}) = 1.88 \times IA(B) \times R^2/W$$

$$C\ (\text{L-sites}) = 1.42 \times IA(L) \times R^2/W$$

in which C is the concentration (mmol/g catalyst); IA(B,L) is the integrated absorbance of the B or L band (cm^{-1}); R is the radius of the catalyst disk (cm); and W is the weight of disk (mg).

NH_3 -TPD analysis was conducted using a BEL-CAT instrument [14]. The morphology of the sulfide phase was characterized using a transmission electron microscope, Tecnai G^2F20S-TWIN [15].

Prior to the TEM analysis, the oxide samples were sulfided. After sulfidation, the temperature was decreased to room temperature under a N_2 atmosphere. Then, sulfide catalysts were suspended in cycloheptane and deposited on a carbon-coated copper grid. The WS_2 slab size distribution was determined by measuring about 700–800 particles through a reference. The average particle length ($\overline{L}$,nm) and stacking number ($\overline{N}$) were calculated according to the first moment of the distribution; the calculation equations are shown as follows:

$$\overline{L} = \frac{\sum_{i=1}^{n} n_i L_i}{\sum_{i=1}^{n} n_i} \tag{10}$$

$$\overline{N} = \frac{\sum_{i=1}^{n} n_i N_i}{\sum_{i=1}^{n} n_i} \tag{11}$$

where L_i represents the length of a single MoS_2 wafer, N_i represents the stacked layers of a single MoS_2 wafer, and n represents the number of MoS_2 wafers.

3.4. *Hydrocracking Reaction*

The tests were conducted in a continuously flowing tubular fixed-bed reactor loaded with 1.0 g of catalyst (~0.25 mm) diluted with the same quantity of quartz particles. Before the reaction, the catalyst was treated with sulfidation feedstock (5 wt% CS_2/cyclohexane solution) at 0.4 mL/min in the presence of 360 mL/min of hydrogen at 4.0 MPa of total pressure. Sulfidation was carried out at 633 K for 2 h, and then the feed was switched to tetralin. The reaction temperature varied from 603 K to 663 K, and the liquid flow rate was 0.1 mL/min. The reaction products were analyzed on an Agilent 6890 GC using a PONA column (50 m × 0.2 mm × 0.5 μm) and an FID detector.

4. Conclusions

The Si/Al ratio is the core property of zeolites, and the acid properties of the zeolites is based on this ratio. Therefore, two groups of molecular sieves were selected from multiple samples. The characterization results showed that in each group of zeolites, the Si/Al ratios were relatively similar, and the acid amounts characterized by NH_3-TPD were similar, while the distribution of different types acid sites characterized by Py-IR were different; meanwhile, their pore properties varied greatly. The sizes of the metal particles in the sulfidation catalysts changed slightly. The evaluation results from tetralin hydrocracking showed that increasing the proportion of V_{micro}/V_{meso} in zeolites was conducive for the production of BTEX. The hydrogenation-cracking reaction is controlled by aromatic saturation thermodynamics, and strong acidity is beneficial to the backward shift of equilibrium temperature.

Author Contributions: Data curation, Y.M.; Methodology, X.L. (Xiang Li); Project administration, Z.W.; Resources, X.L. (Xiangyun Long); Supervision, X.L. (Xiangyun Long); Writing—original draft, C.X.; Writing—review & editing, Z.C. All authors have read and agreed to the published version of the manuscript.

Funding: This research received no external funding.

Institutional Review Board Statement: Not applicable.

Informed Consent Statement: Not applicable.

Data Availability Statement: All data used during the study appear in this article.

Conflicts of Interest: The authors declare no conflict of interest.

References

1. Peng, C.; Zhou, Z.; Cheng, Z.; Fang, X. Upgrading of Light Cycle Oil to High-Octane Gasoline through Selective Hydrocracking over Non-Noble Metal Bifunctional Catalysts. *Energy Fuels* **2019**, *33*, 1090–1097. [CrossRef]
2. Peng, C.; Fang, X.C.; Zeng, R.H.; Guo, R.; Hao, W.Y. Commercial Analysis of Catalytic Hydroprocessing Technologies in Producing Diesel and Gasoline by Light Cycle Oil. *Catal. Today* **2016**, *276*, 11–18. [CrossRef]
3. Laredo, G.; Pérez-Romo, P.; Escobar, J.; Garcia-Gutierrez, J.L.; Vega-Merino, P.M. Light Cycle Oil Upgrading to Benzene, Toluene and Xylenes by Hydrocracking: Studies Using Model Mixtures. *Ind. Eng. Chem. Res.* **2017**, *56*, 10939–10948. [CrossRef]
4. Laredo, G.C.; Vega-Merino, P.M.; Hernández, P.S. Light Cycle Oil Upgrading to High Quality Fuels and Petrochemicals: A Review. *Ind. Eng. Chem. Res.* **2018**, *57*, 7315–7321. [CrossRef]
5. Calemma, V.; Giardino, R.; Ferrari, M. Upgrading of Lco by Partial Hydrogenation of Aromatics and Ring Opening of Naphthenes over Bi-Functional Catalysts. *Fuel Process. Technol.* **2010**, *91*, 770. [CrossRef]
6. Corma, A.; González-Alfaro, V.; Orchillés, A. Decalin and Tetralin as Probe Molecules for Cracking and Hydrotreating the Light Cycle Oil. *J. Catal.* **2001**, *200*, 34–44. [CrossRef]
7. Sato, K.; Iwata, Y.; Miki, Y.; Shimada, H. Hydrocracking of Tetralin over Niw/Usy Zeolite Catalysts: For the Improvement of Heavy-Oil Upgrading Catalysts. *J. Catal.* **1999**, *186*, 45–56. [CrossRef]
8. Upare, P.D.; Park, S.; Kim, M.S.; Kim, J.; Lee, D.; Lee, J.; Chang, H.; Choi, W.; Choi, S.; Jeon, Y.-P.; et al. Cobalt Promoted Mo/Beta Zeolite for Selective Hydrocracking of Tetralin and Pyrolysis Fuel Oil into Monocyclic Aromatic Hydrocarbons. *J. Ind. Eng. Chem.* **2016**, *35*, 99–107. [CrossRef]
9. Laredo, G.C.; Pérez-Romo, P.; Vega-Merino, P.M.; Arzate-Barbosa, E.; García-López, A.; Agueda-Rangel, R.; Martínez-Moreno, V.H. Effect of the Catalytic System and Operating Conditions on Btx Formation Using Tetralin as a Model Molecule. *Appl. Petrochem. Res.* **2019**, *9*, 185–198. [CrossRef]
10. Cao, Z.; Zhang, X.; Xu, C.; Huang, X.; Wu, Z.; Peng, C.; Duan, A. Selective Hydrocracking of Light Cycle Oil into High-Octane Gasoline over Bi-Functional Catalysts. *J. Energy Chem.* **2021**, *52*, 41–50. [CrossRef]
11. Park, J.I.; Lee, J.K.; Jin, M.; Kim, Y.K.; Yoon, S.H.; Mochida, I. Hydro-Conversion of 1-Methyl Naphthalene into (Alkyl)Benzenes over Alumina-Coated Usy Zeolite-Supported Nimos Catalysts. *Fuel* **2011**, *90*, 182–189. [CrossRef]
12. Shin, J.; Youngseok, O.; Yeseul, C.; Lee, J.; Kyoo, J. Design of Selective Hydrocracking Catalysts for Btx Production from Diesel-Boiling-Range Polycyclic Aromatic Hydrocarbons. *Appl. Catal. A* **2017**, *547*, 12–21. [CrossRef]
13. Escalona, G.; Rai, A.; Betancourt, P.; Sinha, A.K. Selective Poly-Aromatics Saturation and Ring Opening During Hydroprocessing of Light Cycle Oil over Sulfided Ni-Mo/SiO_2-Al_2O_3 Catalyst. *Fuel* **2018**, *219*, 270–278. [CrossRef]
14. Karakhanov, E.; Maximov, A.; Kardasheva, Y.; Vinnikova, M.; Kulikov, L. Hydrotreating of Light Cycle Oil over Supported on Porous Aromatic Framework Catalysts. *Catalysts* **2018**, *8*, 397. [CrossRef]
15. Singh, B.K.; Kim, Y.; Kwon, S.; Na, K. Synthesis of Mesoporous Zeolites and Their Opportunities in Heterogeneous Catalysis. *Catalysts* **2021**, *11*, 1541. [CrossRef]
16. Qiao, K.; Li, X.; Yang, Y.; Subhan, F.; Liu, X.; Yan, Z.; Xing, W.; Qin, L.; Dai, B.; Zhang, Z. Modification of Usy Zeolites with Malic-Nitric Acid for Hydrocracking. *Appl. Petrochem. Res.* **2016**, *6*, 353–359. [CrossRef]
17. Kazakov, M.O.; Nadeina, K.A.; Danilova, I.G.; Dik, P.P.; Klimov, O.V.; Pereyma, V.Y.; Paukshtis, E.A.; Golubev, I.S.; Prosvirin, I.P.; Gerasimov, E.Y.; et al. Influence of Usy Zeolite Recrystallization on Physicochemical Properties and Catalytic Performance of Nimo/Usy-Al_2O_3 Hydrocracking Catalysts. *Catal. Today* **2019**, *329*, 108–115. [CrossRef]
18. Nakajima, K.; Suganuma, S.; Tsuji, E.; Katada, N. Mechanism of Tetralin Conversion on Zeolites for the Production of Benzene Derivatives. *React. Chem. Eng.* **2020**, *5*, 1272–1280. [CrossRef]
19. Ren, S.; Meng, B.; Sui, X.; Duan, H.; Gao, X.; Zhang, H.; Zeng, P.; Guo, Q.; Shen, B. Preparation of Mesoporous Zeolite Y by Fluorine-Alkaline Treatment for Hydrocracking Reaction of Naphthalene. *Ind. Eng. Chem. Res.* **2019**, *58*, 7886–7891. [CrossRef]
20. Oh, Y.; Shin, J.; Noh, H.; Kim, C.; Kim, Y.S.; Lee, Y.K.; Lee, J.K. Selective Hydrotreating and Hydrocracking of Fcc Light Cycle Oil into High-Value Light Aromatic Hydrocarbons. *Appl. Catal. A* **2019**, *577*, 86–98. [CrossRef]
21. Jacobs, P.A.; Leuven, K.U. Acid Zeolites: An Attempt to Develop Unifying Concepts (P. H. Emmett Award Address, 1981). *Cat. Rev.—Sci. Eng.* **1982**, *24*, 415–440. [CrossRef]

22. Usman, M.; Li, D.; Razzaq, R.; Yaseen, M.; Li, C.; Zhang, S. Novel Mop/Hy Catalyst for the Selective Conversion of Naphthalene to Tetralin. *J. Ind. Eng. Chem.* **2015**, *23*, 21–26. [CrossRef]
23. Emeis, C.A. Determination of Integrated Molar Extinction Coefficients for Infrared Absorption Bands of Pyridine Adsorbed on Solid Acid Catalysts. *J. Catal.* **1993**, *141*, 347–354. [CrossRef]
24. Pires, J.; Fernandes, R.; Pinto, M.L.; Batista, M. Microporous Volumes from Nitrogen Adsorption at 77 K: When to Use a Different Standard Isotherm? *Catalysts* **2021**, *11*, 1544. [CrossRef]
25. Dik, P.P.; Danilova, I.G.; Golubev, I.S.; Kazakov, M.O.; Nadeina, K.A.; Budukva, S.V.; Pereyma, V.Y.; Klimov, O.V.; Prosvirin, I.P.; Gerasimov, E.Y. Hydrocracking of Vacuum Gas Oil over Nimo/Zeolite-Al_2O_3: Influence of Zeolite Properties. *Fuel* **2019**, *237*, 178–190. [CrossRef]
26. Ferraz, S.; Zotin, F.; Araujo, L.; Zotin, J.L. Influence of Support Acidity of Nimos Catalysts in the Activity for Hydrogenation and Hydrocracking of Tetralin. *Appl. Catal. A* **2010**, *384*, 51–57. [CrossRef]
27. Martinez, A.; Arribas, M.A.; Pergher, S.B. Bifunctional Noble Metal/Zeolite Catalysts for Upgrading Low-Quality Diesel Fractions via Selective Opening of Naphthenic Rings. *Catal. Sci. Technol.* **2016**, *6*, 2528–2542. [CrossRef]
28. Zhou, J.; Fan, W.; Wang, Y.; Xie, Z. The Essential Mass Transfer Step in Hierarchical/Nano Zeolite: Surface Diffusion. *Natl. Sci. Rev.* **2019**, *7*, 1630–1632. [CrossRef]
29. Contreras, M.S.; Bruin, T.D.; Mougin, P.; Toulhoat, H. Thermochemistry of 1-Methylnaphthalene Hydroconversion: Comparison of Group Contribution and Ab Initio Models. *Energy Fuels* **2013**, *27*, 5475–5482. [CrossRef]
30. Gutierrez, A.; Arandes, J.M.; Castano, P.; Aguayo, A.T.; Bilbao, J. Role of Acidity in the Deactivation and Steady Hydroconversion of Light Cycle Oil on Noble Metal Supported Catalysts. *Energy Fuels* **2011**, *25*, 3389–3399. [CrossRef]

catalysts

Article

Interventions to the Spontaneous Fabrication of Hierarchical ZSM-5 Zeolites by Fluorination-Alkaline Treatment

Zifeng Guo [1], Meihua Hong [1], Yonghua Yu [2,3], Guanfeng Liu [1,*], Jiazong Zang [1,*], Dazhi Zhang [2], Huimin Gong [2], Keyu Yang [2,3] and Shengjun Huang [2,*]

1 CenerTech Tianjin Chemical Research and Design Institute Co., Ltd., Tianjin 300131, China
2 Division of Fossil Energy Conversion, Dalian Institute of Chemical Physics, Chinese Academy of Sciences, Dalian 116023, China
3 University of Chinese Academy of Sciences, Beijing 100049, China
* Correspondence: liugf7@cnooc.com.cn (G.L.); zangjzh2@cnooc.com.cn (J.Z.); huangsj@dicp.ac.cn (S.H.)

Abstract: The sequential fluorination-alkaline treatment protocol has been applied for the tailoring of siliceous ZSM-5 zeolite. The original spontaneous growth of mesoporosity in alkaline medium is altered due to the antecedent fluorination step. The outcome is demonstrated by the apparent delay in the mesoporosity growth, whose essential duration for the well-defined mesoporosity is therefore extended from 30 min to 60 min. A low fluorination level decelerates the mesoporosity growth, whereas a high fluorination level enables the achievement of the mesoporosity. These impacts are closely linked with the alteration to the states of Al sites as the function of fluorination level. Compared to the states of Al sites in the pristine and steamed zeolites, the electronic and steric consequences on the environment of Al species by fluorination is proposed for the interplay with the alkaline medium for the mesoporosity growth.

Keywords: hierarchical zeolites; fluorination; siliceous ZSM-5; dealumination

Citation: Guo, Z.; Hong, M.; Yu, Y.; Liu, G.; Zang, J.; Zhang, D.; Gong, H.; Yang, K.; Huang, S. Interventions to the Spontaneous Fabrication of Hierarchical ZSM-5 Zeolites by Fluorination-Alkaline Treatment. *Catalysts* **2022**, *12*, 954. https://doi.org/10.3390/catal12090954

Academic Editors: Maja Milojević-Rakić and Danica Bajuk-Bogdanović

Received: 30 June 2022
Accepted: 17 August 2022
Published: 28 August 2022

1. Introduction

Zeolites are crystalline materials with well-defined microporosity, good hydrothermal stability, tunable compositions and functionalities. These features not only lead to their feasible applications in various industrial petrochemical processes including catalytic cracking, alkylation, aromatization, etc., but also render promising prospects in the productions of fine chemicals [1–4]. Stringent situation of feedstocks compositions for petrochemical processes and appearing versatile "platform" compounds from biomass resource bring forth more requirements for the zeolite-based catalysts: enhanced mass transfer efficiency, higher utilization efficiency of active sites and better resistance to deactivation [2,5]. Nevertheless, the pure micropore networks (<1.2 nm) of microporous zeolites may constrain their performances in these emerging fields [6,7]. Hierarchical zeolites, with the intrinsic micropore and a coupled intra/intercrystalline mesopore networks, have proved their potential in the emerging processes including synthesis of bulky fine chemicals, methanol conversions, pyrolysis and biomass conversions [8–12]. Therefore, synthesis of hierarchical zeolites attracts extensive attentions from basic research and industrial community [13–18].

Alkaline treatment has been proved as a cost-efficient top-down methodology for the construction of hierarchical porosity in various industrial-relevant zeolites with versatile topologies (MFI [19,20], FAU [21], MOR [22,23], BEA [24,25], etc.), which demands a precondition of optimal Si/Al range with 25–50. The underlying process for the achievement of the hierarchical porosities roots from a dynamic equilibrium between dissolution of matrix and reverse surface passivation in alkaline medium, which is regulated by the environments and transformation of Al sites inside zeolite [20,26]. However, in the case of microporous zeolites with starting optimal Si/Al ratios, the fabrication of hierarchical porosity has been displayed as a spontaneous process under the suitable treatment conditions. In other words, the fabrication of hierarchical porosities is confined to the limited alterations by the external parameters including temperature and concentration of alkaline solution [27]. The protocols beyond the external parameters are desirable given the resulting flexibility in porosity tailoring.

In this work, a fluorination-alkaline sequential treatment methodology is applied to the tailoring of MFI zeolite with optimal Si/Al ratio (Si/Al = 34). Accordingly, the initial spontaneous construction of hierarchical porosities is artificially intervened, which results in significant alterations to the achievement of hierarchical porosities. Compared to the usual alkaline treatment, the construction of hierarchical porosities is apparently delayed by the antecedent fluorination step. The resulting porosities still can be tailored and display the dependence on the antecedent fluorination step as the function of fluorination level. The implication of the antecedent fluorination step in the sequential alkaline treatment is proposed as the intervention to the electronic and steric of Al species, upon which their interplay with the alkaline medium is altered for the mesoporosity growth. Our results display the feasibility of beyond the spontaneous construction of hierarchical porosity upon alkaline treatment and therefore enrich the associated mechanistic understandings on the governing interplays for the mesoporosity growth.

2. Results and Discussion

The fluorination-desilication significantly alters the porosity development in the siliceous ZSM-5 zeolite. In line with the reported results, the spontaneous alkaline treatment conditions give rise to a combined type-I and type-IV with a hysteresis in the isotherm of AT(65,30)-Z5, which corresponds to a substantial contribution centered around 10 nm in the pore size distribution (PSD) curves. In contrast, mesoporosity developments in fluorinated Z5 samples are barely observed upon the alkaline treatment at 65 °C for 30 min. As shown in Figures 1 and 2, the isotherms of AT(65,30)-FZ5a and AT(65,30)-FZ5b only display subtle hysteresis loops, and the derived PSD curves evidence the suppressed contributions compared to that of AT(65,30)-Z5. Nevertheless, significant mesoporosities appear again upon longer treatment duration of 60 min. As shown in Figures 1 and 2, pronounced hysteresis loops are observed in the isotherms of AT(65,60)-FZ5a and AT(65,60)-FZ5b, corresponding to significant contributions in the derived PSD curves. The preceding fluorination level has substantial impacts on the final mesoporosity. The mesopore contribution of AT(65,60)-FZ5a shifts to a smaller dimension around 7 nm, whereas the PSD curve of AT(65,60)-FZ5b similarly concentrates on the typical dimension for AT(65,30)-Z5 around 10 nm. Apparently, the antecedent fluorination step is responsible for the alterations to mesoporosity development. Analogously, alteration to the mesopore dimension can be also observed for the sample upon a low-level fluorinated precursor (Figure 1b). The textural parameters derived from the N_2 adsorption-desorption isotherms are shown in Table 1. The low-level fluorination step (e.g., 0.12 wt% for FZ5a) dramatically enhances the mesoporosity development in the subsequent alkaline treatment. Accordingly, the mesoporous surface area (S_{meso}) and mesopore volume (V_{meso}) of AT(65,60)-FZ5a achieve 175 m^2/g and 0.29 cm^3/g, respectively, which are obviously higher than the parameters (S_{meso} 81 m^2/g and V_{meso} 0.18 cm^3/g) of AT(65,30)-Z5. Compared to the microporosities of pristine Z5 and AT(65,30)-Z5, the micropore surface area (S_{micro}, 362 m^2/g and 358 m^2/g) and micropore volume (V_{micro} 0.16 cm^3/g) decreased by approximately 23% and 25%, respectively, for AT(65,60)-FZ5a. The high-level fluorination step (0.72 wt% F) leads to serious reductions in microporosities (S_{micro} 332 m^2/g and V_{micro} 0.15 cm^3/g) for AT(65,60)-FZ5b, but brings about less pronounced enhancement in mesoporosities. The SEM and HRTEM images in Figure 3 show that the pristine zeolite has smooth and characteristic morphology. In comparison to the pristine zeolite with smooth surface, dense pore openings throughout the external surface were obtained by fluorination-alkaline treatment, while the cube-shape morphology is still well preserved (Figure 4a–d). The HRTEM images exhibit the alternately bright and dark contrast areas in Figure 5a–d. The light patches are associated with the presence of mesopores in the obtained hierarchical zeolites. Physical adsorption, SEM and

HRTEM observations confirm the intracrystalline nature of the created mesopores by the fluorination-alkaline treatment.

Table 1. Textural properties of pristine and fluorination-alkaline treated ZSM-5 samples.

Sample	*Yield* [a] [%]	S_{BET} [b] [m^2g^{-1}]	S_{micro} [c] [m^2g^{-1}]	S_{meso} [d] [m^2g^{-1}]	V_{micro} [d] [cm^3g^{-1}]	V_{meso} [e] [cm^3g^{-1}]	V_{pore} [f] [cm^3g^{-1}]
Z5	-	377	362	15	0.16	0.06	0.22
AT(65,30)	65	439	358	81	0.16	0.18	0.34
AT(65,60)-*a*FZ5	72	452	277	175	0.12	0.29	0.41
AT(65,60)-*b*FZ5	84	440	332	108	0.15	0.25	0.40
AT(80,30)	42	446	221	225	0.10	0.34	0.44
AT(80,30)-*a*FZ5	66	475	280	195	0.12	0.35	0.47
AT(80,30)-*b*FZ5	79	445	324	121	0.15	0.29	0.44

[a] Grams of solid after treatment per gram of starting material. [b] BET method. [c] S_{BET}-S_{meso}. [d] *t*-plot method. [e] V_{pore}-V_{micro}. [f] Volume of N_2 adsorbed at $P/P_0 = 0.99$.

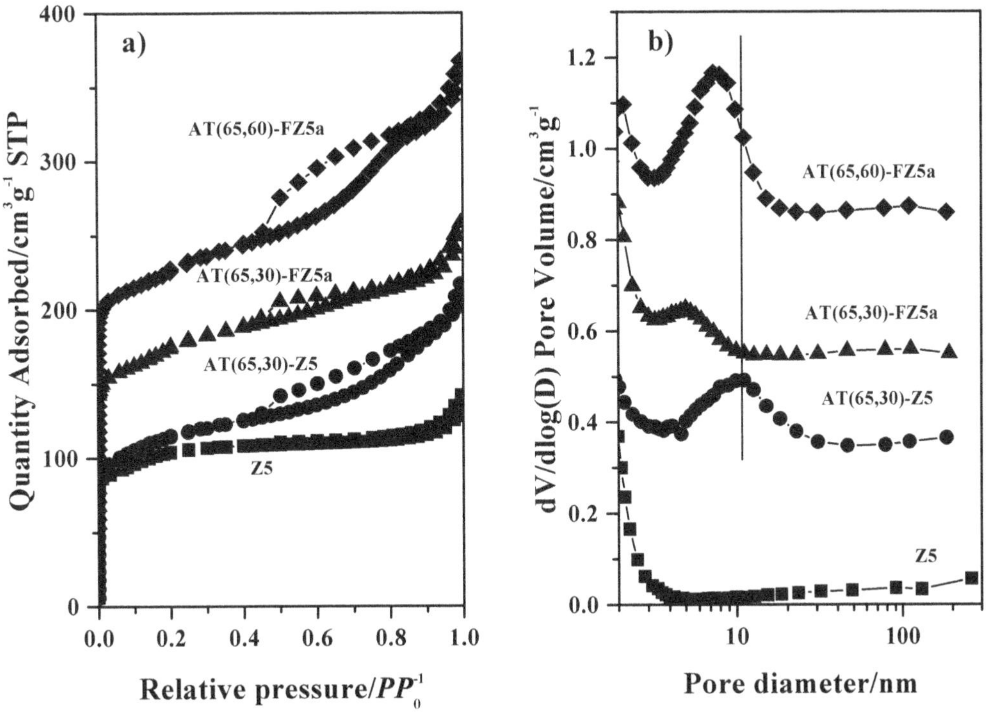

Figure 1. (**a**) N_2 adsorption-desorption isotherms of treated samples. (**b**) BJH pore size distribution derived from the adsorption branches of the isotherms. The isotherms for AT(65,30)-FZ5a and AT(65,60)-FZ5a were offset vertically by 50 and 100 cm^3g^{-1}, respectively. (**b**) Barrett–Joyner–Halenda (BJH) pore size distribution (PSD) data derived from the adsorption branches of the isotherms. The PSD curves of AT(65,30)-Z5, AT(65,30)-FZ5a and AT(65,60)-FZ5a were offset vertically by 0.3, 0.5 and 0.8 cm^3g^{-1}, respectively.

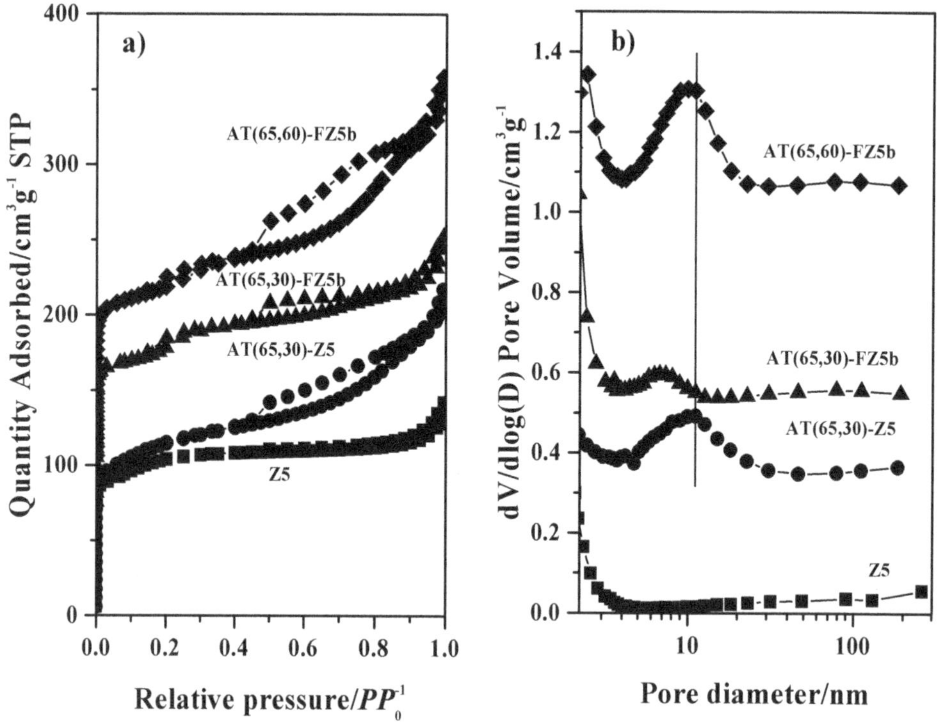

Figure 2. (**a**) N_2 adsorption-desorption isotherms of treated samples at 65 °C. (**b**) BJH pore size distribution derived from the adsorption branches of the isotherms. The isotherms for AT(65,30)-FZ5b and AT(65,60)-FZ5b were offset vertically by 70 and 100 cm^3g^{-1}, respectively. (**b**) Barrett–Joyner– Halenda (BJH) pore size distribution (PSD) data derived from the adsorption branches of the isotherms. The PSD curves of AT(65,30)-Z5,AT(65,30)-FZ5b and AT(65,60)-FZ5b were offset vertically by 0.3, 0.5 and 1.0 cm^3g^{-1}, respectively.

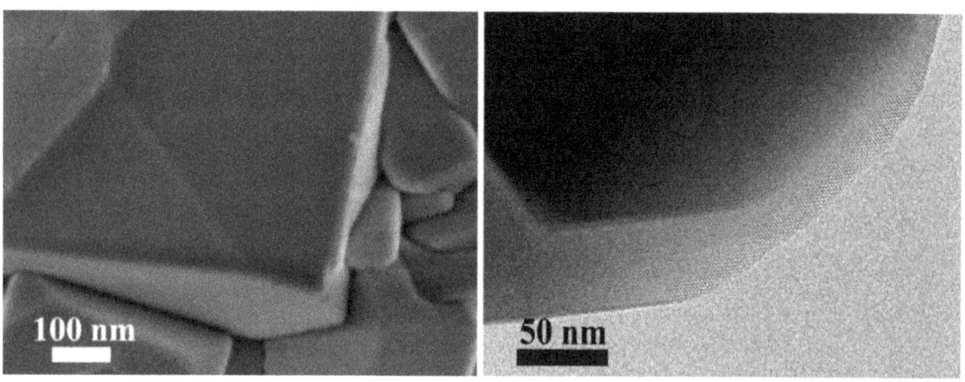

Figure 3. FESEM and HR-TEM micrographs of pristine ZSM-5 zeolite.

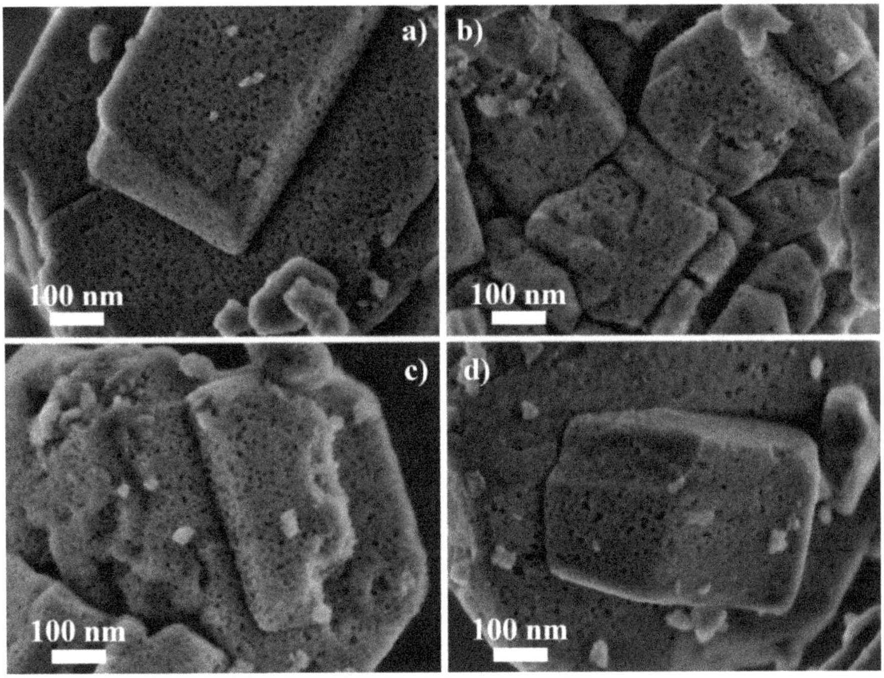

Figure 4. FESEM micrographs of fluorination-alkaline treated samples. (**a**) AT(65,60)-FZ5a, (**b**) AT(65,60)-FZ5b, (**c**) AT(80,30)-FZ5a, (**d**) AT(80,30)-FZ5b.

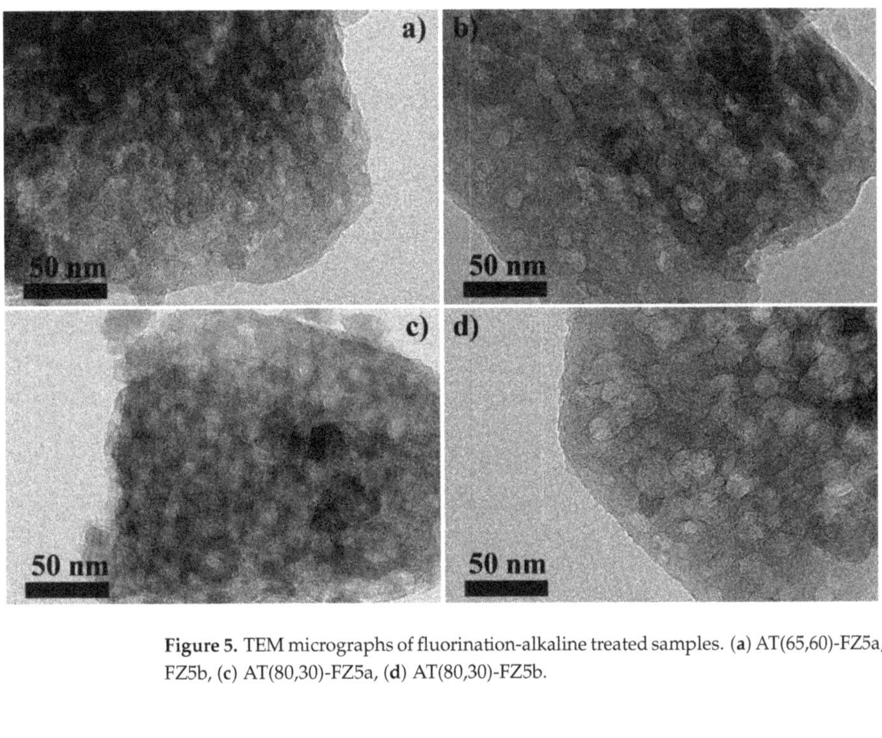

Figure 5. TEM micrographs of fluorination-alkaline treated samples. (**a**) AT(65,60)-FZ5a, (**b**) AT(65,60)-FZ5b, (**c**) AT(80,30)-FZ5a, (**d**) AT(80,30)-FZ5b.

The impact of fluorination step on the state of Al sites is reflected by the ^{27}Al MAS NMR spectroscopy. As shown in Figure 6, the pristine ZSM-5 exhibits a significant resonance at 55 ppm, which is assigned to the four-coordinated framework Al sites.

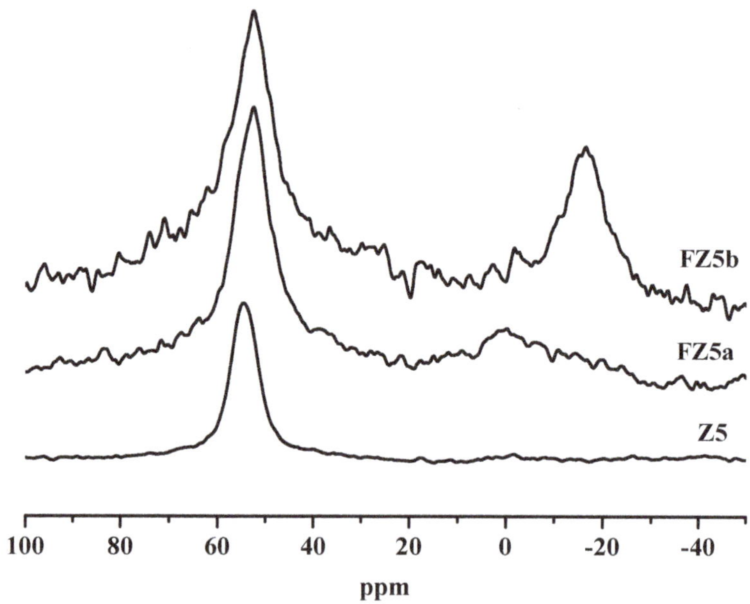

Figure 6. ^{27}Al MAS NMR of pristine ZSM-5 zeolite and fluorinated zeolites.

Upon fluorination treatment, a shift of resonance from 55 ppm to 54 ppm is observed for the fluorinated FZ5a and FZ5b, reflecting the interaction/perturbation of fluoride with the framework Al sites. At low level fluorination, the resonance at 0 ppm appears, indicating the parallel dealumination of framework Al sites to non-framework Al species. However, the high fluorination level leads to the reduction of the resonance at 0 ppm and a significant resonance at −16 ppm, the latter of which is assigned to the AlF_3 phases [28]. Therefore, the fluorination of siliceous ZSM-5 zeolites leads to various alterations to the framework sites as the function of fluorination level: perturbation of the environment of framework Al sites due to the Al-F complexation, dealumination to the octa-coordinated non-framework Al species and formation of the AlF_3 phase. The alterations, in particular the latter two different modes of dislodge of framework Al sites, intervene the sequential development of mesoporosity according to their interplay with the alkaline treatment conditions. Upon the low fluorination level, the generated non-framework Al sites are responsible for poor mesoporosity development in the sequential alkaline treatment at 65 °C for 30 min. It is interesting to note that such a decelerating effect is overcome by the extension of alkaline treatment duration to 60 min as reflected by the appearance of significant mesoporosity in the BJH profile of AT(65,60)-FZ5a. The similar delayed mesoporosity growth is also observed for the alkaline treatment of FZ5b along the transformation to the AlF_3 phase at the higher fluorination level, in which the significant mesoporosity is observed until the longer treatment duration of 60 min. It is worthy of noting that the "decelerating effect" or "delayed effect" imposed by fluorination is intrinsically different to the "inhibitive effect" imposed by the usual dealumination of steaming. The antecedent dealumination of steaming of siliceous ZSM-5 zeolites (Si/Al = 34) has been reported to disable the sequential mesoporosity development by alkaline treatment, which was assigned to the inhibitive effect by the non-framework Al species due to the steaming treatment. Moreover, our comparative steaming-alkaline treatment experiments prove that such inhibitive effects cannot be overcome by the extension of alkaline treatment duration. The corresponding PSD profile in Figure 7 provides evidence of the absence of mesoporosity development in AT(65,60)-ST-Z5 by sequential steaming-alkaline treatment despite the extended alkaline treatment duration of 60 min.

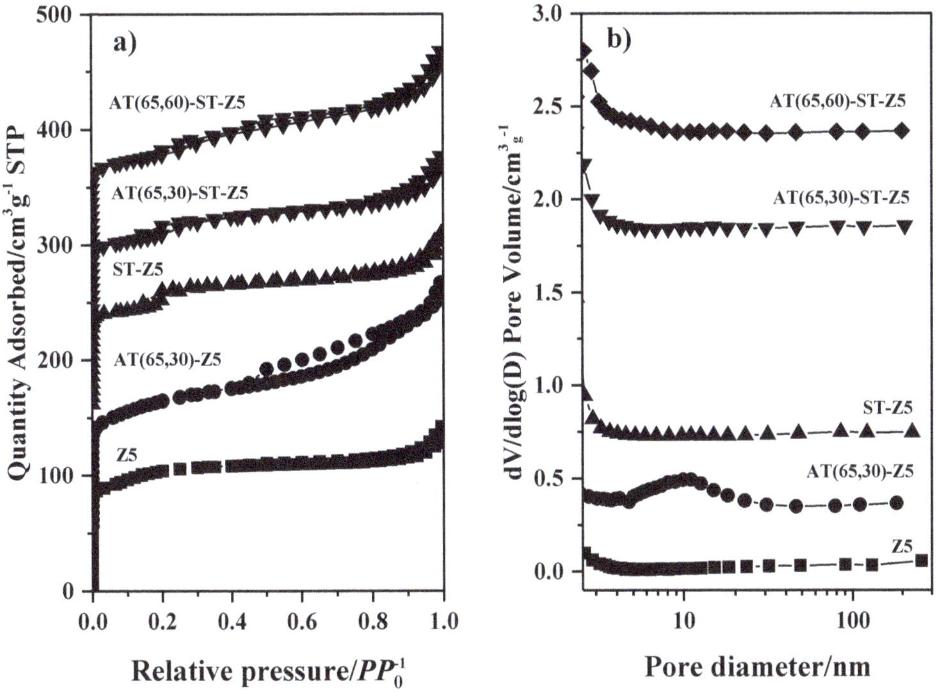

Figure 7. (**a**) N_2 adsorption-desorption isotherms of steaming-alkaline treated samples. (**b**) BJH pore size distribution derived from the adsorption branches of the isotherms. The isotherms for AT(65,30)-Z5, ST-Z5, AT(65,30)-ST-Z5 and AT(65,60)-ST-Z5 were offset vertically by 50, 150, 220 and 270 cm^3g^{-1}, respectively. (**b**) Barrett–Joyner–Halenda (BJH) pore size distribution (PSD) data derived from the adsorption branches of the isotherms. The PSD curves of AT(65,30)-Z5, ST-Z5, AT(65,30)-ST-Z5 and AT(65,60)-ST-Z5 were offset vertically by 0.3, 0.7, 1.8 and 2.3 cm^3g^{-1}, respectively.

Therefore, the above results indicate that the mode of mesoporosity growth can be tailored to an intermediate mode between spontaneous manner and complete inhibitive status. The underlying process of spontaneous manner relies on the framework Al-directing dissolution of microporous zeolite matrix, which is governed by the dynamic transformation of framework Al sites [20]. The apparent outcome of such spontaneous manner is shown as the requirement on an optimal Si/Al ratio range from 25 to 50 for the mesoporosity development [20]. However, the presence of non-framework Al species passivates the mesoporosity development despite their optimal overall Si/Al ratios in this case and other reports [29]. As the intermediate mode, the fluorination step introduces a different response to the alkaline medium, which has close association with the progress of mesoporosity. Based on the ^{27}Al MAS NMR results, the comparison among the non-framework Al species and the fluorinated Al species are depicted in Scheme 1. Upon the mild fluorination level for FZ5a, the perturbation to the $[AlO_4]$ units is proposed to result in the formation of F-bearing tetrahedral Al sites. The high-level fluorination leads to the formation of aluminum fluoride phases including the ideal phases and their imperfect counterparts [28]. Such structural alterations induce a different response to the alkaline medium due to the imposed electronic and steric consequences. The link with the origin of the inhibitive effect of the non-framework species enlightens the understanding of the decelerating effects by the fluorinated phases. As an ideal structural model, the non-framework Al species is characteristic of the unit with an octahedral coordination [30]. Compared to the tetrahedral framework Al sites (illustration a vs. illustration a1 in Scheme 1), the octahedral non-framework Al unit bears a higher density of negative charge density. From the viewpoint of the charge distribution, the creation of such ideal octahedral units assembles the negative charge, and therefore creates harder bastions to the alkaline ions, which are proposed to account for the inhibitive effect for mesoporosity growth due to the electron repulsive between negative charged Al-bearing species and hydroxides ($[OH]^-$). Given the electronically neutral character of the overall zeolite, such impacts refer to the redistribution of negative charges within the zeolite framework. At least, the fluorination step introduces the electronic and steric impacts on the framework and non-framework Al species. As shown in Scheme 1, the fluorine ligand enhances the charge density of the tetrahedral unit due to its smaller radius, which inhibits the interplay with hydroxides. On the other hand, the introduction of the fluorine ligand reduces the symmetry of the tetrahedron and therefore enhances the interplay with the alkaline media. In the case of octahedral state, the introduction of fluorine ligand reduces the negative charge's density along with reducing the size of the unit. The balance of these opposite factors decides the overall interplay with the alkaline hydroxides and the corresponding growth of mesoporosity. Upon the low level of fluorination, the enhanced charge effect overtakes that of the reduced symmetry, which therefore decelerates the mesoporosity. As the comparison with pure O-coordinated non-framework Al species upon steaming treatment, the F-coordinated non-framework Al species ease the strong repulsion to the hydroxides, which enables the mesoporosity growth inside the fluorinated zeolites. Compared to the inhibitive effect by the steaming step, the fluorination step brings the enabling impact for the generation of mesoporosity despite its apparent longer essential duration. It is interesting to note that the essential duration can be eliminated by the harsh alkaline treatment conditions. The enhanced alkaline treatment temperature accelerates the mesoporosity development. Upon alkaline treatment at 80 °C, the pronounced hysteresis loops and significant contributions are observed upon the duration of 30 min as shown by the N_2 adsorption-desorption isotherms and the derived PSD curves of AT(80,30)-*a*FZ5 and AT(80,30)-*b*FZ5, respectively (Figure 8). Meanwhile, upon the harsh alkaline treatment, the fluorination step fully demonstrates its beneficial impacts on the achievement of significant well-defined mesoporosity. Therefore, this suggests the alteration of the interplay between fluorine-bearing species and alkaline medium as the function of fluorination level and treatment conditions.

The sequential fluorination-alkaline treatment protocol has been successfully designed for the creation of hierarchical porosities inside Al-rich ZSM-5 zeolites (Si/Al < 15). In the case of the Al-rich ZSM-5 zeolites, the critical roles of fluorination have been owed to the F-Al complexation effects, which are responsible for the alleviation of initial strong resistance of high-density Al sites to the alkaline medium and the regulated hydrolysis process to sustain the growth of mesoporosity. The application of fluorination-alkaline treatment protocol to the siliceous ZSM-5 uncovers a different side of the fluorination step which cannot be observed in the case of Al-rich ZSM-5 counterpart.

The interplays between non-framework Al species and mesoporosity development still demand fundamental insights, which, however, are confronted with multiple and long-term challenges including the elaboration of the precise structure of non-framework Al species and the mechanistic understandings on the working modes [30]. It should be noted that the schematic descriptions in Scheme 1 are simplified ideal models, and their amending to the complexed non-framework Al species requires further efforts. Our attempt is based on our accumulation on this topic for years, and is beneficial to stimulate more discussion and insights [23,28,31].

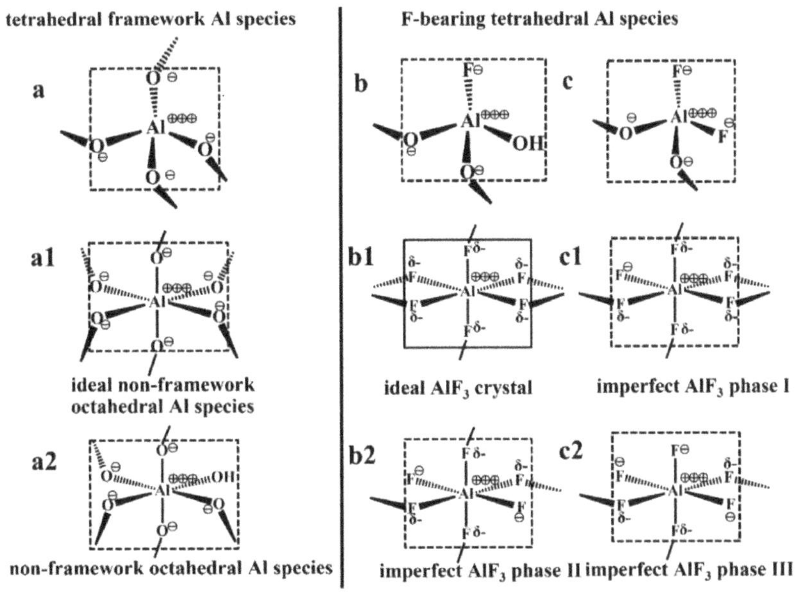

Scheme 1. The Schematic depiction of the comparison among the non-framework Al species (**a**) and the fluorinated Al species (**b**,**c**). ⊖⊕: the separated valence state to the schematic unit; δ^-: the shared valence state to the schematic unit.

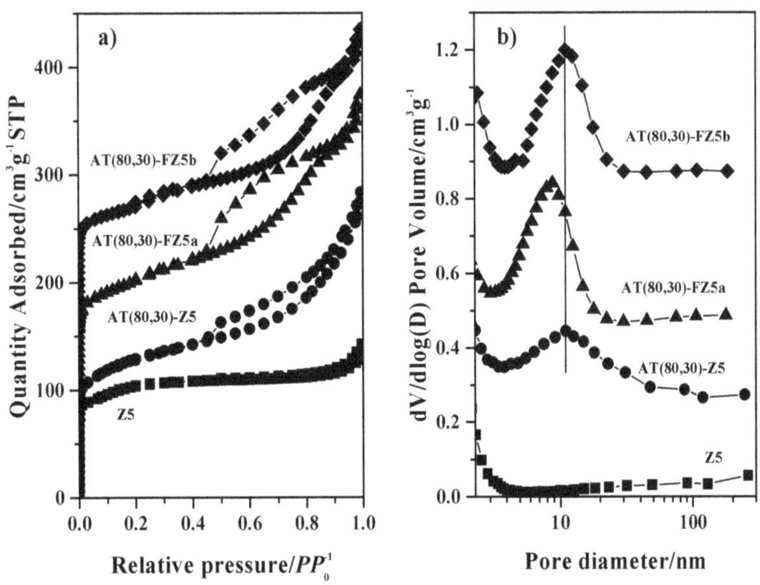

Figure 8. (**a**) N_2 adsorption-desorption isotherms of sequential fluorination-alkaline treated samples. The alkaline treatment temperature is 80 °C. (**b**) BJH pore size distribution derived from the adsorption branches of the isotherms. The isotherms for AT(80,30)-FZ5a and AT(80,30)-FZ5b were offset vertically by 70 and 150 cm^3g^{-1}, respectively. (**b**) Barrett–Joyner–Halenda (BJH) pore size distribution (PSD) data derived from the adsorption branches of the isotherms. The PSD curves of AT(80,30)-Z5, AT(80,30)-FZ5a and AT(80,30)-FZ5b were offset vertically by 0.2, 0.4 and 0.8 cm^3g^{-1}, respectively.

3. Experimental Section

3.1. Sample Preparation

Sequential fluorination-desilication: The commercial ZSM-5 zeolite (Z5, Si/$Al_{nominal}$ = 34, HuangMa Chemical Co., Ltd., Nanjing, China) was used as the starting material. The pristine ZSM-5 was firstly impregnated with the aqueous NH_4F solution according to the impregnation method. Typically, ZSM-5 powder was stirred in the NH_4F solution (0.32 M and 0.60 M, weight/volume ratio for zeolite to solution is 1:3) at room temperature for 8 h. The impregnated samples were dried at 120 °C overnight, and calcined at 550 °C for 3 h. The obtained sample is coded as FZ5a and FZ5b in which *a* (0.12 wt%) and *b* (0.23 wt%) corresponds to the F loading on Z5, respectively. Then, the fluorinated sample was then alkaline treated in a polypropylene flask by an aqueous NaOH solution (0.2 M) at 65 °C or 80 °C for 30 min or 60 min. The alkaline-treated samples were collected by centrifugation (4000 rpm for 5 min), thoroughly washed and dried at 120 °C overnight. Finally, these samples were transformed to protonic-form by a threefold ammonium-exchange (NH_4NO_3 solutions, 0.8 M) and subsequent calcination at 550 °C for 3 h. The final H-form zeolites are designated as AT(x, y)-FZ5a and AT(x, y)-FZ5b, where x and y represent the alkaline treatment temperature (°C) and alkaline treatment time (minutes), respectively.

Comparative spontaneous alkaline treatment: The spontaneous alkaline treatment of the same pristine ZSM-5 zeolite was conducted as the same procedures as described for the alkaline treatment of fluorinated zeolites. After the same ammonia-exchange and calcination steps, the obtained zeolite is designated as AT(x, y)-Z5,where x and y represent the alkaline treatment temperature (°C) and alkaline treatment time (minutes), respectively.

Comparative steaming-alkaline treatment: Pristine ZSM-5 zeolite powder was firstly pressed into pellets under 3.0 MPa for 1 min, and then crushed into granules of around 5 mm to minimize the pressure-drop during the steaming process. Approximately 10 g of zeolite granules were loaded in a fixed quartz reactor and heated to 600 °C with a ramp rate of 5.6 °C min^{-1} in dry air flow under ambient pressure. Afterwards, water vapor (0.8 mL min^{-1}) was charged and the dry air flow was stopped. After 3 h of steaming treatment, the water injection was stopped, and the steamed zeolite was cooled and dried under air flow. The obtained zeolites are denoted as ST-Z5. These steamed samples were subsequently subjected to the same alkaline treatment as the above same procedures. After the same ammonia-exchange and calcination steps, the obtained zeolite is designated as AT(60, y)-ST-Z5,where y represents the alkaline duration in minutes.

3.2. Characterization

Nitrogen adsorption measurements were carried out at −196 °C on a Micromeritics ASAP-2020 analyzer. Prior to analysis, each sample was evacuated at 350 °C for 10 h. Field emission scanning electron microscope (FESEM) measurement was performed with JEOL JSM-7800F at an accelerating voltage of 5 kV. High resolution transmission electron microscopy (HRTEM) measurement was performed with a JEOL JEM-2100 at an accelerating voltage of 120 KV. Solid NMR spectra were recorded on Bruker AVANCE III 500 MHz spectrometer equipped with a 4 mm MAS probe. ^{27}Al MAS NMR spectra were recorded at a resonance frequency of 130.3 MHz with a spinning rate of 12 KHz. Chemical shifts were referenced to 1 M of $Al(NO_3)_3$ at 0 ppm as a reference. The spectra were accumulated for 1000 scans with $\pi/12$ flip angle and a 2 s pulse delay.

4. Conclusions

The sequential fluorination-alkaline treatment is committed to the tailoring of the siliceous ZSM-5 zeolite (Si/Al = 34). The spontaneous growth of mesoporosity inside siliceous ZSM-5 zeolite is thereby intervened by the antecedent fluorination step. As a consequence, the essential duration for the achievement of well-defined mesoporosity is extended from 30 min for pristine zeolite to 60 min for the fluorinated zeolite. The fluorination leads to the various alterations to the framework sites as the function of fluorination level: perturbation of the environment of framework Al sites due to the Al-

F complexation, dealumination to the octa-coordinated non-framework Al species and formation of octa-coordinated aluminum fluoride phases. At least, the introduction of fluorine-ligand gives simultaneous electronic and steric impacts on the interplay between fluorinated zeolite and alkaline medium. Upon the low fluorination level, the charge effect overtakes that steric effect, which intensifies the electronic repulsion to the alkaline medium and therefore decelerates the mesoporosity growth. On the other hand, the steric effect overtakes the charge effect at the high fluorination level for F-coordinated non-framework Al species and eases the strong repulsion to the hydroxides. Compared to the inhibitive effect imposed by steaming, high level fluorination brings an enabling effect for mesoporosity growth. The impacts of fluorination-alkaline treatment on the mesoporosity growth of siliceous zeolite uncover more sides of the interplay between framework/non-framework Al species and alkaline medium, and provide insights into the understanding of the tailoring of porosities.

Author Contributions: Conceptualization, G.L., J.Z. and S.H.; Methodology, Z.G., M.H. and D.Z.; Experiments, Z.G., M.H., Y.Y., H.G. and K.Y.; Project administration, G.L. and J.Z., Writing—original draft preparation, Z.G., M.H. and Y.Y.; Writing—review and editing, Z.G., Y.Y., K.Y. and S.H.; Supervision and funding acquisition, S.H. All authors have read and agreed to the published version of the manuscript.

Funding: Dalian Bureau of Science and Technology (No. 2020JJ25CY007) and National Natural Science Foundation of China (No. 21978282).

Data Availability Statement: All the data generated or analyzed within the present investigation are included in this manuscript.

Acknowledgments: This work was supported by CenerTech Tianjin Chemical Research and Design Institute (Co., Ltd.), Dalian Bureau of Science and Technology (No. 2020JJ25CY007) and National Natural Science Foundation of China (No. 21978282).

Conflicts of Interest: The authors declare no conflict of interest in publishing the results.

References

1. Vermeiren, W.; Gilson, J.-P. Impact of Zeolites on the Petroleum and Petrochemical Industry. *Top. Catal.* **2009**, *52*, 1131–1161. [CrossRef]
2. Primo, A.; Garcia, H. Zeolites as catalysts in oil refining. *Chem. Soc. Rev.* **2014**, *43*, 7548–7561. [CrossRef] [PubMed]
3. Choudary, N.V.; Newalkar, B.L. Use of zeolites in petroleum refining and petrochemical processes: Recent advances. *J. Porous Mater.* **2011**, *18*, 685–692. [CrossRef]
4. Dai, W.J.; Zhang, L.N.; Liu, R.Z.; Wu, G.J.; Guan, N.J.; Li, L.D. Plate-Like ZSM-5 Zeolites as Robust Catalysts for the Cracking of Hydrocarbons. *ACS Appl. Mater. Interfaces* **2022**, *14*, 11415–11424. [CrossRef] [PubMed]
5. Jacobs, P.A.; Dusselier, M.; Sels, B.F. Will Zeolite-Based Catalysis be as Relevant in Future Biorefineries as in Crude Oil Refineries? *Angew. Chem. Int. Ed.* **2014**, *53*, 8621–8626. [CrossRef]
6. Davis, M.E. Ordered porous materials for emerging applications. *Nature* **2002**, *417*, 813–821. [CrossRef]
7. Perego, C.; Millini, R. Porous materials in catalysis: Challenges for mesoporous materials. *Chem. Soc. Rev.* **2013**, *42*, 3956–3976. [CrossRef]
8. Nuttens, N.; Verboekend, D.; Deneyer, A.; Van Aelst, J.; Sels, B.F. Potential of Sustainable Hierarchical Zeolites in the Valorization of α-Pinene. *Chemsuschem* **2015**, *8*, 1197–1205. [CrossRef]
9. Dapsens, P.Y.; Mondelli, C.; Pérez-Ramírez, J. Biobased Chemicals from Conception toward Industrial Reality: Lessons Learned and To Be Learned. *ACS Catal.* **2012**, *2*, 1487–1499. [CrossRef]
10. Smith, K.; El-Hiti, G.A. Use of zeolites for greener and more *para*-selective electrophilic aromatic substitution reactions. *Green Chem.* **2011**, *13*, 1579–1608. [CrossRef]
11. Liu, Y.; Zhang, Q.; Li, J.; Wang, X.; Terasaki, O.; Xu, J.; Yu, J. Protozeolite-Seeded Synthesis of Single-Crystalline Hierarchical Zeolites with Facet-Shaped Mesopores and Their Catalytic Application in Methanol-to-Propylene Conversion. *Angew. Chem. Int. Ed.* **2022**, *134*, e202205716. [CrossRef]
12. Mardiana, S.; Azhari, N.J.; Ilmi, T.; Kadja, G.T.M. Hierarchical zeolite for biomass conversion to biofuel: A review. *Fuel* **2022**, *309*, 122119. [CrossRef]
13. Verboekend, D.; Pérez-Ramírez, J. Design of hierarchical zeolite catalysts by desilication. *Catal. Sci. Technol.* **2011**, *1*, 879–890. [CrossRef]

14. Pérez-Ramírez, J.; Mitchell, S.; Verboekend, D.; Milina, M.; Michels, N.-L.; Krumeich, F.; Marti, N.; Erdmann, M. Expanding the Horizons of Hierarchical Zeolites: Beyond Laboratory Curiosity towards Industrial Realization. *ChemCatChem* **2011**, *3*, 1731–1734. [CrossRef]
15. Kerstens, D.; Smeyers, B.; Waeyenberg, J.V.; Zhang, Q.; Yu, J.; Sels, B.F. State of the Art and Perspectives of Hierarchical Zeolites: Practical Overview of Synthesis Methods and Use in Catalysis. *Adv. Mater.* **2020**, *32*, 2004690. [CrossRef]
16. Fernandez, S.; Ostraat, M.L.; Zhang, K. Toward rational design of hierarchical beta zeolites: An overview and beyond. *AIChE J.* **2020**, *66*, e16943. [CrossRef]
17. Wang, Z.; Zhang, R.; Wang, J.; Yu, Z.; Xiang, Y.; Kong, L.; Liu, H.; Ma, A. Hierarchical zeolites obtained by alkaline treatment for enhanced n-pentane catalytic cracking. *Fuel* **2022**, *313*, 122669. [CrossRef]
18. Abdulridha, S.; Zhang, R.; Xu, S.; Tedstone, A.; Ou, X.; Gong, J.; Mao, B.; Frogley, M.; Bawn, C.; Zhou, Z.; et al. An efficient microwave-assisted chelation (MWAC) post-synthetic modification method to produce hierarchical Y zeolites. *Microporous Mesoporous Mater.* **2021**, *311*, 110715. [CrossRef]
19. Ogura, M.; Shinomiya, S.-Y.; Tateno, J.; Nara, Y.; Nomura, M.; Kikuchi, E.; Matsukata, M. Alkali-treatment technique—New method for modification of structural and acid-catalytic properties of ZSM-5 zeolites. *Appl. Catal. A Gen.* **2001**, *219*, 33–43. [CrossRef]
20. Groen, J.C.; Peffer, L.A.A.; Moulijn, J.A.; Pérez-Ramírez, J. Mechanism of Hierarchical Porosity Development in MFI Zeolites by Desilication: The Role of Aluminium as a Pore-Directing Agent. *Chem. Eur. J.* **2005**, *11*, 4983–4994. [CrossRef]
21. de Jong, K.P.; Zecevic, J.; Friedrich, H.; de Jongh, P.E.; Bulut, M.; van Donk, S.; Kenmogne, R.; Finiels, A.; Hulea, V.; Fajula, F. Zeolite Y Crystals with Trimodal Porosity as Ideal Hydrocracking Catalysts. *Angew. Chem. Int. Ed.* **2010**, *49*, 10074–10078. [CrossRef] [PubMed]
22. van Laak, A.N.C.; Sagala, S.L.; Zecevic, J.; Friedrich, H.; de Jongh, P.E.; de Jong, K.P. Mesoporous mordenites obtained by sequential acid and alkaline treatments—Catalysts for cumene production with enhanced accessibility. *J. Catal.* **2010**, *276*, 170–180. [CrossRef]
23. Huang, S.; Liu, X.; Yu, L.; Miao, S.; Liu, Z.; Zhang, S.; Xie, S.; Xu, L. Preparation of hierarchical mordenite zeolites by sequential steaming-acid leaching-alkaline treatment. *Microporous Mesoporous Mater.* **2014**, *191*, 18–26. [CrossRef]
24. Sammoury, H.; Toufaily, J.; Cherry, K.; Hamieh, T.; Pouilloux, Y.; Pinard, L. Impact of desilication of * BEA zeolites on the catalytic performance in hydroisomerization of n-C_{10}. *Appl. Catal. A Gen.* **2018**, *551*, 1–12. [CrossRef]
25. van Laak, A.N.C.; Zhang, L.; Parvulescu, A.N.; Bruijnincx, P.C.A.; Weckhuysen, B.M.; de Jong, K.P.; de Jongh, P.E. Alkaline treatment of template containing zeolites: Introducing mesoporosity while preserving acidity. *Catal. Today* **2011**, *168*, 48–56. [CrossRef]
26. Zhang, D.; Jin, C.; Zou, M.; Huang, S. Mesopore Engineering for Well-Defined Mesoporosity in Al-Rich Aluminosilicate Zeolites. *Chem. Eur. J.* **2019**, *25*, 2675–2683. [CrossRef]
27. Groen, J.C.; Moulijn, J.A.; Pérez-Ramírez, J. Alkaline Posttreatment of MFI Zeolites. From Accelerated Screening to Scale-up. *Ind. Eng. Chem. Res.* **2007**, *46*, 4193–4201. [CrossRef]
28. Yu, L.; Xie, S.; Huang, S.; Xu, L. A Facile Top-Down Protocol for Postsynthesis Modification of Hierarchical Aluminum-Rich MFI Zeolites. *Chem. Eur. J.* **2015**, *21*, 1048–1054. [CrossRef]
29. Groen, J.C.; Moulijn, J.A.; Pérez-Ramírez, J. Decoupling mesoporosity formation and acidity modification in ZSM-5 zeolites by sequential desilication-dealumination. *Microporous Mesoporous Mater.* **2005**, *87*, 153–161. [CrossRef]
30. Silaghi, M.C.; Chizallet, C.; Raybaud, P. Challenges on molecular aspects of dealumination and desilication of zeolites. *Microporous Mesoporous Mater.* **2014**, *191*, 82–96. [CrossRef]
31. Yang, S.; Yu, C.; Yu, L.; Miao, S.; Zou, M.; Jin, C.; Zhang, D.; Xu, L.; Huang, S. Bridging Dealumination and Desilication for the Synthesis of Hierarchical MFI Zeolites. *Angew. Chem. Int. Ed.* **2017**, *56*, 12553–12556. [CrossRef] [PubMed]

catalysts MDPI

Review

A Comparative Review of Binder-Containing Extrusion and Alternative Shaping Techniques for Structuring of Zeolites into Different Geometrical Bodies

Zahra Asgar Pour [1,*], Marwan M. Abduljawad [2], Yasser A. Alassmy [2], Ludwig Cardon [3], Paul H. M. Van Steenberge [4] and Khaled O. Sebakhy [4,*]

1 Department of Chemical Engineering, Engineering and Technology Institute Groningen (ENTEG), University of Groningen, Nijenborgh 4, 9747AG Groningen, The Netherlands
2 Refining and Petrochemicals Technologies Institute, Center of Excellence for Petrochemicals (Oxford), King Abdulaziz City for Science and Technology (KACST), Riyadh 11442, Saudi Arabia
3 Centre for Polymer and Material Technologies (CPMT), Department of Materials, Textiles and Chemical Engineering, Ghent University, Technologiepark 130 (zone c3), 9052 Ghent, Belgium
4 Centre for Polymer and Material Technologies (CPMT), Laboratory for Chemical Technology (LCT), Department of Materials, Textiles and Chemical Engineering, Ghent University, Technologiepark 125, 9052 Ghent, Belgium
* Correspondence: z.asgar.pour@rug.nl (Z.A.P.); khaled.sebakhy@ugent.be (K.O.S.); Tel.: +31-(0)6-2992-8542 (K.O.S.)

Abstract: Zeolites are crystalline metallosilicates displaying unique physicochemical properties with widespread applications in catalysis, adsorption, and separation. They are generally obtained by a multi-step process that starts with primary mixture aging, followed by hydrothermal crystallization, washing, drying, and, finally, a calcination step. However, the zeolites obtained are in the powder form and because of generating a pressure drop in industrial fixed bed reactors, not applicable for industrial purposes. To overcome such drawbacks, zeolites are shaped into appropriate geometries and desired size (a few centimeters) using extrusion, where zeolite powders are mixed with binders (e.g., mineral clays or inorganic oxides). The presence of binders provides good mechanical strength against crushing in shaped zeolites, but binders may have adverse impacts on zeolite catalytic and sorption properties, such as active site dilution and pore blockage. The latter is more pronounced when the binder has a smaller particle size, which makes the zeolite internal active sites mainly inaccessible. In addition to the shaping requirements, a hierarchical structure with different levels of porosity (micro-, meso-, and macropores) and an interconnected network are essential to decrease the diffusion limitation inside the zeolite micropores as well as to increase the mass transfer because of the presence of larger auxiliary pores. Thus, the generation of hierarchical structure and its preservation during the shaping step is of great importance. The aim of this review is to provide a comprehensive survey and detailed overview on the binder-containing extrusion technique compared to alternative shaping technologies with improved mass transfer properties. An emphasis is allocated to those techniques that have been less discussed in detail in the literature.

Keywords: zeolites shaping technology; binder-containing extrusion; zeolitic monoliths; binder effects; binder-free shaping alternatives; extrusion

Citation: Asgar Pour, Z.; Abduljawad, M.M.; Alassmy, Y.A.; Cardon, L.; Van Steenberge, P.H.M.; Sebakhy, K.O. A Comparative Review of Binder-Containing Extrusion and Alternative Shaping Techniques for Structuring of Zeolites into Different Geometrical Bodies. *Catalysts* **2023**, *13*, 656. https://doi.org/10.3390/catal13040656

Academic Editors: Maja Milojević Rakić and Danica Bajuk-Bogdanović

Received: 26 December 2022
Revised: 29 January 2023
Accepted: 23 March 2023
Published: 27 March 2023

1. Introduction

In recent decades, rapidly growing rates in chemical demands from the market and the need for greener considerations for chemical production have accelerated the production and utilization of heterogeneous catalysts and adsorbents in industry. Large-scale application is always dictated by several pre-requisites that have to be fulfilled in order to make these materials suitable for commercial application. Among these requirements, shaping is one of the most important steps, since bulk zeolites in powder form will increase the pressure drop in packed bed reactors [1,2]. Conversely, shaped bodies with larger particle sizes (several mm) and a high degree of mechanical resistance to abrasion and dust formation are appropriate to minimize the pressure drop in a fixed-bed setup [3], and, thus, are good candidates for high throughput processes as well. Each shaped catalyst displays a specific flow pattern, which is more influenced by flow direction (e.g., up- or down-flow mode) [4], while a powder-loaded bed is a compact structure where the lack of enough void space complicates reactant feed in, permeability, and well-mixing across the bed, and, consequently, impedes the establishment of a steady-state flow regime inside the bed. Furthermore, powder application increases the volume of the bed for a given reactant flow rate, which, in turn, gives rise to the fabrication costs as well.

Zeolites as crystalline metallosilicates, conventionally in the form of aluminosilicates, have well-defined microporous networks containing regular cavities and channels [5]. Up to now, various excellent features have been reported for zeolites, such as molecular sieving, shape selectivity, high surface area, pore size tuning, high thermal and hydrothermal stability, tunable acidity, crystallinity, and ion-exchange property [6–12]. However, despite the outstanding features exhibited by zeolites, they are inherently microporous materials (d_{pore} < 1.5 nm) and, thus, suffer from intense diffusion limitations imposed by their micropores, particularly, in contact with bulky molecules [12]. In brief, molecules larger than the diameter of micropores are not able to diffuse through the narrow openings of pores and, thus, have a low chance of accessing the zeolite internal active sites. To overcome the microporosity limitations, many strategies such as synthesis of nano-size zeolites, structuring large pore size zeolites, and the design of hierarchical zeolite have been exploited to enhance the molecular diffusion [13–15]. For the sake of clarity, nano-size zeolites can diminish the diffusion limitations drawback of conventional micron-size zeolites due to their larger external surface area, which can be utilized for molecular adsorption and catalysis of bulkier species which cannot diffuse into the micropores. In addition, reduced mean diffusion path in nano-size zeolites can enhance the molecular diffusion rate by means of the shorter path length.

In addition, application of zeolite in industrial scale is strongly interrelated to the shaping technologies applied for structuring zeolite powder, where their fundamental microscopic physico-chemical properties (e.g., acidity, porosity, surface area) are highly dependent on the engineering features of macroscopic catalytic bodies (e.g., transport phenomena and pressure drop in packed bed reactors). Moreover, the utmost preservation of intrinsic catalytic properties is imperative during the shaping step, which provides adequate mechanical strength to the shaped bodies. Thanks to their resistance against breakage or abrasion as well as preventing high pressure drops, structured bodies are applicable for large scale operation conditions (e.g., high pressures and temperatures). Structuring is possible into final bodies with several different geometries (e.g., pellets, cylinders, honeycomb monoliths, hollow alternatives, etc.) [16,17]. Basically, the type of reactor dictates what geometry is more suitable. For instance, large shaped bodies are suitable for packed and moving beds, which are mainly prepared using extrusion, pelletizing, or granulation (with approximate pellet size in the range of 1 mm up to 3 cm). However, for slurry three phase reactors and fluidized beds, powders with a particle size in the range of 20 to 100 µm (prepared via spray drying) are more applicable [18]. In the subject of shaping, some researchers have tested lab-made zeolite bodies in larger pilot equipment [19–21], but less focus has been methodically directed toward shaping techniques their pros and cons, and their effects on the zeolite physicochemical properties. On the basis of what we have found so far, macroscopic shaping technologies of zeolites have been never completely collected and discussed in one study in detail. This motivated us to prepare this review from academic research and industrial patents, with the impetus of elucidation of the influence of several shaping methods, such as binder-containing extrusion technique and binder-free alternatives on the physico-chemical properties and performance of zeolites bodies. Moreover, the future outlook is depicted for more practical shaping techniques.

2. Zeolites Shaping Using Extrusion Technique and by Means of Inorganic and/or Organic Binders

The extrusion technique is frequently carried out in industry to shape powders with various geometries (e.g., cylinders, hollow cylinders, monoliths, etc.) to form extrudates with an average length range of 0.5–2 cm [22–24]. A schematic illustration of the extrusion process and some examples of commercial extrudates are shown in Figure 1A,B, respectively. Generally, there are some parameters that need to be smartly tuned to provide sufficient pastiness and rheological integrity, such as water mass ratio [25], which is determined on the basis of empirical trials, mixing and aging times [18], and binder content [26,27] as well as drying and calcination temperature. The mechanical strength of extrudates is mainly obtained by the adhesive interaction and partial interconnection of zeolite terminal hydroxyl groups with the adjacent binder particles, or by a fusion between contacting surfaces of binder and zeolites particles that is more pronounced after the calcination step [28]. The applied quantities of binders are varied and even values higher than 50 wt% were reported [29], but the more preferable amounts of binders are in the range of 10 to 20 wt% [29]. Binders not only provide good mechanical stability to zeolites but also some types of binders, such as oxides (e.g., alumina), improve the activity of silicate materials with low or no acidity [30]. The type of binder is normally selected on the basis of shaping necessities and the relevant application requirements [25]. In general, two types of common binders are used for zeolite shaping: inorganic binders (e.g., clays, silica, alumina, titania, zirconia, or a combination of them) and organic binders (e.g., water-soluble cellulose products, polyethylene glycol, and polyvinyl alcohol.) [31]. Organic binders are more applicable as thickening additives, lubricants, wetting agents, and plasticizers [26,28,32]. A list of selected organic/inorganic binders is provided in Table 1.

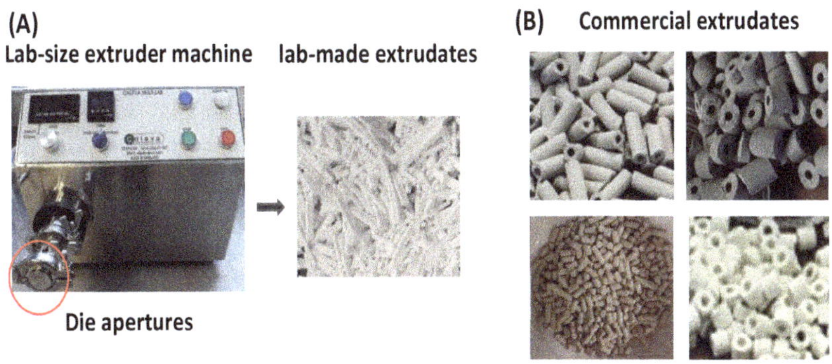

Figure 1. (**A**) Schematic illustration of zeolite shaping using extrusion technique, (**B**) Commercial extrudates with different geometrical structures.

In addition to binders and the above-mentioned additives, peptizing agents such as nitric acid can be utilized to stabilize particles and prevent their agglomeration and aid in better dispersion of binder particles in the zeolite-binder matrix [3,33]. Peptizing agents are also used to set the pH far as a point zero charge in order to prevent agglomeration of the paste [3]. In fact, large zeta potentials are known to stabilize the dispersion of particles by creating repulsion forces between positively charged particles and granting the paste mixture long-term suspension stability, consequently decreasing the viscosity of the paste [3,34]. To obtain a highly homogenized paste mixture, high degrees of mixing and stirring have been also recommended [34]. During the extrusion, the rheological behavior of primary paste is a crucial parameter impacting the ultimate mechanical stability. While a dry paste is highly viscous and can increase the possibility of the formation of cracks in the extrudates, a very wet initial paste mixture provides a very poor mechanical stability for extrudates [3,25]. To improve the rheological properties of paste parallel to final textural properties of extrudates, different additives such as plasticizers (e.g., starch, sugars, cellulose derivatives, and mineral clays), water soluble polymers (e.g., polyethylene glycol, poly vinyl alcohol), organic porogens (e.g., carbon black, starch, or organic fillers), and lubricants and peptizing additives (e.g., dilute nitric or acetic acids) are added to produce a low viscous paste displaying a lower frictional behavior and, thus, less crack in the final extrudates [3,34–37]. Calcination, which is the last step for obtaining final extrudates from green extrudates (i.e., before thermal treatment at high temperatures), can enhance the strength of extrudates and make them more mechanically stable through agglomeration of dealuminated sections inside the extrudates [18,38]. Poor mechanical strength leads to catalyst malfunction and significant pressure drop in packed beds, owing to dust formation. Several techniques for the mechanical strength measurement of shaped bodies have been developed, such as crushing, knife edge cutting, and three-point ending [39]. However, the single pellet crushing strength test has been generally accepted as a standard technique for this measurement [40]. By testing different geometries such as spheres, cylindrical tablets, and cylindrical and trilobe extrudates that have been subjected to mechanical strength measurement with the above-mentioned methods, it was reported that the geometry of shaped bodies is strongly determinative for the selection of measurement techniques to obtain a reliable assessment of mechanical stability. For instance, both crushing and cutting tests are appropriate for tablets [39], while for extrudates cutting and bending, and for spherical shaped bodies crushing test is more practical [39]. More details can be found in a previously cited reference [41].

Table 1. A list of selected organic/inorganic binders.

Entry	Zeolite Type	Organic Binder	Inorganic Binder	Ref.
1	3A	Copolymer (ethylene/butyle acrylate, 17 wt% and 30 wt% copolymer)	-	[31]
2	3A	Polyethylene	-	[31]
3	Y	Cellulose fibers	-	[32]
4	TiO_2 monolith	Polyethylene and methyl-hydroxyl-cellulose	Bentonite, glass fiber	[42]
5	ZSM-5		Aluminum phosphate γ-Al_2O_3	[28]
6	X, Y, ZSM-12, Mordenite, zeolite A, P, ZSM-5, MCM-41	-	Portland cement, High Al cement Sulphoaluminate cements, phosphate, bonded cements, blast furnace slag cements, Ca_3SiO_5, (Ca_2SiO_4), alumino-ferrite, tricalcium aluminate, calcium aluminates such as monocalcium aluminate ($Ca_3Al_2O_6$), and calcium hexoaluminate ($CaAl_2O_4$), used alone or as a mixture	[29]
7	ZSM-5	-	α-Alumina monohydrate	[30]
8	ZSM-5	Cellulose	Catapal D (Boehmite alumina)	[34]
9	5A	carboxymethycellulose	Kaolin	[43]
10	ZSM-5	-	Silica/Alumina	[36]
11	ZSM-5, 11, 23, 35, 38, 48, Beta, X, Y, L	-	Silica	[44]

Although other shaping technologies such as granulation, spray drying, wet or dry pressing, and pelletizing are well-known industrial techniques [35,45], extrusion is the most applicable method at the industrial level due to its high capacity and low preparation costs as well as other advantages such as narrow distribution of particle size and diversity in geometries [3,34,35]. In contrast to shape diversity in extrusion, only spherical bodies are produced by granulation or spray drying where their average particle size is in the range between 1 mm to 20 mm for granulation and 10 μm to 100 μm for spray drying [34,45]. For instance, high Resolution Scanning Electron Microscopy (HRSEM) and Focused Ion Beam Scanning Electron Microscopy (FIB-SEM) are two imaging techniques used for the structural study of extrudates and proved this uniformity, which has been attributed to two different agglomeration mechanisms taking place during extrusion or granulation [46].

2.1. The Important Parameters in Rheological Behavior of Extrudates

The rheological behavior of primary paste is determined on the empirical basis, and it aims to produce low viscous paste displaying lower frictional behavior and, thus, lesser cracks in the final extrudates. Peptizing agents are usually used for setting the pH to the target point of zero charge and prevent agglomeration of the paste [34]. Another important parameter is zeta potential (i.e., isoelectric potential in colloidal chemistry), and it is known that large zeta potentials stabilize the dispersion of particles by creating repulsive forces between positively charged particles and decrease the viscosity of the paste mixture [3,34]. Herein, zeta potential is an indicator of long-term suspension stability [3].

As mentioned before, adjusting water content during paste preparation is a critical factor to provide low viscous but sufficiently adhesive paste, which is easily shaped into noodles when passing through the die apertures. This occurs as a result of high shear stress in the extrusion chamber, as depicted in Figure 2. The shear stress is described by Equation (1):

$$\tau = \tau_0 + k \cdot \gamma^n \quad (1)$$

in which, τ is the shear stress (Pa), τ_0 is the minimum stress (the quantity that a substance can easily flow above this point), k is the consistency coefficient (Pa.sn), γ is the shear rate (Pa/s), and n is the flow index. For $n = 1$, Newtonian fluid properties are observed, while for $n < 1$, fluid displays Thixotropic behavior, and, for $n > 1$, shows dilatants behavior.

In addition, viscosity can be measured based on Equation (2) [30],

$$\eta = k \cdot \gamma^{n-1} \quad (2)$$

which can be reformulated in the form of Equation (3) as follows [3]:

$$\eta = \frac{\tau_0}{\gamma} + k \cdot \gamma^{n-1} \quad (3)$$

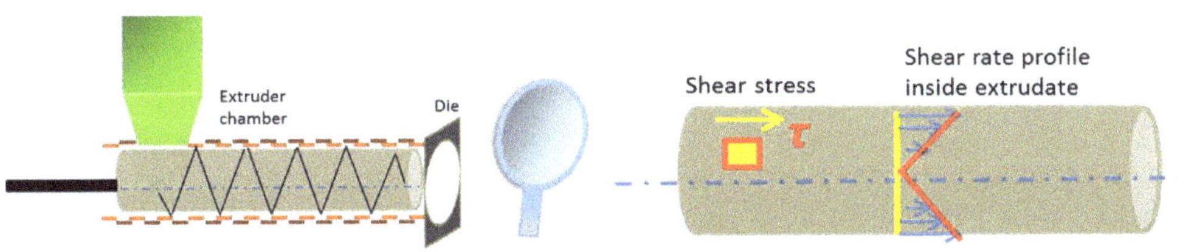

Figure 2. Shear stress (τ) applied to the extrudates during extrusion and shear rate profile on the extrudates.

Based on the Herschel–Bulky model, both shear stress and the viscosity of the paste passing through the extrusion chamber will depend on the shear rate imposed on the paste mixture [28]. A flow index close to zero shows plug flow properties, which are known as the ideal flow regime, but the flow behavior of the paste is close to 0.7, which fundamentally indicates non-Newtonian fluid behavior [3,36,42]. When the paste is passing through the die, different radial velocity profiles are generated. If the velocity index is significant in the radial axis, extruded noodles will possess macro-defects, particularly macro-cracks [36]. Furthermore, inside the extrusion chamber, appropriate zeta potentials separate the moving phase of the paste from the stationary phase, which is attached to the surfaces of the chamber and allows the moving phase to flow freely [42]. The zeta potential can be obtained from Henry's equation as follows [47]:

$$\zeta = \frac{3\eta U_e}{2\varepsilon f(Ka)} \tag{4}$$

in which ζ is zeta potential, η is viscosity of paste, U_e is electrophoretic mobility, ε is the dielectric constant, and $f(K_a)$ is the Henry's function with values of 1.0 or 1.5 [47]. Finally, the relation between shear yield stress and zeta potential is shown in the following equation [36,48]:

$$\tau_0 = \tau_{0,max} + k_0 \tau_{0,max} \zeta^2 \tag{5}$$

where $\tau_{0,max}$ is the maximum shear stress at zeta potential of zero and k_0 is a specific constant for the system. More details about binder content, its influence on shear yield stress, and the role of additives such as poly vinyl alcohol to adjust proper zeta potential can be found elsewhere [49]. Setting the zeta potential at a value where the best dispersion for the paste mixture is obtained dedicates the mechanical stability, reduces defects at the end of the process, and prevents the formation of agglomerated areas inside the final extrudates. Modification of zeta potential so that the paste will behave close to a Newtonian fluid should be prevented as it leads to lesser shape integrity [3].

2.2. *Binder Effects on Zeolite Physicochemical Properties*

Binders have different physicochemical properties compared with pure zeolites and, thus, the presence of a binder can adversely affect the zeolite textural, adsorption, and catalytic properties. It has been reported that commercial binder-containing 5A zeolite has a lower adsorption capacity for n-paraffins compared to the pure powder of 5A zeolite [50]. Like zeolite, the binder can also adsorb molecules and change the polar interaction between adsorbate and zeolite surface, leading to the reduction of adsorption heat and a higher coalescing factor between adsorbed molecules and active sites [43,50]. From a textural point of view, the porosity and surface area of shaped zeolites is affected by the binder materials, particularly those that have different particle size and, remarkably, a lower surface area than zeolite, which, in turn, results in pore blockage and the reduction of the surface area. The generation of larger meso- and macropores in the extrudates depends on the size of the binder particles (which is around 10 nm in the case of silica or alumina binders) [28]. Furthermore, the presence of additional porosity inside the extrudates is highly dependent on the degree of the porosity of the binder. A more closely connected network in the final extrudates is created using binders with smaller particle sizes (e.g., silica, alumina, or boehmite) while the larger degrees of interparticle porosities are generated by binders with large-size particles, such as plate-like kaolin [25,51]. From a chemical alteration point of view, dilution of acidic centers of zeolite, variation in their sorption/catalytic properties, and migration of undesired cations or impurities from binder to zeolite might also occur in the course of the extrusion step. For example, the ZSM-5 extrudates prepared with different binders such as boehmite, attapulgite, and silica displayed a lesser amount of the Brønsted acidity as proved by FTIR spectroscopy analysis [25], but kaolin did not show any chemical interaction with ZSM-5, and the kaolin containing ZSM-5 displayed acidic properties close to pure ZSM-5 [25]. Similar observation has been cited for H-gallosilicate and ascribed to the consequence of zeolite partial neutralization through ion-exchange of active sites of zeolite with mobile ions (e.g., alkali or alkaline earth cations such as Na^+ or Mg^{2+}) of binders, and dealumination was also proved by ^{27}Al MAS NMR analysis [52]. In contrast, insertion of alumina as a binder can produce extra acid sites inside the zeolite extrudates [30]. For instance, in a study, the effect of alumina binder was investigated on the zeolite acidic properties using UV/Vis and confocal fluorescence microspectroscopy and demonstrated the migration of Al species from the alumina binder into the ZSM-5 extrudates [53]. These Al species are tetrahedrally coordinated into the zeolite framework and thus increase the amount of zeolite Brønsted acidity [53]. Higher catalytic activities for ZSM-5 extrudates have been previously reported when alumina was used as a binder in acid-catalyzed reactions (e.g., n-hexane cracking, propylene oligomerization, lube oil dewaxing, and methanol conversion to hydrocarbons) [44]. In addition, the influence of the alumina binder in the enhancement of the catalytic activity of ZSM-5 extrudates have been formerly admitted in several acid-catalyzed reactions [44]. However, the catalytic activity of zeolite/alumina matrix did not display an increasing trend when they were dry-mixed in the absence of water proved that the wet zeolite/alumina phase is prone to the generation of additional Brønsted acid centers in the final composite [44]. In addition, several studies have been performed to determine the effect of different binders on zeolite acidity, and more details can be found elsewhere [54–62]. This evidence demonstrates that the choice of proper materials as binder agents is not identical for different cases and highly depends on the binder interaction with zeolites as well as its effects on the physicochemical properties of zeolite for a given reaction [44,63,64]. The non-uniform distribution of binders inside the extrusion mixture can generate chemically different zones. While rich zeolite zones (e.g., those parts with lesser binder content) inside the extrudates increase the reaction rate, rich binder-containing zones show much lower catalytic activity, which leads to lower reproducibility in the catalytic/adsorption performance of different batches of extrudates. Owing to the different textural and chemical structures in the two inhomogeneous zones, the reaction rate is controlled by mass transfer rather than by kinetics [56]. However, the solid state ion-exchange, due to the close contact of zeolite and the binder, has displayed two-sided effects, including both positive and negative outcomes such as neutralization of active sites, Si and Al migration, and poison trapping [57].

3. Zeolites Shaping Using Alternative Binder-Free Techniques

Because of the possible negative impacts of binders on zeolite properties, as explained previously, many studies have been conducted to develop binder-free alternative methods for zeolite structuring. Although some early studies have been conducted by Universal Oil Products Company (UOP) [65,66], more recently, different and new methods such as hard templating technique, hydrothermal transformation, pulsed current [67], and some other novel processes [68] have been tested for structuring zeolite particles using binder-free technologies. Amongst them, the hard templating technique and hydrothermal transformation are discussed in detail in the coming sections of this paper. The ultimate shape and geometry of binder-less zeolitic structures is almost identical to binder-containing analogous (e.g., spheres, cylinders, hollow cylinders, honeycombs, etc.) but they are prepared in the absence of any binder agent. The additional additives utilized during the shaping step (e.g., organic materials) are combusted and removed by thermal treatments, therefore, final structured zeolite bodies can contain up to 100% pure zeolite particles.

3.1. Macroscopic Hard Templating Casting

Macroscopic-size and porous hard templates are generally applied to generate both larger porosity in zeolite as well as to structure them simultaneously. Indeed, the hard templating technique is a casting method used to produce porous zeolitic bodies by immersing a solid template in the primary mixture of zeolite initiators. The zeolite particles are crystallized inside the pores of the template, and after solidification by hydrothermal treatment and following drying, the template is removed by calcination at an elevated temperature and a zeolitic body that is texturally a replica of the original template remains. A wide range of hard templates, such as polystyrene spheres [69], latex beads [70] and carbon materials (e.g., monoliths, fibers or disks) [71], have been applied as hard templates to shape the zeolites. For instance, binder-free silicalite-1 spheres were synthesized using ion-exchange resin beads as macroscopic hard templates [72]. After resin removal by calcination, hierarchically porous and spheres with the particle diameter in a range between 0.4 and 0.85 mm remain (Figure 3).

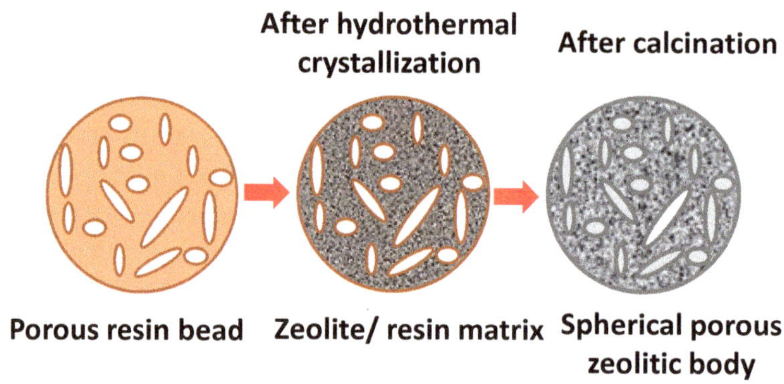

Figure 3. Synthesis of bead shape silicalite-1 using resin beads as hard templates.

Afterwards a series of zeolites in bead format were synthesized using this technique (e.g., zeolite Beta and ZSM-5) and metal-containing zeolite beads (e.g., Cr- and Pd-Beta) have also been synthesized in bead format [73–76]. Other porous solids such as organic aerogel, Polyurethane foams (PUFs) [77], carbon materials [78], and natural bio-based materials [79] have also been used as hard templates to shape zeolites. Some examples are: spongy silicalite obtained using starch and hierarchical silicalite-1 and Beta synthesized using aerogels as hard templates [77–79]. A list of different zeolitic frameworks in the spherical format as discussed above is summarized in Table 2.

Table 2. Summary of properties of synthesized zeolite beads.

Entry	Type of Zeolite Framework	Commercial Name of Resin	Si/Al Ratio	Surface Area (m^2/g)	Application	Ref.
1	Silicalite-1/MFI	MSA-1	∞	930	Catalysis/Molecular Sieve	[72]
2	Silicalite-1/MFI	WBA	∞	559	Catalysis/Molecular Sieve	[72]
3	Beta/BEA	MSA-1	50	640	Catalysis	[73]
4	ZSM-5/MFI	MSA-1	50	300–850	Catalysis	[74]
5	Cr-Beta/BEA	MSA-1	50	651	Catalysis	[75]
6	Pd-Beta/BEA	MSA-1	50	144	Catalysis	[76]

3.2. Hydrothermal Transformation

By means of hydrothermal transformation, a temporary binder mixed with zeolite powder, is converted to the zeolitic phase in the course of hydrothermal treatment. In addition to this, non-zeolitic materials (e.g., different types of clays or silica) can be transformed into zeolitic frameworks. The latter transformation needs an aqueous medium and, practically, a partial autogenic hydrothermal condition. By means of this method and in the presence of organic fillers such as cellulose, zeolites Y, X and A have been obtained by binder transformation inside the pre-fabricated bodies [80]. In addition, some inorganic materials such as silica or mineral clays such as kaolin and metakaolin have been converted into different zeolite frameworks (e.g., MFI, BEA, FAU) by aging them in an alkaline medium (OH/SiO_2 molar ratio up to 1.2), and they showed acceptable mechanical stability [81]. Clay minerals such as kaolin or metakaolin and silica consist of similar elements that are present in zeolites (e.g., Si, Al, Na), and they can thus be utilized as the resource of starting materials to initiate zeolite formation and following crystallization propagation. It has been reported that by means of a three-step hydrothermal conversion, kaolin extrudates can be transformed to silica X [80]. In addition, zeolite X, A, and mordenite have been obtained through the conversion of metakaolin (the activated form of kaolin prepared by kaolin thermal activation at 973 K) [81,82]. Next, using a mixture of NaOH, KOH, distilled water, and metakaolin, agitated for a maximum of 10 days, a 100% crystalline zeolite phase was obtained [81]. The alkaline solution is used as a mineralizer to assist the conversion of non-zeolitic materials to binder-free zeolite bodies. More details can be found in the previously cited literature [83]. Another class of pre-fabricated materials are glassy substances and, in this context, porous glass beads and glassy granules were transformed to MFI beads and to ZSM-5 granules, respectively [84,85]. Likewise, zeolite A tubes have been also synthesized, but, in this case, crystalline structures with large macropores in the range of 3 μm were obtained [86]. In addition, silicalite-1 disk has been synthesized by transformation of a kanemite disk (a single-layer silicate) using triphenylamine (TPA) cation by phase separation under a dry treatment (or so-called dry solid-state transformation) [87]. This technique is involved with gradual dissolution of bulk material (so-called BMD) resulting in zeolite monolith formation [88]. For instance, tubular zeolite monoliths with MFI topology were obtained by direct conversion of porous glass materials [88]. The BMD technique is based on the direct conversion of raw materials into zeolites under static conditions. This technique can be also conducted under a stirring condition, which is known as dynamic bulk material dissolution (DBMD). By employing the BMD technique and adjusting the solubility of precursors, zeolite tubular bodies have been obtained (Figure 4).

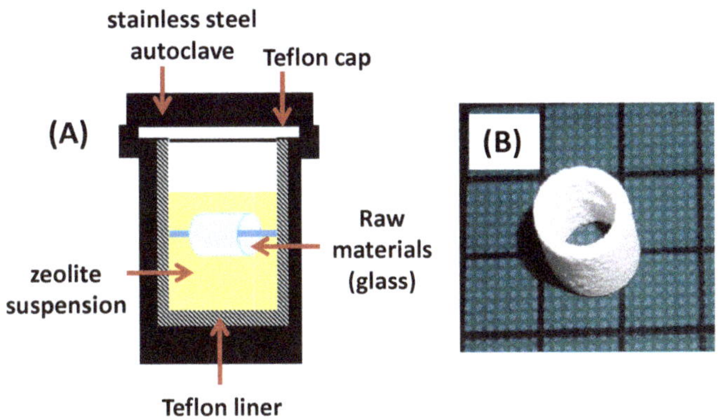

Figure 4. (**A**): Schematic illustration of the synthesis route for bulk material dissolution technique under static condition; (**B**): obtained monolith [87].

By means of this technique, a tubular and porous quartz glass was gradually converted to a MFI framework. The transformation reaction took place by inserting quartz glass in an autoclave equipped with polytetrafluoroethylene (PTFE) liner. In addition, other reagents such as tetrapropylammonium hydroxide (TPAOH), hydrogen fluoride (HF), and about 50 mg of MFI zeolite seeds were added to this mixture and stirred under 10 rpm at 200 °C for 61 days. From the point of view of the formation mechanism, a gradual formation of the MFI phase took place and was followed by a very moderate growth of MFI crystals, which propagated from the surface of the quartz toward the inside of the quartz body and gradually converted the glassy SiO_2 species to the MFI phase, and this led to that zeolite entirely replaced the quartz [88]. Here, seeds of MFI as initiators form new nuclei to trigger the formation of primary MFI building blocks and accelerate the crystallization process under a very slow stirring rate. Based on the selected conditions of this technique, partially transformed fibers and fabrics have also been obtained [88]. The drawbacks of this technique are the need for a large amount of time for the conversion of inorganic materials into zeolite, and, in some cases, the formation of a non-homogeneous composite consisting of the zeolite phase and other raw material phases has been observed [88].

Other techniques such as dry sol-gel transformation and solid-state transformation have been also studied in order to synthesize binder-free zeolitic bodies including binder-free ZSM-11 and mordenite pellets obtained in a low water medium [89] as well as ZSM-5 disks obtained by the conversion of amorphous aluminosilicate fibers using solid-state transformation [90]. Other structures such as MFI disks and films have been fabricated from primary aluminosilicate glasses using a solid-state transformation followed by the phase separation under hydrothermal treatment [86,90].

One interesting alternative is the pseudomorphic transformation of raw materials with the possibility to optimize the morphology of shaped zeolites. Using this technique, spherical Alginate/zeolite and alginate/zeolite-bentonite composite have been obtained, as depicted in Figure 5 [91]. In addition, pre-shaped zeolite Y/alginate spheres have been also transformed into pure zeolite Y with the same geometrical appearance (i.e., pre-shaped spheres) [91]. Following calcination at 600 °C for 44 h and decomposition of alginate, the binder-free zeolite granules were obtained [91].

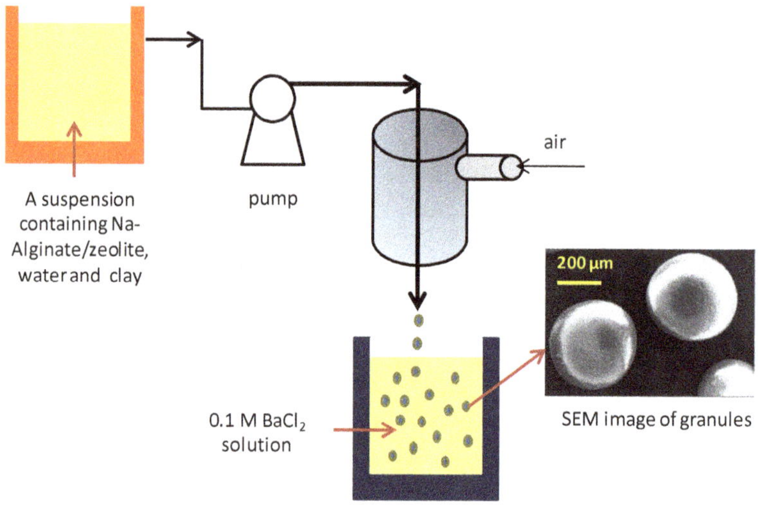

Figure 5. Schematic representation of synthesis granular and binder-free zeolite Y [91].

Other frameworks such as sodalite (SOD), Linde Type A (LTA) and faujasite X (FAU-X) have also been prepared by pseudomorphic transformation [92]. A type of hierarchical zeolite Beta monoliths have also been obtained through a layer-by-layer process using silica gel transformation [93], for which more details about the properties and synthetic routes can be found in previously cited literatures [92,93]. All these transformation techniques have a multi-step synthetic procedure for obtaining zeolite bodies. Recently, a one-step in situ hydrothermal synthesis has been developed in which silica sol was suspended in an oil/water system (immiscible liquids system) and granular zeolite A has been obtained by silica sol transformation (Figure 6) [94]. In this synthesis, heptane was used as the oil phase while the aqueous phase was a mixture of sodium aluminate, NaOH, and deionized (DI) water. Silica sol was gradually dropped into the oil phase and, over time, Al species slowly diffused to the spherical silica particles. Consequently, more silica precursors were gently dissolved in the reaction medium to release more Si species and push forward the nucleation step. Indeed, Al species were attached to the Si species via oxygen bridges in a way that these species produced the zeolite primary building blocks. This synthesis rout was involved with the sol suspension inside of two immiscible phases and further followed by the next gelation step. The full conversion of silica to zeolite was completed during the hydrothermal crystallization inside the alkaline solution. As mentioned above, clay extrudates can be also converted in an alkaline pH into zeolites. Zeolite A, X and MFI have also been obtained using this technique and alkaline pH [94]. The average diameter of the shaped particles is in the millimeter range, but a wide particle size distribution was observed.

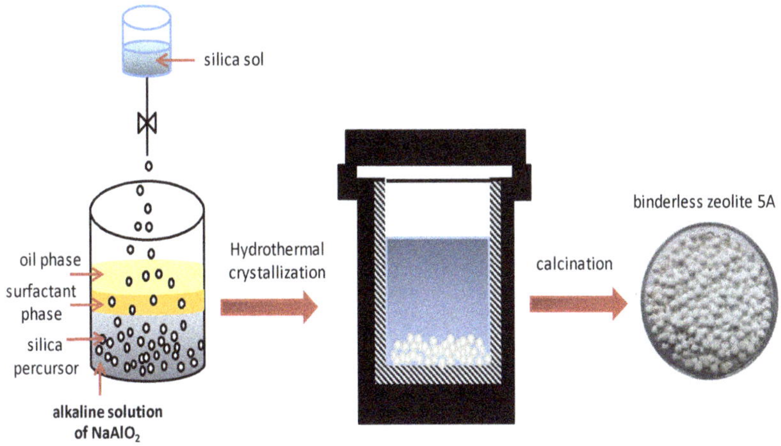

Figure 6. Schematic illustration of one-step synthesis protocol for preparing binder-free granules of zeolite A [94].

As the last case study, a one-pot template- and a binder-free synthesis rout for obtaining shaped mordenite under an acid-hydrolysis condition is reviewed here. By means of this technique, primary species were converted into the hierarchical and mechanically strong mordenite bodies during the hydrothermal crystallization [95]. In a typical synthesis, TEOS (tetraethyl orthosilicate as silica precursor) was dissolved in demineralized water (DI), and, next, HCl was slowly added to the mixture, which further followed by stirring for 24 h at 20 °C to thoroughly fulfill the hydrolysis step. The Al precursor was obtained by mixing aluminum sulfate ($Al_2(SO_4)_3$), sodium hydroxide (NaOH), and DI water with pre-defined molar ratios. Silica mixture was added to Al solution, entirely mixed, and aged for 24 h at room temperature. Finally, the resultant mixture was transferred to a Teflon-lined stainless steel autoclave and hydrothermally heated for preselected times at 170 °C (Figure 7) [95]. The highest degree of crystallinity and mechanical stability was observed after 144 h hydrothermal heating. At the end, mordenite monoliths with cubic or cylindrical structures with high mechanical strengths were obtained without any shrinkage by calcination at 550 °C. The obtained structured mordenite were used as catalysts in a Friedel–Crafts benzylation reaction of benzene with benzyl alcohol, and showed higher activity compared with the powder analogue. From a mechanical strength point of view, these monoliths were reported to be close to industrially available counterparts, and thereby, they show promising performance for industrial applications.

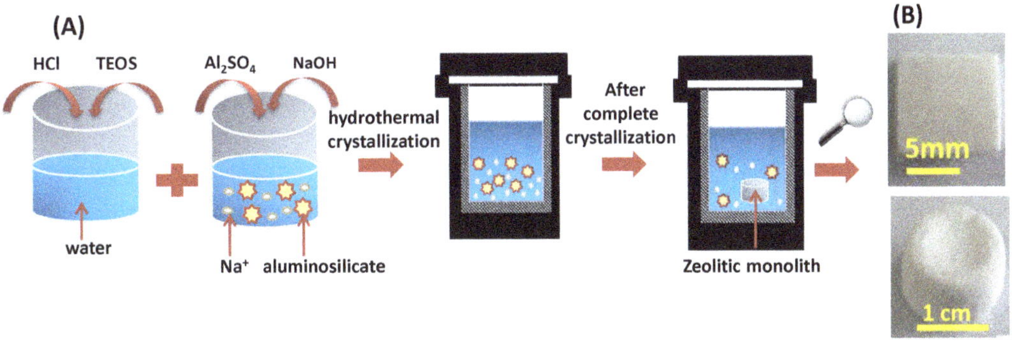

Figure 7. (**A**): Schematic illustration of one-step template- and binder-free synthesis rout for preparing mordenite monoliths, (**B**): obtained monoliths with different shape geometries [95].

4. Conclusions, Recommendations and Future Perspectives

The shaping of zeolites is a complex task that entails several considerations such as mass transfer prerequisites, pressure drop prevention, cost, and stability investigations [96–98]. The present review provides an outline of different shaping techniques as well as their possible effect on the physicochemical properties of structured bodies, as highlighted in different sections. Each technique has its own impact on the original physicochemical properties of zeolite, which are discussed thoroughly in the context of this review. Extrusion is a common industrial technique in which binders (e.g., silica, alumina, or clays) are employed to increase mechanical stability. Hence, they might have their own so-called side-effects on basic features of zeolites, so that some of these effects are undesired or detrimental. Some of them, as mentioned before, involve a change in the adsorption properties of binder-containing zeolites, which may reduce relative to the increase in binder quantity. Other major drawbacks frequently observed for several binders are neutralization of active sites and partial pore blockage, and, consequently, a decrease in the surface area. Regarding the probable disadvantages of binders, alternative methods have been investigated and showed a possible capacity to diminish binder drawbacks by higher degrees of preservation of the inherent physicochemical properties of the zeolites. However, the downside of these techniques is the lower mechanical strength, which is more pronounced for the hard templating approach. It is noteworthy that binder selection is a function of different parameters, such as specific applications and their requirements, and, thus, is a trade-off between inhibiting and promoting results in different cases. With a critical look at how shaping techniques have been developed in recent decades, a forward progress trend is obvious by several numbers of patents and research delivered each year to the public. However, looking to the future, there is still a need for the development and refinement of available techniques, in particular, with a combination of new computational techniques with experimental results. Evidently, more accurate and realistic criteria should be defined for preserving original properties of zeolites after shaping, and might need to seek more adapted technologies. Despite the limitations and reported downsides, binder-free alternatives proved to be an applicable method for retaining the original properties in shaped zeolites, but the scale-up feasibility is still ongoing and requires comprehensive study.

Author Contributions: Conceptualization, Z.A.P. and K.O.S.; methodology, Z.A.P., L.C. and K.O.S.; formal analysis, Z.A.P., Y.A.A. and K.O.S.; investigation, Z.A.P., M.M.A. and K.O.S.; writing—original draft preparation, Z.A.P. and K.O.S.; writing—review and editing, Z.A.P., P.H.M.V.S. and K.O.S.; visualization, Z.A.P. and K.O.S.; supervision, K.O.S. All authors have read and agreed to the published version of the manuscript.

Funding: This research received no external funding.

Data Availability Statement: Not applicable.

Acknowledgments: The authors would like to sincerely thank both the University of Groningen, Netherlands and the Laboratory for Chemical Technology (LCT), Ghent University, Belgium.

Conflicts of Interest: The authors declare no conflict of interest.

References

1. Ergun, S. Fluid flow through packed columns. *Chem. Eng. Prog.* **1952**, *48*, 89–94.
2. Foumeny, E.A.; Kulkarni, A.; Roshani, S. Elucidation of pressure drop in packed-bed systems. *J. Appl. Therm. Eng.* **1996**, *16*, 195–202. [CrossRef]
3. Devyatkov, S.; Kuzichkin, N.V.; Murzin, D.Y. On Comprehensive Understanding of Catalyst Shaping by Extrusion. *Chim. Oggi-Chem. Today* **2015**, *33*, 57–64.
4. De Wind, M.; Plantenga, F.L.; Heinerman, J.J.L.; Free, H.W.H. Upflow versus Downflow Testing of Hydrotreating Catalysts. *Appl. Catal.* **1988**, *43*, 239–252. [CrossRef]
5. Corma, A.; Martinez, A. Zeolites and Zeotypes as Catalysts. *Adv. Mater.* **1995**, *7*, 137–144. [CrossRef]
6. Asgar Pour, Z.; Sebakhy, K.O. A Review on the Effects of Organic Structure-Directing Agents on the Hydrothermal Synthesis and Physicochemical Properties of Zeolites. *Chemistry* **2022**, *4*, 431–446. [CrossRef]

7. Rabo, J.A.; Schoonover, M.W. Early Discoveries in Zeolite Chemistry and Catalysis at Union Carbide, and Follow-up in Industrial Catalysis. *Appl. Catal. A Gen.* **2001**, *222*, 261–275. [CrossRef]
8. Tanabe, K.; Hölderich, W.F. Industrial Application of Solid Acid–Base Catalysts. *Appl. Catal. A Gen.* **1999**, *181*, 399–434. [CrossRef]
9. Dapsens, P.Y.; Mondelli, C.; Pérez-Ramírez, J. Design of Lewis-Acid Centres in Zeolitic Matrices for the Conversion of Renewables. *Chem. Soc. Rev.* **2015**, *44*, 7025–7043. [CrossRef]
10. Ch Deka, R. *Acidity in Zeolites and Their Characterization by Different Spectroscopic Methods*, 5th ed.; Indian Journal of Technology: New Delhi, India, 1998.
11. Moshoeshoe, M.N.; Nadiye-Tabbiruka, M.S.; Obuseng, V.C. A Review of the Chemistry, Structure, Properties and Applications of Zeolites. *Am. J. Mater. Sci.* **2017**, *7*, 196–221.
12. Primo, A.; Garcia, H. Zeolites as Catalysts in Oil Refining. *Chem. Soc. Rev.* **2014**, *43*, 7548–7561. [CrossRef]
13. Davis, M.E. Ordered Porous Materials for Emerging Applications. *Nature* **2002**, *417*, 813–821. [CrossRef]
14. Corma, A. From Microporous to Mesoporous Molecular Sieve Materials and Their Use in Catalysis. *Chem. Rev.* **1997**, *97*, 2373–2420. [CrossRef]
15. Mintova, S.; Gilson, J.-P.; Valtchev, V. Advances in Nanosized Zeolites. *Nanoscale* **2013**, *5*, 6693–6703. [CrossRef]
16. Lakiss, L.; Gilson, J.P.; Valtchev, V.; Mintova, S.; Vicente, A.; Vimont, A.; Bedard, R.; Abdo, S.; Bricker, J. Zeolites in a good shape: Catalyst forming by extrusion modifies their performances. *Microporous Mesoporous Mater.* **2020**, *299*, 110114. [CrossRef]
17. Wang, Y.; Chang, Y.; Liu, M.; Zhang, A.; Guo, X. A facile strategy to prepare shaped ZSM-5 catalysts with enhanced para-xylene selectivity and stability for toluene methylation: The effect of in situ modification by attapulgite. *Molecules* **2019**, *24*, 3462. [CrossRef]
18. Perego, C.; Villa, P. Catalyst Preparation Methods. *Catal. Today* **1997**, *34*, 281–305. [CrossRef]
19. Pérez-Ramírez, J.; Mitchell, S.; Verboekend, D.; Milina, M.; Michels, N.-L.; Krumeich, F.; Marti, N.; Erdmann, M. Expanding the Horizons of Hierarchical Zeolites: Beyond Laboratory Curiosity towards Industrial Realization. *ChemCatChem* **2011**, *3*, 1731–1734. [CrossRef]
20. Verboekend, D.; Mitchell, S.; Pérez-Ramírez, J. Hierarchical Zeolites Overcome All Obstacles: Next Stop Industrial Implementation. *Chimia* **2013**, *67*, 327. [CrossRef]
21. Michels, N.-L.; Mitchell, S.; Milina, M.; Kunze, K.; Krumeich, F.; Marone, F.; Erdmann, M.; Marti, N.; Pérez-Ramírez, J. Hierarchically Structured Zeolite Bodies: Assembling Micro-, Meso-, and Macroporosity Levels in Complex Materials with Enhanced Properties. *Adv. Funct. Mater.* **2012**, *22*, 2509–2518. [CrossRef]
22. Beeckman, J.W.L.; Fassbender, N.A.; Datz, T.E. Length to diameter ratio of extrudates in catalysis technology I. Bending Strength versus impulsive forces. *AICHE J.* **2016**, *62*, 2658–2669. [CrossRef]
23. Gustafson, W.R.; Conn, T. Shaped Catalyst Particle. US Patent 3,966,644, 29 June 1967.
24. Hagen, J. Catalyst Shapes and Production of Heterogeneous Catalysts. *Ind. Catal.* **2015**, 211–238. [CrossRef]
25. Michels, N.-L.; Mitchell, S.; Pérez-Ramírez, J. Effects of Binders on the Performance of Shaped Hierarchical MFI Zeolites in Methanol-to-Hydrocarbons. *ACS Catal.* **2014**, *4*, 2409–2417. [CrossRef]
26. Jasra, R.V.; Tyagi, B.; Badheka, Y.M.; Choudary, V.N.; Bhat, T.S.G. Effect of Clay Binder on Sorption and Catalytic Properties of Zeolite Pellets. *Ind. Eng. Chem. Res.* **2003**, *42*, 3263–3272. [CrossRef]
27. Tischer, R.E. Preparation of Bimodal Aluminas and Molybdena/Alumina Extrudates. *J. Catal.* **1981**, *72*, 255–265. [CrossRef]
28. Freiding, J.; Patcas, F.-C.; Kraushaar-Czarnetzki, B. Extrusion of Zeolites: Properties of Catalysts with a Novel Aluminium Phosphate Sintermatrix. *Appl. Catal. A General.* **2007**, *328*, 210–218. [CrossRef]
29. Bazer-Bachi, D.; Harbuzaru, B.; Lecolier, E. Zeolite Formed by Extrsion and Pelleting with a Hydraulic Binder Having Improved Mechanical Properties and Process and Preparing Same. U.S. Patent 20,160,288,109A1, 6 October 2016.
30. Shihabi, D.S.; Garwood, W.E.; Chu, P.; Miale, J.N.; Lago, R.M.; Chu, C.T.-W.; Chang, C.D. Aluminum Insertion into High-Silica Zeolite Frameworks: II. Binder Activation of High-Silica ZSM-5. *J. Catal.* **1985**, *93*, 471–474. [CrossRef]
31. Bouvier, L.; Nicolas, S.; Medevielle, A.; Alex, P. Zeolite Adsorbent Having an Organic Binder. U.S. Patent 8,932,386B2, 13 January 2015.
32. Akhtar, F.; Andersson, L.; Ogunwumi, S.; Hedin, N.; Bergström, L. Structuring Adsorbents and Catalysts by Processing of Porous Powders. *J. Eur. Ceram. Soc.* **2014**, *34*, 1643–1666. [CrossRef]
33. Lynn, M. Method of Making a High Strength Catalys, Catalyst Support or Adsorber. U.S. Patent 5,633,217, 27 May 1997.
34. Whiting, G.T.; Chung, S.-H.; Stosic, D.; Chowdhury, A.D.; van der Wal, L.I.; Fu, D.; Zecevic, J.; Travert, A.; Houben, K.; Baldus, M.; et al. Multiscale Mechanistic Insights of Shaped Catalyst Body Formulations and Their Impact on Catalytic Properties. *ACS Catal.* **2019**, *9*, 4792–4803. [CrossRef]
35. Baldovino-Medrano, V.G.; Le, M.T.; Van Driessche, I.; Bruneel, E.; Alcázar, C.; Colomer, M.T.; Moreno, R.; Florencie, A.; Farin, B.; Gaigneaux, E.M. Role of Shaping in the Preparation of Heterogeneous Catalysts: Tableting and Slip-Casting of Oxidation Catalysts. *Catal. Today* **2015**, *246*, 81–91. [CrossRef]
36. Devyatkov, S.Y.; Al Zinnurova, A.; Aho, A.; Kronlund, D.; Peltonen, J.; Kuzichkin, N.V.; Lisitsyn, N.V.; Murzin, D.Y. Shaping of Sulfated Zirconia Catalysts by Extrusion: Understanding the Role of Binders. *Ind. Eng. Chem. Res.* **2016**, *55*, 6595–6606. [CrossRef]
37. Absi-Halabi, M.; Stanislaus, A.; Al-Zaid, H. Effect of Acidic and Basic Vapors on Pore Size Distribution of Alumina under Hydrothermal Conditions. *Appl. Catal. A Gen.* **1993**, *101*, 117–128. [CrossRef]
38. Mitchell, S.; Michels, N.L.; Pérez-Ramírez, J. From Powder to Technical Body: The Undervalued Science of Catalyst Scale Up. *Chem. Soc. Rev.* **2013**, *42*, 6094–6112. [CrossRef]

39. Li, Y.; Wu, D.; Zhang, J.; Chang, L.; Wu, D.; Fang, Z.; Shi, Y. Measurement and Statistics of Single Pellet Mechanical Strength of Differently Shaped Catalysts. *Powder Technol.* **2000**, *113*, 176–184. [CrossRef]
40. *ASTM Committee D-32*; 1985 Annual Book of ASTM Standards, vol. 5.03. American Society for Testing and Materials: New York, NY, USA, 1984.
41. Rhee, Y.W.; Guin, J.A. Preparation of Alumina Catalyst Supports and NiMo/Al_2O_3 Catalysts. *Korean J. Chem. Eng.* **1993**, *10*, 112–123. [CrossRef]
42. Forzatti, P.; Orsenigo, C.; Ballardini, D.; Berti, F. On the Relations Between the Rheology of Tio2-Based Ceramic Pastes and the Morphological and Mechanical Properties of the Extruded Catalysts. In *Studies in Surface Science and Catalysis*; Delmon, B., Jacobs, P.A., Maggi, R., Martens, J.A., Grange, P., Poncelet, G., Eds.; Elsevier: Amsterdam, The Netherlands, 1998; Volume 118, pp. 787–796, ISBN 0167-2991.
43. Shams, K.; Mirmohammadi, S.J. Preparation of 5A Zeolite Monolith Granular Extrudates Using Kaolin: Investigation of the Effect of Binder on Sieving/Adsorption Properties Using a Mixture of Linear and Branched Paraffin Hydrocarbons. *Microporous Mesoporous Mater.* **2007**, *106*, 268–277. [CrossRef]
44. Bowes, E.; Amwell, E. Extrusion of Silica-Rich Solids. U.S. Patent 4,582,815, 15 April 1986.
45. Gleichmann, K.; Unger, B.; Brandt, A. Industrial Zeolite Molecular Sieves. *Zeolites Useful. Miner.* **2016**, *5*, 89–108. [CrossRef]
46. Mitchell, S.; Michels, N.L.; Kunze, K.; Pérez-Ramírez, J. Visualization of Hierarchically Structured Zeolite Bodies from Macro to Nano Length Scales. *Nat. Chem.* **2012**, *4*, 825–831. [CrossRef]
47. Koutsoukos, P.K.; Klepetsanis, P.G.; Spanos, N.; Somasundaran, P. *Calculation of Zeta-Potentials from Electrokinetic Data, Encyclopedia of Surface and Colloid Science*; CRC Press: New York, NY, USA, 2006.
48. Foundas, M.; Britcher, L.G.; Fornasiero, D.; Morris, G.E. Boehmite Suspension Behaviour upon Adsorption of Methacrylate–Phosphonate Copolymers. *Powder Technol.* **2015**, *269*, 385–391. [CrossRef]
49. Catalyst Fundamentals. In *Catalytic Air Pollution Control*; Wiley: Hoboken, NJ, USA, 2009; pp. 1–23, ISBN 978-1-118-39774-9.
50. Sun, H.; Shen, B.; Liu, J. N-Paraffins Adsorption with 5A Zeolites: The Effect of Binder on Adsorption Equilibria. *Sep. Purif. Technol.* **2008**, *64*, 135–139. [CrossRef]
51. Luna-Murillo, B.; Pala, M.; Paioni, A.L.; Baldus, M.; Ronsse, F.; Prins, W.; Bruijnincx, P.C.A.; Weckhuysen, B.M. Catalytic fast pyrolysis of biomass: Catalyst characterization reveals the feed-dependent deactivation of a technical ZSM-5-based catalyst. *ACS Sustain. Chem. Eng.* **2020**, *9*, 291–304. [CrossRef]
52. Devadas, P.; Kinage, A.K.; Choudhary, V.R. Effect of Silica Binder on Acidity, Catalytic Activity and Deactivation Due to Coking in Propane Aromatization over H-Gallosilicate (MFI). In *Studies in Surface Science and Catalysis*; Rao, T.S.R.P., Dhar, G.M., Eds.; Elsevier: Amsterdam, The Netherlands, 1998; Volume 113, pp. 425–432. ISBN 0167-2991.
53. Whiting, G.T.; Meirer, F.; Mertens, M.M.; Bons, A.-J.; Weiss, B.M.; Stevens, P.A.; de Smit, E.; Weckhuysen, B.M. Binder Effects in SiO^{2-} and Al2O3-Bound Zeolite ZSM-5-Based Extrudates as Studied by Microspectroscopy. *ChemCatChem* **2015**, *7*, 1312–1321. [CrossRef] [PubMed]
54. Mitra, B.; Kunzru, D. Washcoating of Different Zeolites on Cordierite Monoliths. *J. Am. Ceram. Soc.* **2008**, *91*, 64–70. [CrossRef]
55. Lee, H.J.; Kim, J.H.; Park, D.-W.; Cho, S.J. Effect of Base Binder, Flash Calcined Hydrotalcite, in MFI Zeolite Granule: Catalytic Activity over 1-Butene Isomerization and MTO Reaction. *Appl. Catal. A Gen.* **2015**, *502*, 42–47. [CrossRef]
56. Du, X.; Kong, X.; Chen, L. Influence of Binder on Catalytic Performance of Ni/HZSM-5 for Hydrodeoxygenation of Cyclohexanone. *Catal. Commun.* **2014**, *45*, 109–113. [CrossRef]
57. Kasture, M.W.; Niphadkar, P.S.; Bokade, V.V.; Joshi, P.N. On the Catalytic Performance in Isopropylation of Benzene over H/β Zeolite Catalysts: Influence of Binder. *Catal. Commun.* **2007**, *8*, 1003–1008. [CrossRef]
58. Mehlhorn, D.; Valiullin, R.; Kärger, J.; Schumann, K.; Brandt, A.; Unger, B. Transport Enhancement in Binderless Zeolite X- and A-Type Molecular Sieves Revealed by PFG NMR Diffusometry. *Microporous Mesoporous Mater.* **2014**, *188*, 126–132. [CrossRef]
59. Kong, X.; Liu, J. Influence of Alumina Binder Content on Catalytic Performance of Ni/HZSM-5 for Hydrodeoxygenation of Cyclohexanone. *PLoS ONE* **2014**, *9*, e101744. [CrossRef]
60. Sánchez, P.; Dorado, F.; Fúnez, A.; Jiménez, V.; Ramos, M.J.; Valverde, J.L. Effect of the Binder Content on the Catalytic Performance of Beta-Based Catalysts. *J. Mol. Catal. A Chem.* **2007**, *273*, 109–113. [CrossRef]
61. Uguina, M.A.; Sotelo, J.L.; Serrano, D.P. Toluene Disproportionation over ZSM-5 Zeolite: Effects of Crystal Size, Silicon-to-Aluminum Ratio, Activation Method and Pelletization. *Appl. Catal.* **1991**, *76*, 183–198. [CrossRef]
62. Yang, K.; Zhang, D.; Zou, M.; Yu, L.; Huang, S. The Known and Overlooked Sides of Zeolite-Extrudate Catalysts. *ChemCatChem* **2021**, *13*, 1414–1423. [CrossRef]
63. Dorado, F.; Romero, R.; Cañizares, P. Hydroisomerization of N-Butane over Pd/HZSM-5 and Pd/Hβ with and without Binder. *Appl. Catal. A Gen.* **2002**, *236*, 235–243. [CrossRef]
64. de Lucas, A.; Valverde, J.L.; Sánchez, P.; Dorado, F.; Ramos, M.J. Influence of the Binder on the N-Octane Hydroisomerization over Palladium-Containing Zeolite Catalysts. *Ind. Eng. Chem. Res.* **2004**, *43*, 8217–8225. [CrossRef]
65. Michaiko, E. Method for Producing Molecular Sieves Zeolite Particles. U.S. Patent 3,348,911A, 24 December 1967.
66. Michaiko, E. Preparation of Crystalline Zeolite Particles. U.S. Patent 3,359,068A, 19 December 1967.
67. Vasiliev, P.; Akhtar, F.; Grins, J.; Mouzon, J.; Andersson, C.; Hedlund, J.; Bergstrom, L. Strong hierarchically porous monoliths by pulsed current processing of zeolite powder assemblies. *ACS Appl. Mater. Interfaces* **2010**, *2*, 732–737. [CrossRef]

68. Lu, S.; Han, R.; Wang, H.; Song, C.; Ji, N.; Lu, X.; Ma, D.; Liu, Q. Three birds with one stone: Designing a novel binder-free monolithic zeolite pellet for wet VOC gas adsorption. *Chem. Eng. J.* **2022**, *448*, 137629. [CrossRef]
69. Holland, B.T.; Abrams, L.; Stein, A. Dual Templating of Macroporous Silicates with Zeolitic Microporous Frameworks. *J. Am. Chem. Soc.* **1999**, *121*, 4308–4309. [CrossRef]
70. Dong, A.; Wang, Y.; Tang, Y.; Zhang, Y.; Ren, N.; Gao, Z. Mechanically Stable Zeolite Monoliths with Three-Dimensional Ordered Macropores by the Transformation of Mesoporous Silica Spheres. *Adv. Mater.* **2002**, *14*, 1506–1510. [CrossRef]
71. García-Martínez, J.; Cazorla-Amorós, D.; Linares-Solano, A.; Lin, Y.S. Synthesis and Characterisation of MFI-Type Zeolites Supported on Carbon Materials. *Microporous Mesoporous Mater.* **2001**, *42*, 255–268. [CrossRef]
72. Tosheva, L.; Valtchev, V.; Sterte, J. Silicalite-1 Containing Microspheres Prepared Using Shape-Directing Macro-Templates. *Microporous Mesoporous Mater.* **2000**, *35–36*, 621–629. [CrossRef]
73. Tosheva, L.; Mihailova, B.; Valtchev, V.; Sterte, J. Zeolite Beta Spheres. *Microporous Mesoporous Mater.* **2001**, *48*, 31–37. [CrossRef]
74. Tosheva, L.; Sterte, J. ZSM-5 Spheres Prepared by Resin Templating. In *Studies in Surface Science and Catalysis*; Aiello, R., Giordano, G., Testa, F., Eds.; Elsevier: Amsterdam, The Netherlands, 2002; Volume 142, pp. 183–190. ISBN 0167-2991.
75. Naydenov, V.; Tosheva, L.; Sterte, J. Chromium Containing Zeolite Beta Macrostructures. In *Studies in Surface Science and Catalysis*; Aiello, R., Giordano, G., Testa, F., Eds.; Elsevier: Amsterdam, The Netherlands, 2002; Volume 142, pp. 1449–1455, ISBN 0167-2991.
76. Naydenov, V.; Tosheva, L.; Sterte, J. Palladium-Containing Zeolite Beta Macrostructures Prepared by Resin Macrotemplating. *Chem. Mater.* **2002**, *14*, 4881–4885. [CrossRef]
77. Lee, Y.-J.; Lee, J.S.; Park, Y.S.; Yoon, K.B. Synthesis of Large Monolithic Zeolite Foams with Variable Macropore Architectures. *Adv. Mater.* **2001**, *13*, 1259–1263. [CrossRef]
78. Li, W.-C.; Lu, A.-H.; Palkovits, R.; Schmidt, W.; Spliethoff, B.; Schüth, F. Hierarchically Structured Monolithic Silicalite-1 Consisting of Crystallized Nanoparticles and Its Performance in the Beckmann Rearrangement of Cyclohexanone Oxime. *J. Am. Chem. Soc.* **2005**, *127*, 12595–12600. [CrossRef] [PubMed]
79. Tong, Y.; Zhao, T.; Li, F.; Wang, Y. Synthesis of Monolithic Zeolite Beta with Hierarchical Porosity Using Carbon as a Transitional Template. *Chem. Mater.* **2006**, *18*, 4218–4220. [CrossRef]
80. Flank, W.H.; Fethke, W.P.; Marte, J. Process for Preparing Molecular Sieve Bodies. U.S. Patent 4,818,508A, 4 April 1989.
81. Verduijn, J.P. Process for Producing Substantially Binder-Free Zeolites. U.S. Patent 5,460,796A, 24 October 1995.
82. Pavlov, M.L.; Basimova, R.A. Improvement of Synthesis Methods of Powdery Mordenite Type Zeolite. *J. Oil Gas. Bus.* **2012**, *2*, 459–469.
83. Akolekar, D.; Chaffee, A.; Howe, R.F. The Transformation of Kaolin to Low-Silica X Zeolite. *Zeolites* **1997**, *19*, 359–365. [CrossRef]
84. Rauscher, M.; Selvam, T.; Schwieger, W.; Freude, D. Hydrothermal Transformation of Porous Glass Granules into ZSM-5 Granules. *Microporous Mesoporous Mater.* **2004**, *75*, 195–202. [CrossRef]
85. Scheffler, F.; Schwieger, W.; Freude, D.; Liu, H.; Heyer, W.; Janowski, F. Transformation of Porous Glass Beads into MFI-Type Containing Beads. *Microporous Mesoporous Mater.* **2002**, *55*, 181–191. [CrossRef]
86. Özcan, A.; Kalıpçılar, H. Preparation of Zeolite A Tubes from Amorphous Aluminosilicate Extrudates. *Ind. Eng. Chem. Res.* **2006**, *45*, 4977–4984. [CrossRef]
87. Shimizu, S.; Kiyozumi, Y.; Maeda, K.; Mizukami, F.; Pál-Borbély, G.; Mihályi, R.M.; Beyer, H.K. Transformation of Intercalated Layered Silicates to Zeolites in the Solid State. *Adv. Mater.* **1996**, *8*, 759–762. [CrossRef]
88. Shimizu, S.; Hamada, H. Direct Conversion of Bulk Materials into MFI Zeolites by a Bulk-Material Dissolution Technique. *Adv. Mater.* **2000**, *12*, 1332–1335. [CrossRef]
89. De Luca, P.; Crea, F.; Fonseca, A.; Nagy, J.B. Direct Formation of Self-Bonded Pellets during the Synthesis of Mordenite and ZSM-11 Zeolites from Low Water Content Systems. *Microporous Mesoporous Mater.* **2001**, *42*, 37–48. [CrossRef]
90. Madhusoodana, C.D.; Kameshima, Y.; Yasumori, A.; Okada, K. Preparation of Fiber-Reinforced Binderless Zeolite Disks in Solid State. *J. Mater. Sci. Lett.* **2003**, *22*, 553–556. [CrossRef]
91. Charkhi, A.; Kazemeini, M.; Ahmadi, S.J.; Kazemian, H. Fabrication of Granulated NaY Zeolite Nanoparticles Using a New Method and Study the Adsorption Properties. *Powder Technol.* **2012**, *231*, 1–6. [CrossRef]
92. Mańko, M.; Vittenet, J.; Rodriguez, J.; Cot, D.; Mendret, J.; Brosillon, S.; Makowski, W.; Galarneau, A. Synthesis of Binderless Zeolite Aggregates (SOD, LTA, FAU) Beads of 10, 70μm and 1mm by Direct Pseudomorphic Transformation. *Microporous Mesoporous Mater.* **2013**, *176*, 145–154. [CrossRef]
93. Lei, Q.; Zhao, T.; Li, F.; Wang, Y.F.; Hou, L. Zeolite Beta Monoliths with Hierarchical Porosity by the Transformation of Bimodal Pore Silica Gel. *J. Porous Mater.* **2008**, *15*, 643–646. [CrossRef]
94. Sun, H.; Sun, Z.; Shen, B.; Liu, J.; Li, G.; Wu, D.; Zhang, Y. One-Pot Synthesis of Binderless Zeolite A Spheres via in Situ Hydrothermal Conversion of Silica Gel Precursors. *AIChE J.* **2018**, *64*, 4027–4038. [CrossRef]
95. Zhang, J.; Mao, Y.; Li, J.; Wang, X.; Xie, J.; Zhou, Y.; Wang, J. Ultrahigh Mechanically Stable Hierarchical Mordenite Zeolite Monolith: Direct Binder-/Template-Free Hydrothermal Synthesis. *Chem. Eng. Sci.* **2015**, *138*, 473–481. [CrossRef]
96. Sebakhy, K.O.; Vitale, G.; Pereira-Almao, P. Production of Highly Dispersed Ni within Nickel Silicate Materials with the MFI Structure for the Selective Hydrogenation of Olefins. *Ind. Eng. Chem. Res.* **2019**, *58*, 8597–8611. [CrossRef]

97. Sebakhy, K.O.; Vitale, G.; Pereira-Almao, P. Dispersed Ni-doped Aegirine Nanocatalysts for the Selective Hydrogenation of Olefinic Molecules. *ACS. Appl. Nano Mater.* **2018**, *1*, 6269–6280. [CrossRef]
98. Asgar Pour, Z.; Koelewijn, R.; El Hariri El Nokab, M.; van der Wel, P.C.A.; Sebakhy, K.O.; Pescarmona, P.P. Binder-free Zeolite Beta Beads with Hierarchical Porosity: Synthesis and Application as Heterogeneous Catalysts for Anisole Acylation. *ChemCatChem* **2022**, *14*, e202200518. [CrossRef]

catalysts

Article

Tuning the Structure and Acidity of Pt/Hierarchical SSZ-32 Catalysts to Boost the Selective Hydroisomerization of n-Hexadecane

Xinyue Yang [1,2], Wenli Zhao [1,2], Linlin Liu [1,2], Xiaopo Niu [1,2] and Qingfa Wang [1,2,3,*]

1 Key Laboratory for Green Chemical Technology of Ministry of Education, School of Chemical Engineering and Technology, Tianjin University, Tianjin 300072, China
2 Collaborative Innovation Center of Chemical Science and Engineering (Tianjin), Tianjin University, Tianjin 300072, China
3 Zhejiang Institute of Tianjin University, Ningbo 315201, China
* Correspondence: qfwang@tju.edu.cn; Tel.: +86-22-27892340

Abstract: Developing highly selective and efficient bifunctional catalysts is an important issue for the hydroisomerization of long-chain *n*-alkanes. It is vital to tailor the balance of isomerization and cracking reactions in hydroisomerization. Herein, a bifunctional Pt/hierarchical SSZ-32 catalyst was fabricated with a sequential desilication–dealumination treatment to boost the selective hydroisomerization of *n*-hexadecane (C_{16}). The pore structure and acid sites of SSZ-32 zeolite were tailored. More mesopore and Brønsted acid sites were generated, and the ratio of weak to strong Brønsted acidity (B_w/B_s) was increased by the sequential desilication–dealumination. The generated hierarchical structure had little effect on the selectivity of the reaction pathways of hydroisomerization versus cracking. The ratio of isomers/cracking products increased almost linearly with the increase in the B_w/B_s ratios. Meanwhile, the synergetic effect of the hierarchical structure and acidity regulation promoted the selectivity of monobranched *i*-C16 products. Therefore, the resulting Pt/SSZ-0.6AS exhibited the highest activity with a total isomer yield of 71.5% at 255 °C and the enhanced formation mechanism of monobranched isomers occurred via the pore mouth.

Keywords: hierarchical SSZ-32 zeolites; bifunctional catalysts; *n*-hexadecane hydroisomerization; Brønsted acidity distribution; reaction pathways

Citation: Yang, X.; Zhao, W.; Liu, L.; Niu, X.; Wang, Q. Tuning the Structure and Acidity of Pt/Hierarchical SSZ-32 Catalysts to Boost the Selective Hydroisomerization of n-Hexadecane. *Catalysts* **2023**, *13*, 702. https://doi.org/10.3390/catal13040702

Academic Editors: Maja Milojević-Rakić and Danica Bajuk-Bogdanović

Received: 26 February 2023
Revised: 31 March 2023
Accepted: 3 April 2023
Published: 5 April 2023

1. Introduction

The hydroisomerization of long-chain *n*-alkanes plays an important role in the petroleum industry, in which heavy distillate and residue are converted into value-added products, such as gasoline, jet fuel, and lubricant oils [1]. The branching of *n*-alkanes through hydroisomerization is one strategy to improve the octane number of gasoline and to enhance the performance of diesel or lubricating oils at low temperatures [2,3]. It is agreed that the hydroisomerization reaction is always accompanied by a side reaction of hydrocracking, resulting in a much lower selectivity and yield of isomers than 100%, especially for the long-chain alkanes, which maintain a high tendency towards cracking [4]. Therefore, boosting the selectivity of hydroisomerization by controlling the cracking reaction is a very necessary and severe challenge. How to construct efficient and excellent hydrogen isomerization catalysts for n-alkanes to maintain high selectivity of isomeric alkanes and low cracking rate has always been the focus of research. Many works have been carried out to achieve a better balance of hydroisomerization and cracking reactions by regulating their reaction pathways. Unfortunately, there is still a lack of comprehensive understanding on how to tailor the reaction pathways in the hydroisomerization of *n*-alkanes.

Traditionally, metal–acid bifunctional catalysts are used for the hydroisomerization of *n*-alkanes, which undergo hydrogenation/dehydrogenation on metallic sites (usually Pt or

Pd) and protonation/deprotonation and skeleton rearrangement on acidic sites [5–9]. In the hydroisomerization of n-C_8 to n-C_{16} alkanes, silicoaluminophosphate and aluminosilicate molecular sieves, such as SAPO-11 [10], ZSM-22 [11–14], ZSM-23 [15,16], ZSM-48 [17], and SSZ-32 [18,19], are commonly used as the support of the bifunctional catalysts for shape-selective hydroisomerization [20–22]. Recently, SSZ-32 (0.45 nm × 0.52 nm) zeolite with a unique MTT structure and one-dimensional pores has attracted more attention in the hydroisomerization of long-chain n-alkanes because of its intrinsic microporous framework structure, suitable acid sites, high hydrothermal stability, environmental friendliness, and good shape selectivity [20]. In addition, compared with ZSM-23 zeolite, SSZ-32 has been considered to be superior for the hydrodewaxing of long-chain alkanes with good selectivity and smaller crystal size, which has many advantages in applied catalysis fields [23], such as producing lubricant base oils from Fischer–Tropsch wax [24]. The activity and selectivity of hydroisomerization are directly affected by the textural properties and acidity of supports, particle dispersion and reductivity of metal centers, and the metal–acid balance [6]. Many approaches have been developed to alter pore structure and acidity distribution. For example, a series of SAPO-11 molecular sieves were synthesized in an alcohol–water concentrated gel system with improved acidity and mesoporous volume and were used as the support for the hydroisomerization of n-hexadecane [25]. Liang et al. synthesized ZSM-23 zeolites with dual-structure directing agents and Fe-substituted ZSM-23 for the hydroisomerization of n-hexadecane [26,27]. In our previous work, we developed a partial detemplation method to investigate the influence of pore structure and acidity of Pt/ZSM-22 catalysts on the hydroisomerization of n-dodecane [20]. Post-synthetic treatments, such as desilication, dealumination, and sequential post-synthetic treatment, are an effective approach, and many zeolites have been widely used (ZSM-5, Hβ, HY, USY, SAPO-11, ZSM-48, H-ZSM-22, ZSM-23, and so on) for hydrogenation and hydroisomerization [28–31]. The alkali-treated and acid-treated ZSM-23-supported Pt bifunctional catalysts exhibited an outstanding n-C_{16} hydroisomerization performance, with the highest i-C16 yield of 64 wt.% at 340 °C [32]. The alkali–acid-treated hierarchical Pt/ZSM-22 zeolites with 0.3 M NaOH and with 6.0 M HCl and 0.1 M H_2SiF_6 exhibited 81.1% of the highest n-heptane conversion and 76.4% of total i-heptane yield at 260 °C [33]. Zhou et al. post-treated ZSM-5 zeolite with ammonium hexafluorosilicate (AHFS) and potassium carbonate (K_2CO_3) to control pore structure and acidity. They revealed that the use of AHFS and K_2CO_3 post-treatment resulted in the production of more mesopores to boost the diffusion of aromatic products and reduced the strength of strong acid sites in ZSM-5 to prevent excessive cracking reactions [34]. Wang et al. prepared hierarchical HY molecular sieves with good hierarchical structure by alkali treatment, which efficiently promoted the directional alkylation of naphthalene and cyclohexene catalytic reaction. The poriness and acidity of HY zeolite were optimized to increase the accessibility to active acid sites by regulating the concentration of NaOH, resulting in the hierarchical 0.5 HY zeolite with good catalytic activity and cycle stability [35]. However, studies reporting the roles of pore structure and acidity of SSZ-32 zeolite on the hydroisomerization of alkanes are rare. Introducing the secondary mesoporous networks into SSZ-32 molecular sieves with unique pore structures can effectively enhance the diffusion efficiency, expose more active sites, and improve the accessibility of acid sites, which is critical to explore the influence of textural properties and acidity on the reaction pathways of isomerization/cracking.

In this context, we tailored the textural properties and acidity of hierarchical SSZ-32 zeolites by desilication and sequential desilication–dealumination treatments. The catalytic hydroisomerization of long-chain n-alkanes over these post-treated SSZ-32-supported Pt catalysts was systematically carried out. The aim was to reveal the influence of the hierarchical structure as well as acid strength and distribution on the reaction pathways of isomerization/cracking and the formation mechanism of monobranched isomers.

2. Results and Discussion

2.1. Structure and Morphology of Pt/Hierarchical SSZ-32 Catalysts

The skeleton structure of the parent and post-treated SSZ-32 zeolites were determined by XRD characterization. As shown in Figure 1, all the samples showed a typical MTT topological structure with six distinct characteristic diffraction peaks at 2θ = 8.1°, 11.3°, 19.7°, 20.9°, 22.9°, and 24.0° (PDF# 37-0411). The Si/Al ratio decreased from 20.6 to 17.4 for SSZ32-0.3A and 14.3 for SSZ32-0.6A due to the desilication by alkali treatment, while after acid treatment, the Si/Al ratio of the alkali–acid-treated samples increased significantly, ascribed to the removal of extra-framework species. The relative crystallinity (RC) of all samples was calculated and the results are listed in Table 1. The relative crystallinity of the desilicated SSZ-32 samples decreased significantly with the increase in alkali concentration due to the partial collapse of the skeleton structure and the formation of the amorphous phase. However, the relative crystallinity of the SSZ32-0.3AS and SSZ32-0.6AS samples was partially recovered owing to the removal of the amorphous phase by the sequential acid leaching. This was confirmed by TEM characterization. The parent SSZ-32 zeolite showed a smooth external surface and complete rod-like morphology with a nanorod diameter around 20 nm. Meanwhile, certain intercrystalline voids were observed because of the accumulation of SSZ-32 nanorods. After being treated by weak desilication, the morphology and structure of SSZ32-0.3A and SSZ32-0.3AS samples remained nearly intact. A small number of mesopores were observed at the edge of the rod-like crystals in these samples. Compared with the SSZ32-0.3A sample, no crystal fragments were found on the surface of the SSZ32-0.3AS sample, confirming that the crystal fragments generated during the alkali treatment were removed by acid leaching. For the SSZ32-0.6A sample, the amount and size of mesopores increased significantly, and more crystal fragments were generated, while for the SSZ32-0.6AS sample, most of the crystal fragments were removed, and more and larger mesopores were formed. Mavrodinova et al. believed that extra lattices could be removed from the molecular sieve structure by acid reflux [36], thus, improving the accessibility of the inherent micropores of the molecular sieve. These results indicated that the mesopore was generated after the desilication treatment, and the sequential alkali–acid treatment could help to remove crystal fragments and amorphous phases, which further promoted the diffusion of reactants and improved the accessibility of acid sites in zeolite.

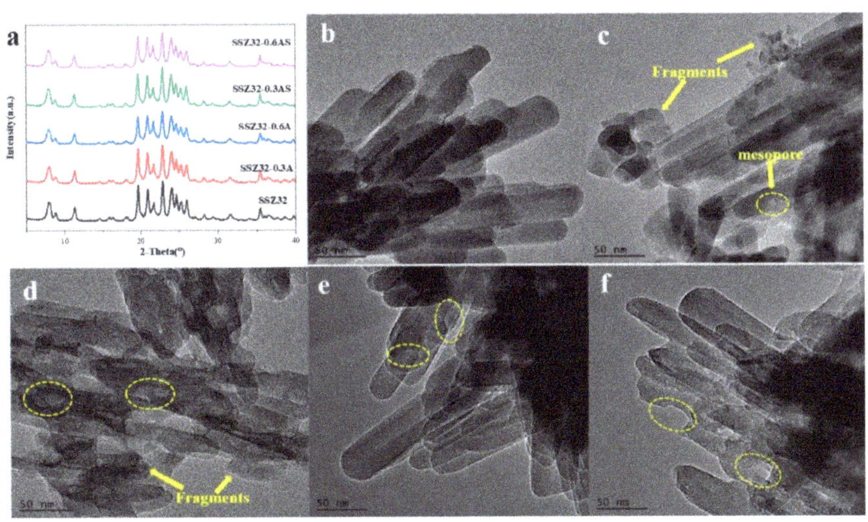

Figure 1. XRD patterns (**a**) and TEM images of parent SSZ32 (**b**) and post-treated SSZ32-0.3A (**c**), SSZ32-0.6A (**d**), SSZ32-0.3AS (**e**), and SSZ32-0.6AS (**f**).

Table 1. Textural properties for parent and hierarchical SSZ-32-supported Pt catalysts.

Samples	RC [a] (%)	Specific Surface Area (m^2/g)			Volume (cm^3/g)		
		S_{BET}	S_{micro}	S_{meso}	V_{total}	V_{micro}	V_{meso}
Pt/SSZ32	100	103.73	71.65	32.08	0.158	0.0400	0.118
Pt/SSZ32-0.3A	95.5	92.92	42.05	50.87	0.147	0.0203	0.127
Pt/SSZ32-0.6A	84.7	98.50	33.48	65.01	0.162	0.0184	0.144
Pt/SSZ32-0.3AS	99.7	134.85	62.30	72.55	0.178	0.0271	0.151
Pt/SSZ32-0.6AS	88.6	160.75	68.63	92.12	0.218	0.0342	0.184

[a] Determined by XRD.

To further understand the textural structure of the parent and alkali-/acid-treated catalysts, the pore size distribution was determined by conducting the N_2 isothermal adsorption–desorption curve. As shown in Figure 2, the parent SSZ32 zeolite showed a typical type-I isotherm profile with a rapid uptake at very low relative pressure of $p/p_0 < 0.02$ and an adsorption saturation platform at high relative pressure, which exhibited the characteristic of a microporous structure. Additionally, an obvious uptake at the relative pressure of $p/p_0 = 0.9$–0.99 and a faint hysteresis was also shown in the parent sample, indicating the existence of mesopores caused by the stacked gaps or voids [18,37]. Compared with the parent SSZ-32, all the alkali-/acid-treated catalysts showed a typical type-VI isotherm with a hysteresis in the relative pressure range of $p/p_0 = 0.4$–0.99, suggesting the formation of mesopores. The hysteresis became more and more obvious with the increase in the desilication degree. This indicated that micro-mesopore hierarchical structures were formed and more intracrystalline mesopores were generated by high-concentration alkali treatment. In addition, the uptake at the relative pressure of $p/p_0 < 0.1$ decreased gradually with the increase in desilication degree. The Pt/SSZ32-0.6A catalyst showed the lowest uptake at low relative pressure ($p/p_0 < 0.1$) and more obvious hysteresis at high relative pressure, indicating that more mesopores were generated with the decrease in micropores. After acid leaching, the uptake of alkali-treated samples at $p/p_0 < 0.1$ was significantly enhanced and a bigger hysteresis was obtained, which indicated that more micropores and mesopores blocked by the dissolved fragment were recovered. The pore size distribution derived from the BJH adsorption branch is plotted in Figure 2b. The intercrystalline mesopore in the Pt/SSZ32 was centered at 25 nm. After alkali treatment, the average mesopore size significantly increased up to 30 nm for Pt/SSZ32-0.3A and 40 nm for Pt/SSZ32-0.6A because of the generation of a large number of mesopores. With additional acid treatment, the mesopore size further increase slightly by about 3 nm due to the removal of leached extra-framework oxides. Additionally, the signal strength also increased significantly with the increase in desilication degree from alkali treatment, and further increased after acid treatment, especially for the Pt/SSZ32-0.6AS. This indicated that the content of mesopore increased significantly. As listed in Table 1, the parent SSZ32-supported Pt catalyst (Pt/SSZ32) showed a BET-specific surface area (S_{BET}) of 103.73 m^2/g with a S_{micro} value of 71.65 m^2/g, indicating the microporous characteristic of a SSZ-32 support. With the increase in alkali concentration, the S_{micro} value of alkali-treated catalysts decreased significantly with an increase in S_{meso} value. The S_{BET} value first showed a decrease for Pt/SSZ32-0.3A and then an increase for Pt/SSZ32-0.6A because of the significant increase in S_{meso} value. Additionally, the mesopore volume significantly increased from 0.118 cm^3/g for Pt/SSZ32 to 0.127 cm^3/g for Pt/SSZ32-0.3A and 0.144 cm^3/g for Pt/SSZ32-0.6A. After further acid treatment, the total S_{BET} value increased significantly with the increase in S_{micro} and S_{meso} values, especially for Pt/SSZ32-0.6AS with a high S_{micro} value (68.63 m^2/g) and S_{meso} value (92.12 m^2/g). Consequently, the mesopore volume of Pt/SSZ32-0.6AS increased significantly, up to 0.184 cm^3/g, derived from the quick increase of generated mesopores. In the alkali treatment, internal Si-OH groups of the zeolite were partially dissociated, and part of the micropores were etched to form the mesopores located on the external surface of the crystal structure or in the defect regions. Within the molecular sieve framework, the reaction occurred between the hydroxide anions (OH^-) and the silanol groups (defect sites).

The alkali metal cation stabilized the extracted silicate anions, leading to voids within the skeleton. The Na+ ion is the best leaching agent for silicon atoms because of its excellent ability to stabilize silicate anions [38]. In the meantime, partial framework Al species are usually removed and deposited on the external surface of the zeolite in the form of amorphous etched oxides to block part of the pore mouth, which resulted in the significant decrease in S_{micro} and V_{micro} values. Simultaneously, the V_{micro} values of Pt/SSZ32-0.3AS and Pt/SSZ32-0.6AS were still smaller than that of the Pt/SSZ32 because of the transformation of micropores into mesopores. These results indicated that the acid treatment of desilicated SSZ-32 in sequential alkali–acid treatment plays a vital role in recovering the pore surface area and volume by significantly removing the etched oxide species.

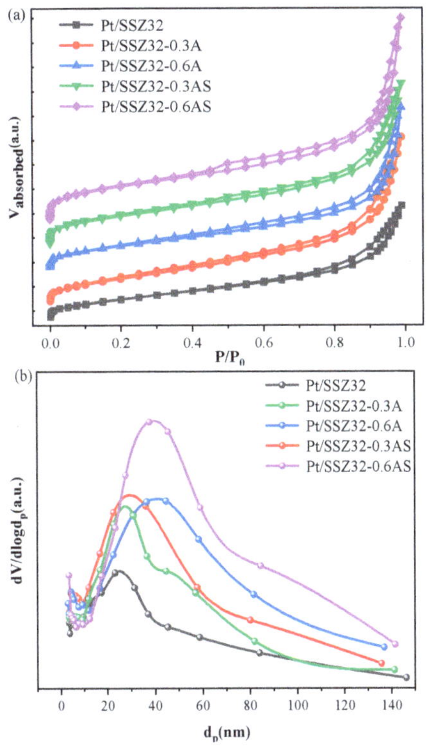

Figure 2. N_2 adsorption–desorption isotherm (**a**) and pore size distribution (**b**) of parent and hierarchical SSZ-32-supported Pt catalysts.

The chemical environment and coordination of Al species in all the samples were characterized by ^{27}Al MAS NMR. As shown in Figure 3, a sharp peak at ca. 57 ppm was observed in all the catalysts, which could be assigned to framework tetrahedral coordination Al species (FAl). A weak peak at 0 ppm assigned to extra-framework Al (EFAl) species was also observed. In the meantime, the framework Al signal at ≈ 57 ppm of the Pt/SSZ32-xA(S) samples showed a broad shoulder resonance extending to the upfield. This broadening might be associated with the distorted tetrahedral Al atoms (40–50 ppm region, Figure 3) [39–41]. The coordination environmental distribution of Al species was further calculated and listed in Table 2. The Pt/SSZ32 had a FAl content of 83.6% with a small number of distorted FAl species (6.1%) and EFAl species (10.3%). The EFAl species in this sample were probably generated from the conversion of the tetrahedrally coordinated framework Al at a high calcination temperature [42]. With the increase in alkali concentration, the framework and distorted framework Al species of the desilicated samples decreased with the significant increase in extra-framework Al species, up to 14% for Pt/SSZ32-0.3 and 24.8% for Pt/SSZ32-0.6A. The increased extra-framework Al species might have resulted in the channels in the zeolites becoming blocked [43]. After further acid treatment, the extra-framework Al species was obviously removed. The content of EFAl species decreased to 10.8% for the Pt/SSZ32-0.3AS sample. However, for the Pt/SSZ32-0.6AS sample, only 5.2% EFAl species was observed, indicating the removal of most extra-framework Al species. Part of the distorted framework Al species was converted into the framework Al species in the Pt/SSZ32-0.3AS sample. In contrast, part of the framework Al species was converted into the distorted framework Al species in the Pt/SSZ32-0.6AS sample. This indicated that in the process of the sequential alkali–acid treatment, acid treatment could not only remove the fragment and the generated extra-framework Al species in the alkali treatment, but also alter the distribution of framework and distorted framework Al species.

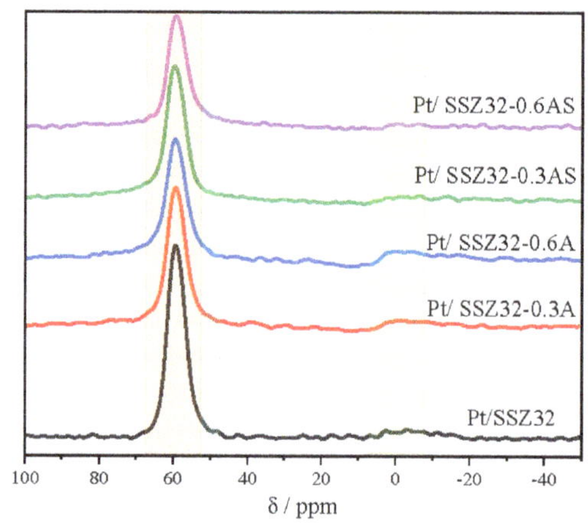

Figure 3. ^{27}Al MAS NMR spectra of parent and hierarchical SSZ−32−supported Pt catalysts.

Table 2. Quantitative analysis of Al proportions in different coordination states based on ^{27}Al MAS NMR spectra.

Catalysts	The Proportion of Different Al Species			Si/Al
	FAl % (57 ppm)	Distorted FAl(~52 ppm)	EFAl % (0 ppm)	
Pt/SSZ32	83.6	6.1	10.3	20.6
Pt/SSZ32-0.3A	80.6	5.4	14.0	17.4
Pt/SSZ32-0.6A	71.5	3.7	24.8	14.3
Pt/SSZ32-0.3AS	85.4	3.8	10.8	22.5
Pt/SSZ32-0.6AS	85.9	8.9	5.2	19.6

Acidity Properties

The acidity of different catalysts was determined by Py-FTIR and NH_3-TPD. To quantify the acid sites of zeolites after alkali treatment and sequential alkali–acid treatment, the Py-FTIR spectra at different temperature were obtained. As shown in Figure 4, the characteristic peaks at 1545 cm^{-1} and 1450 cm^{-1} belonged to the absorption vibration of pyridine adsorbed on the Brønsted acid sites (BAS) and Lewis acid sites (LAS). The quantitative analysis of Brønsted acidity (B) and Lewis acidity (L) was conducted according to the spectra measured at 200 °C and 350 °C, and the results are listed in Table 3. The Pt/SSZ32 catalyst showed a high B/L ratio (1.85) with a total acidity of 133.8 μmmol/g and a high ratio of strong BAS to strong LAS, indicating the dominance of BAS. As for the alkali-treated samples, the Brønsted and total acidity decreased with the increased Lewis acidity, especially the strong Lewis acidity. In addition, it can be observed that the decrease in total Brønsted acidity mainly came from the reduction in strong Brønsted acidity for Pt/SSZ32-0.3A. However, for Pt/SSZ32-0.6A, the decrease in total Brønsted acidity came from the decrease in weak Brønsted acidity. The change in acid sites was attributed to the partial destruction of the SSZ-32 microporous framework and the formation of mesoporous structure, generating the more accessible extra-framework Al and Si species. The generated extra-framework Al species was the precursor of LAS, which promoted the formation of acid sites directly and gave rise to Lewis acidity [44,45], and the B acidity was derived from the framework Al species [45]. As a result, the total acidity and the B/L ratio for the alkali-treated samples decreased, especially the ratio of strong B acid sites to strong L acid sites. After further acid treatment, more BAS sites were generated on the alkali–acid-treated catalyst with fewer LAS sites. This could be attributed to the formation of mesopores exposing more framework acid sites. Additionally, the Brønsted acidity and total acidity were increased and reached a maximum on the Pt/SSA32-0.6AS, and the increase in Brønsted acidity mainly derived from the strong Brønsted acidity for the Pt/SSA32-0.3AS but switched from the weak Brønsted acidity for Pt/SSA32-0.6AS. The Lewis acidity did not show an obvious change, probably due to the similar total extra-framework and distorted FAl species, both of which were the LAS [44]. Moreover, the B/L ratios significantly increased up to 2.44 for Pt/SSA32-0.3AS and 3.54 for Pt/SSA32-0.6AS, elucidating that Brønsted acidity contributed more to the total acidity for the alkali–acid-treated catalysts.

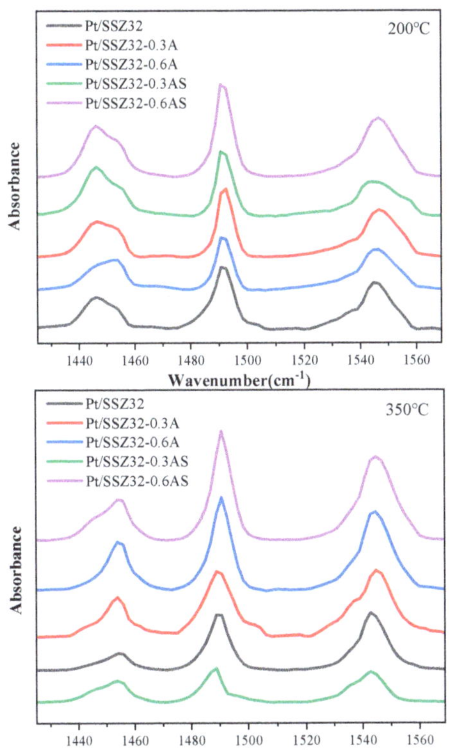

Figure 4. FTIR spectra of pyridine adsorption on parent and hierarchical SSZ−32−supported Pt catalysts.

Table 3. Acidity of parent and hierarchical SSZ-32-supported Pt catalysts determined by Py-FTIR spectra.

Samples	Brønsted Acidity (μmol g^{-1})		Lewis Acidity (μmol g^{-1})		Total Acidity (μmol g^{-1})		B/L Ratio	
	200 °C	350 °C	200 °C	350 °C	200 °C	350 °C	200 °C	350 °C
Pt/SSZ32	86.8	40.8	47.0	13.1	133.8	53.9	1.85	3.12
Pt/SSZ32-0.3A	83.0	35.4	47.8	29.6	130.8	65.0	1.74	1.20
Pt/SSZ32-0.6A	72.5	34.8	53.7	35.4	126.3	70.2	1.35	0.98
Pt/SSZ32-0.3AS	95.5	43.8	39.1	27.9	134.6	71.7	2.44	1.57
Pt/SSZ32-0.6AS	116.1	50.0	32.8	20.2	149.0	70.2	3.54	2.47

The acid properties of different catalysts after alkali and sequential alkali–acid treatment were further characterized by NH_3-TPD. As shown in Figure 5, there were two peaks of NH_3 desorption at 210–230 °C and 370–440 °C, which are usually assigned to the weak and strong acid sites, respectively [19]. Compared with the Pt/SSZ32 catalyst, the peak at high temperature shifted to a lower temperature for the alkali-treated catalysts, implying the decreased acid strength of the alkali-treated catalysts, especially for Pt/SSZ32-0.6A. However, after the alkali–acid treatment, both the NH_3 desorption peaks shifted toward a higher temperature. This suggested the enhanced overall acid strength of the sequential alkali–acid-treated catalysts, especially for Pt/SSZ32-0.6AS. The acid strength change in these catalysts was ascribed to the reduced framework Al species with the increased extra-framework Al species after NaOH treatment, more exposed framework Al species in the formed mesopore, and the regenerated channels by the removal of extra-framework Al species after further acid treatment. These resulted in the increased acid sites per unit of catalyst mass (Table 3). These results indicated that the Pt/SSZ32-0.6AS catalyst had a stronger acidity strength than other catalysts, which was consistent with the results of Py-FTIR, textual properties, and XRF analysis. The acidity variation changed by the alkali–acid treatment would subsequently have a significant direct impact on the performance of the *n*-hexadecane hydroisomerization process.

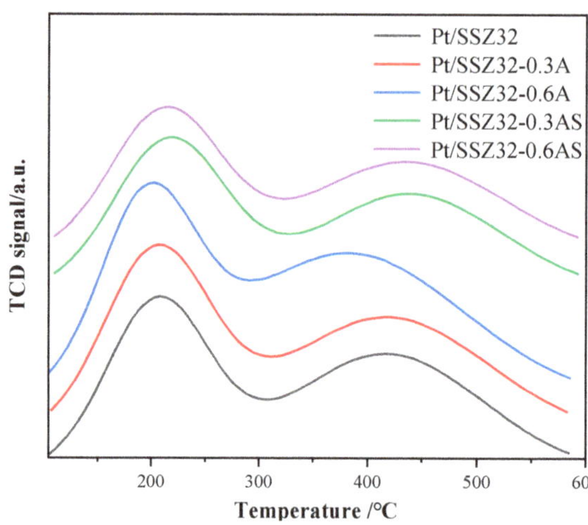

Figure 5. NH_3-TPD profiles of parent and hierarchical SSZ-32-supported Pt catalysts.

2.2. *Catalytic Hydroisomerization of n-Hexadecane over Different Catalysts*

The influence of post-treated SSZ-32-zeolite-supported Pt catalysts on the hydroisomerization reaction process was further investigated. The hydroisomerization performance of different catalysts was evaluated using *n*-hexadecane as the model compound in a fixed-bed reactor. As shown in Figure 6, the conversion of *n*-hexadecane over different catalysts gradually increased with the elevating temperature. The Pt/SSZ32 catalyst showed a high activity for the n-hexadecane hydroisomerization with the conversion of ca. 100% at the temperature of 280 °C. Compared with the Pt/SSZ32 catalyst, the conversion of n-hexadecane over the alkali-treated catalysts at the same reaction temperature became lower and decreased in the order Pt/SSZ32 > Pt/SSZ32-0.3A > Pt/SSZ32-0.6A. On the contrary, the activity of alkali–acid-treated catalysts was significantly enhanced and became higher than that of the Pt/SSZ32 catalyst. The highest activity was achieved over the Pt/SSZ32-0.6AS catalyst with a relatively low temperature at 260 °C to reach the conversion of 100%. Although the apparent hierarchical structure was generated in the alkali-treated catalysts, the catalytic activity was not enhanced. Obviously, the catalytic activity depended on the total B acidity (Table 3). The Pt/SSZ32-0.3A catalyst showed a similar B acidity with the Pt/SSZ32 catalyst. The similar curve of conversion versus temperature was also obtained over these two catalysts. A much lower activity was obtained on the Pt/SSZ32-0.6A catalyst because of the lowest B acidity, while for the Pt/SSZ32-0.3AS and Pt/SSZ32-0.6AS catalysts, the total B acidity increased more than that of Pt/SSZ32, and, accordingly, the activity became much higher. These results indicated that the total B acidity played the vital role in the hydroisomerization of n-alkanes on the bifunctional catalysts, which was consistent with the literature [46,47]. This further proved that the B acid sites participate in the reaction of alkane hydroisomerization, and more B acid sites could enhance n-hexadecane protonation to boost the catalytic performance according to the bifunctional catalytic mechanism [32,47]. Of course, the formed mesopore also contributed to providing more accessible acid sites and reducing the diffusion restriction of reactants and/or intermediates, which was also beneficial to improving the catalytic activity.

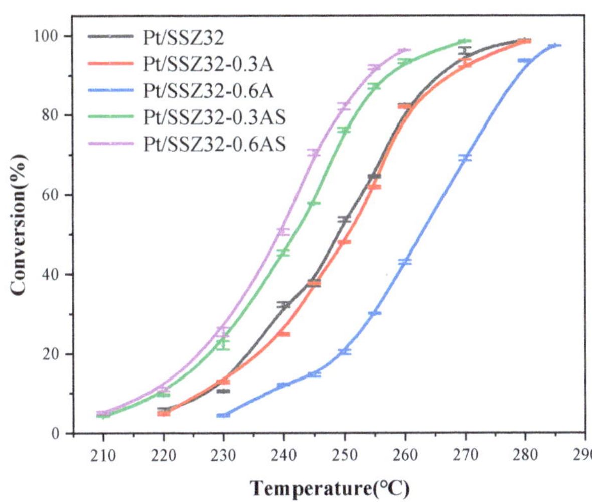

Figure 6. n-Hexadecane conversion over parent and hierarchical SSZ-32-supported Pt catalysts.

The yields of total C_{16} isomers on different catalysts are shown in Figure 7a. The similar yield curves were obtained on all catalysts. As the reaction temperature increased, the yield of i-C_{16} products increased then decreased at high conversion. The highest yield of i-C_{16} products (71.5%) was achieved on the Pt/SSZ32-0.6AS catalyst at 255 °C, while the maximum yield of i-C_{16} products was only 65~68% over other catalysts. In the hydroisomerization of n-alkanes, the rearrangement of carbenium ions is the rate-determining step [47]. The rearrangement of carbenium ions can be accelerated with an increase in reaction temperature, resulting in an increase in isomer yield. However, at higher temperature, the hydrocracking of isomers occurred via two types of scission (type A tertiary alkylcarbenium ions intervene, and type B, β-scission) and became severe. Therefore, the yield of isomers decreased. n-Hexadecane underwent hydroisomerization and hydrocracking reactions simultaneously over the bifunctional catalysts, which were competitive reactions. The regulation of the isomerization-cracking equilibrium mainly focuses on the control of the probability of the intermediates meeting with acid sites and the retention time for intermediates over acid sites. The post-treated catalysts showed different catalytic selectivity for the reactions of hydroisomerization and hydrocracking. As shown in Figure 7b, the selectivity of *i*-C_{16} products decreased gradually with the increase in conversion. As the conversion exceeded 90%, the selectivity of *i*-C_{16} products dramatically decreased due to the rapid increase in cracking products (Figure 7c). All the catalysts except Pt/SSZ32-0.6A showed a high selectivity of *i*-C_{16} products over 85%, with the conversion below 80%. The Pt/SSZ32-0.6A catalyst exhibited the lowest selectivity of ca. 80% with the highest cracking product selectivity (ca. 20%). In the conversion of 35~85%, the selectivity of *i*-C_{16} products decreased over the catalysts as follows: Pt/SSZ32-0.3A > Pt/SSZ32-0.6AS > Pt/SSZ32-0.3AS > Pt/SSZ32 > Pt/SSZ32-0.6A. The yield ratio of isomers to cracking products (*iso*/C ratio) was further obtained to compare the influence of post-treatment on the selectivity of two reaction pathways. As shown in Figure 7d, the *iso*/C ratio decreased as follows: Pt/SSZ32-0.3A > Pt/SSZ32-0.6AS > Pt/SSZ32-0.3AS > Pt/SSZ32 > Pt/SSZ32-0.6A. These results indicated that the post-treated catalysts, except Pt/SSZ32-0.6A, significantly enhanced the hydroisomerization reaction pathway rather than the cracking reaction pathway. This could be attributed to the variation in Brønsted acidity, especially the weak B acidity. As for the hydroisomerization over different zeolites, a detailed understanding of the reaction mechanism in the presence and/or absence of shape selectivity has been developed by extensive studies with both large-pore zeolites and with medium-pore zeolites [48–52]. According to the reported mechanism of alkane hydroisomerization, the consecutive reaction steps occur on an ideal bifunctional catalyst: *n*-hexadecane ↔ monobranched isohexadecane ↔ multibranched isohexadecane ↔ cracked products. Furthermore, the weak Brønsted acidity is beneficial to the hydroisomerization reaction, while the strong B acidity mainly promotes the excessive isomerism reaction, and the generated multibranched isomeric products are likely to crack, thus, contributing to the cracking reaction [53]. It was speculated that the variation in *iso*/C ratios was relative to the change in Brønsted acidity distribution. As listed in Table 3, the alkali treatment caused the decrease of Brønsted acidity for Pt/SSZ32-0.3A and Pt/SSZ32-0.6A. Although more mesopores were generated in Pt/SSZ32-0.6A than in Pt/SSZ32-0.3A, the ratio of weak to strong Brønsted acidity (B_w/B_s) for Pt/SSZ32-0.6A (1.09) became much lower than that for Pt/SSZ32-0.3A (1.35) due to the partial coverage of BAS in the formed hierarchical structure. Moreover, the B_w/B_s ratio for Pt/SSZ32-0.3A was also much higher than that of Pt/SSZ32 (1.13). As shown in Figure 7c, the Pt/SSZ32-0.3A catalyst showed a much higher hydroisomerization selectivity than the Pt/SSZ32 and Pt/SSZ32-0.6A catalysts. This indicated that the reaction pathway selectivity of hydroisomerization and cracking strongly depended on the weak Brønsted acidity distribution of the SSZ-32 zeolite. Compared with Pt/SSZ32-0.3A, the acid-treated Pt/SSZ32-0.3AS catalyst exhibited more and larger mesopores but a lower B_w/B_s ratio (1.18) because of the removal of desilicated Al fragments and the destruction of some framework Al species. In contrary, the Pt/SSZ32-0.6AS showed more mesopores and a higher B_w/B_s ratio (1.32) than that of Pt/SSZ32-0.6A due to the removal of extra-framework Al species. These two catalysts showed higher *iso*/C values than the Pt/SSZ32 catalyst. Interestingly, the *iso*/C values almost showed a linear increase with the increase in B_w/B_s ratio, as shown in Figure 7d. These results confirmed that the selectivity of the hydroisomerization reaction strongly depended on the B_w/B_s ratio. Different from the reported results, the generated hierarchical structure in the post-treatment had no direct influence on the reaction pathways of hydroisomerization and cracking but could regulate the acidity distribution.

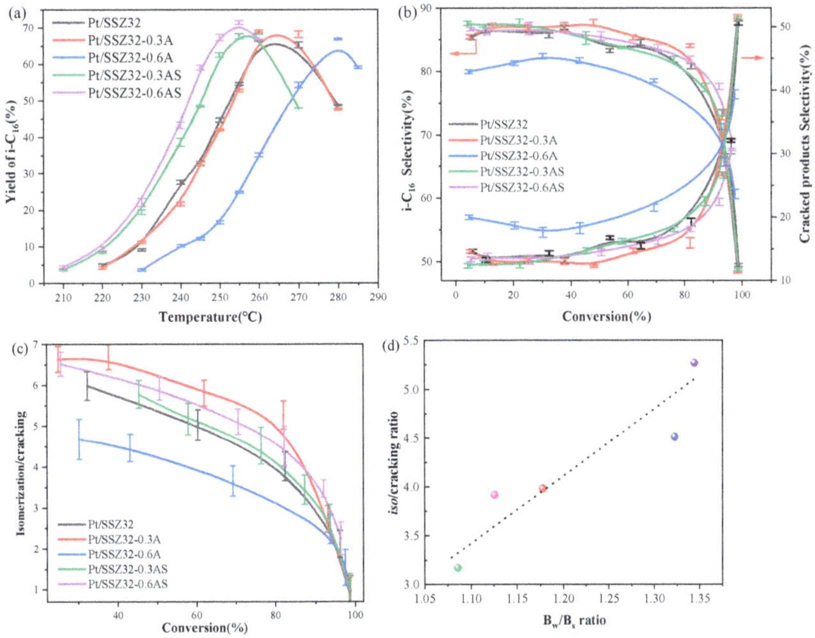

Figure 7. The reaction pathways' selectivity of hydroisomerization and hydrocracking over parent and hierarchical SSZ-32-supported Pt catalysts. (**a**) Yield of *i*-C16 products; (**b**) the selectivity of isomers and cracking products; (**c**) the ratio of isomerization/cracking reaction pathways versus conversion; (**d**) the iso/cracking ratios at ca. 80% conversion versus B_w/B_s ratios.

As per the aforementioned reaction mechanism, the hydroisomerization of *n*-hexadecane could convert it into the monobranched *i*-C_{16} products and multibranched *i*-C_{16} products. The isomer distribution over different catalysts is plotted in Figure 8. All the catalysts showed similar variations in monobranched and multibranched *i*-C_{16} products. As shown in Figure 8c, the ratio of mono-/multibranched *i*-C_{16} products was much higher than 1 and decreased linearly with the increase in *n*-C_{16} conversion. This indicated that the hydrocracking of multibranched isomer occurred similarly on all the catalysts, and the dibranched and tribranched alkanes could not be accommodated by sterical restrictions; thus, monobranched isomers were exclusively formed on these bifunctional catalysts [44]. Compared with the Pt/SSZ32 catalyst, the alkali- and alkali–acid-treated catalysts showed a higher ratio of mono-/multibranched *i*-C_{16} products, and the ratio deceased in the following order: Pt/SSZ32-0.3A > Pt/SSZ32-0.6A ≈ Pt/SSZ32-0.6AS > Pt/SSZ32-0.3AS > Pt/SSZ32. This result suggests that the formed hierarchical structure dominantly enhanced the selectivity of monobranched *i*-C_{16} products by enhancing the diffusion of the reactants and intermediates in the bifunctional catalysts. However, comparing the different hierarchical catalysts, it could be inferred that the synergetic effect of hierarchical structure and acidity regulation promoted the selectivity of monobranched *i*-C_{16} products.

The product distribution of monobranched *i*-C_{16} products was further analyzed to investigate the reaction mechanisms involved. Figure 9 shows the distribution of 2-,3-,4-, 5- and 6/7/8-methyl isomers. All the catalysts showed the highest selectivity of 2-methyl isomer, probably because this skeletal isomer possessed a minimum diffusion limit [54]. Additionally, high central position branching, such as 6/7/8-methyl isomers, were obtained on all the catalysts. The selectivity of 2-methyl isomer over the Pt/SSZ32, Pt/SSZ-0.3A, and Pt/SSZ-0.6A catalysts decreased gradually from the initial 30% to 20% at the *n*-C_{16} conversion of 80%. However, for the Pt/SSZ-0.3AS and Pt/SSZ-0.6AS catalysts, the selectivity of 2-methyl isomer decreased gradually from the initial 35% to 25%. At high conversion (80~100%), the selectivity of 2-methyl isomer showed a rapid decrease to ca. 15%. The total selectivity of 6/7/8-methyl isomers showed an almost linear increase from ca. 32% to 42% with the increase in n-C_{16} conversion up to 100% at the same time. The selectivity of 3-, 4- and 5-methyl isomers showed almost the same change over the different catalysts. All the catalysts showed a selectivity of 3-methyl isomers of about 15%. About 12% selectivity of 4-methyl isomers was obtained at the conversion below 80% and then gradually increased by about 3% at 100% conversion, while the selectivity of 5-methyl isomers slowly increased at below 80% conversion and then also gradually increased by about 3% at 100% conversion. As reported, the isomer product distributions could be explained by the pore mouth and key–lock modes of physisorption of long n-alkanes in pore openings of the zeolite [54]. The 2- and 3-methyl-branched isomers were more favorably formed via the pore mouth mechanism, and key–lock modes favored the central branching of the chain, such as 6/7/8-methylbranched isomers along the carbon chain [19,54]. To further investigate the influence of post-treatment on the isomerization modes, the typical product distribution at 80% conversion was obtained and listed in Table 4. It can be observed that the yield ratio of 2-/3-methylbranched to 6/7/8-methylbranched isomers (P/K ratio) changed over the different post-treated catalysts. The alkali-treated Pt/SSZ32-0.3A and Pt/SSZ32-0.6A showed almost the same P/K ratios (~0.96) with the Pt/SSZ32 catalyst, indicating that the key–lock mode dominated the formation of monobranched isomers over these catalysts. However, the alkali–acid-treated catalysts exhibited a higher P/K ratio (1.009 for Pt/SSZ32-0.3AS and 1.063 for Pt/SSZ32-0.6AS), indicating that the pore mouth mode became the dominant reaction pathway in the formation of monobranched isomers. This suggests that only alkali-treated catalysts had no influence on the isomerization pathways, but the sequential alkali–acid-treated catalysts could tailor the adsorption mode of isomerization intermediates, probably due to the pore widening and more exposed pore mouth, as well as the more exposed acid sites in the pore mouth.

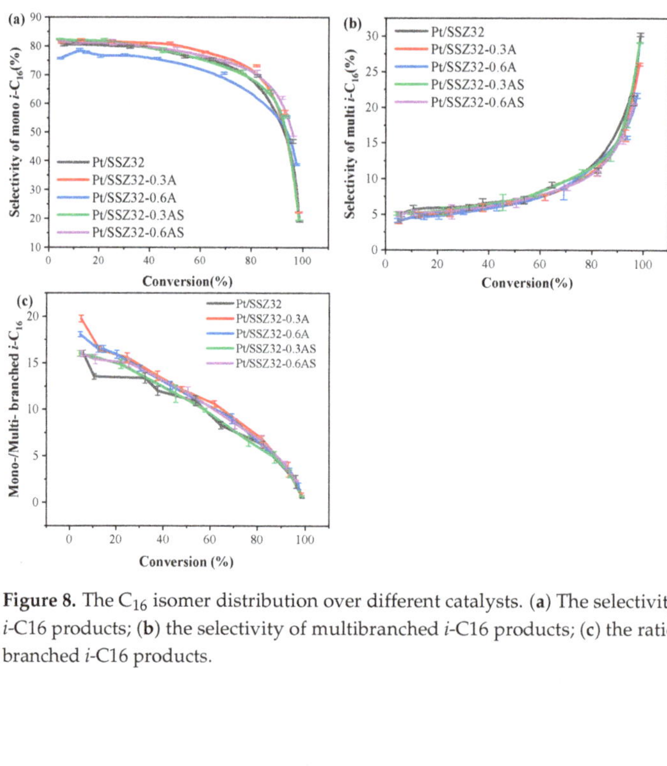

Figure 8. The C_{16} isomer distribution over different catalysts. (**a**) The selectivity of monobranched i-C16 products; (**b**) the selectivity of multibranched i-C16 products; (**c**) the ratio of mono- to multi-branched i-C16 products.

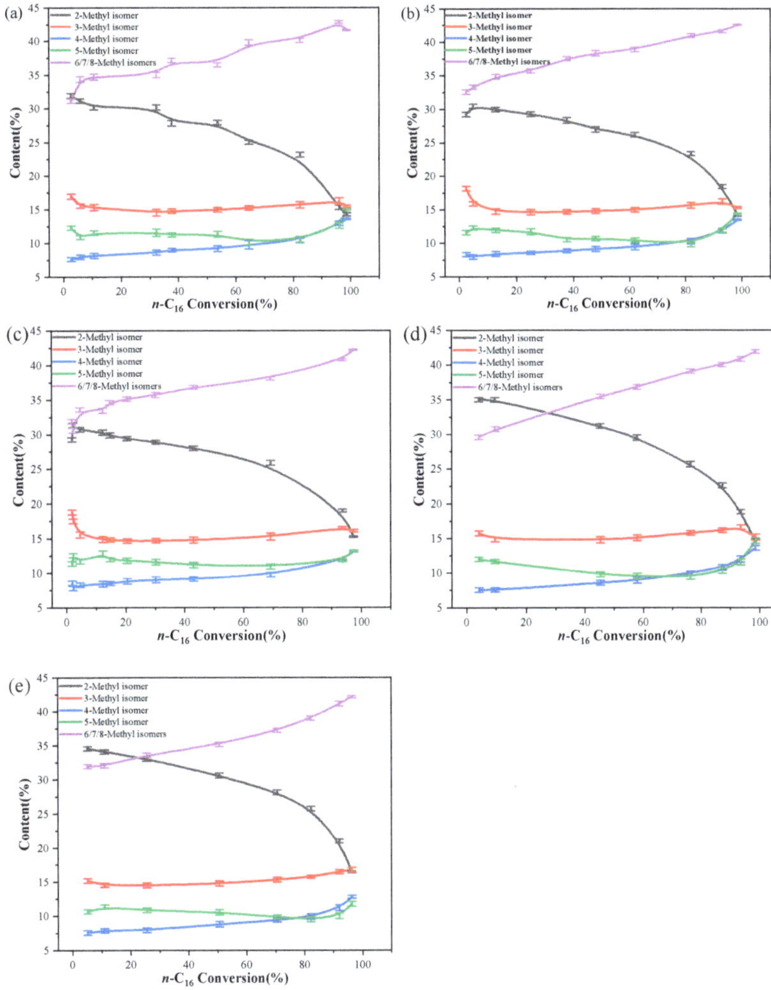

Figure 9. The selectivity of methyl-branched i-C_{16} isomers versus n-C_{16} conversion over different catalysts. (**a**) Pt/SSZ32; (**b**) Pt/SSZ32-0.3A; (**c**) Pt/SSZ32-0.6A; (**d**) Pt/SSZ32-0.3AS; (**e**) Pt/SSZ32-0.6AS.

Table 4. The product distribution of n-C16 hydroisomerization on different catalysts at the conversion rate of 80%.

i-C_{16} Products	Yield (%)				
	Pt/SSZ32	Pt/SSZ32-0.3A	Pt/SSZ32-0.6A	Pt/SSZ32-0.3AS	Pt/SSZ32-0.6AS
2-Methyl isomer (pore mouth)	13.30	14.03	11.19	13.27	15.04
3-Methyl isomer (pore mouth)	9.02	9.40	8.07	8.85	9.19
4-Methyl isomer	5.97	6.18	5.52	5.77	5.82
5-Methyl isomer	6.08	5.90	5.78	5.48	5.57
6/7/8-Methyl isomers (key–lock)	23.03	24.62	20.02	21.92	22.80
Pore mouth/key–lock	0.969	0.952	0.962	1.009	1.063

3. Experimental Section

3.1. Catalysts Preparation

3.1.1. Synthesis of Parent SSZ-32

The parent SSZ-32 zeolite used in this work was synthesized by a hydrothermal method according to the literature [18]. First, 1.2 g of N′N-diisopropylimidazolium chloride (Acros) was dissolved in 17.6 g of deionized water, and 9.35 g of 1.0 M KOH aqueous solution was added to adjust the pH value. After stirring for 15 min, 5.93 g of Ludox AS- 30 (Sigma-Aldrich, Shanghai, China) was added, followed by dropping 1.87 g of Nalco's 1056 colloid sol (Sigma-Aldrich, Shanghai, China) into the mixture. Then, 0.7 g of isobutylamine was added and further stirred for an additional 15 min. The obtained gel was transferred into a Teflon-lined steel autoclave for dynamic crystallization (120 rpm) at 175 °C for 9 days. The product was filtered, washed, and dried at 120 °C for 12 h, and then finally calcined at 595 °C for 5 h.

3.1.2. Synthesis of Hierarchical SSZ-32

The parent SSZ-32 was calcined at 595 °C in air for 5 h before the following treatment. The desilication of SSZ-32 was first carried out. In a typical run, 1 g of SSZ-32 zeolite was mixed with 10 mL NaOH solution (0.3 M, 0.6 M) and heated at 80 °C for 1.5 h under stirring. Subsequently, the obtained slurry was cooled to room temperature, filtered, washed with deionized water to pH = 7.0, and then dried overnight at 120 °C. The resultant zeolites were subjected to continuous ion exchange three times with a 1.0 M NH_4Cl solution at 80 °C for 4 h, filtered, dried, and finally calcined at 550 °C for 4 h. The obtained samples were denoted as SSZ32-*n*A, where *n* represents the concentration of NaOH alkaline solution in desilication.

Then, the sequential dealumination of desilicated SSZ-32 zeolites was further carried out. Part of the above desilicated zeolite was subjected to acid treatment in 0.1 M HNO_3 solution at 80 °C for 6 h with a solid concentration of 20 g L^{-1}. The obtained slurry was cooled to room temperature, filtered, washed with deionized water to be neutral and dried overnight at 120 °C. Then, the resultant zeolites were ion-exchanged three times with NH_4Cl (1.0 M) solution at 80 °C for 4 h and calcined at 550 °C for 4 h to prepare the protonated samples. The obtained samples were denoted as SSZ32-*n*AS.

The parent and as-prepared hierarchical SSZ-32-zeolite-supported 0.5 wt.% Pt catalysts were prepared by an incipient wetness impregnation method using H_2PtCl_6 aqueous solution. The samples were dried overnight at room temperature. Then, the samples were calcined at 450 °C in air for 4 h. Finally, the obtained catalysts loaded with Pt were named Pt/SSZ32, Pt/SSZ32-0.3A, Pt/SSZ32-0.6A, Pt/SSZ32-0.3AS, and Pt/SSZ32-0.6AS, respectively.

3.2. Catalyst Characterization

The phase purity and relative crystallinity of the samples were characterized by X-ray diffraction (XRD) on a D8 advance X-ray diffractometer with (Cu Kα) radiation at 40 kV and 140 mA. The morphology and structure of the samples were characterized by field emission scanning electron microscopy (SEM, Hitachi S-4800) and transmission electron microscopy (TEM, Tecnai G2 F20). The pore structure and specific surface area of the catalysts were measured using a Micromeritics Model ASAP 2460 Version 3.01 volumetric instrument. Prior to the test, the samples were degassed at 300 °C for 12 h under vacuum. The Brunauer–Emmett–Teller (BET) method was used to calculate the total specific surface area. The total pore volume of the pores was obtained at p/p_o = 0.95 and the *t*-plot method was used to determine the micropore volume and micropore area. The mesopore size distribution was obtained from the adsorption branch of the isotherm using the Barrett–Joyner–Halenda (BJH) model. The Si and Al content of each sample were determined with an X-ray fluorescence spectrometer (XRF, Supermini200). Pyridine adsorption Fourier transform infrared spectra (Py-IR) was collected on a VERTEX 70 spectrometer (Bruker) in the scanning range of 400–4000 cm^{-1} to make a distinction between the Lewis (LAS) and Brønsted (BAS) acid sites. The densities of the Lewis and Brønsted acid sites were calculated by integrating the absorbance at 1545 and 1455 cm^{-1}, respectively, in the py-IR spectra at 200 °C and 350 °C. Temperature-programmed desorption of ammonia (NH_3-TPD) was carried out with an AMI-300 chemisorption analyzer to measure the total number of acid sites of the samples. Solid-state ^{27}Al magic-angle spinning (MAS) nuclear magnetic resonance spectra (^{27}Al-NMR) were acquired on a Varian Infinity Plus 300 MHz solid-state NMR spectrometer.

3.3. Catalytic Hydroisomerization of n-Hexadecane

The hydroisomerization of *n*-C16 was carried out in a down-flow fixed bed (1.0 cm i.d. and 45 cm in length). The reaction temperature was monitored with a thermocouple in the catalyst bed and controlled by three thermocouples on the reactor. Typically, 2 g of the catalysts in 20–40 mesh were loaded into the reactor. Before the reaction, the catalysts were reduced in situ with hydrogen at 400 °C and 3 MPa for 4 h. Then, the reaction was carried out under 260–330 °C at 3 MPa with a H_2/n-C16 volume ratio of 1000 and weight hourly space velocity (WHSV) of 2.0 h^{-1}. After the reaction became stable, the products were cooled with a heat exchanger and were separated into gas and liquid fractions in a separator. Different liquid samples at an interval of 30 min were collected. The products were qualitatively and quantitatively analyzed by an Agilent 6890 N gas chromatography/5975 N mass spectrometer (GC/MS) and a gas chromatographer (Agilent 7890A GC). The conversion of *n*-hexadecane(*n*-C16) and selectivity of isomers (*i*-C16) were calculated according to [18]. The average results from the three replicated experiments were used.

4. Conclusions

In summary, the hierarchical SSZ-32-zeolite-supported Pt catalysts were prepared by alkali treatment and sequential alkali–acid treatment for *n*-hexadecane hydroisomerization. The hierarchical structure and acidity of support SSZ-32 zeolite were tailored. The sequential acid treatment after alkali treatment generated more mesopore and more Brønsted acidity and increased the B_w/B_s ratio. Compared with Pt/SSZ32(1.13), the Pt/SSZ32-0.3AS and Pt/SSZ32-0.6AS BW/BS ratios prepared by sequential alkali–acid treatment were 1.18 and 1.32, respectively. The mesopores generated by post-treatment had nearly no effect on the reaction pathway selectivity of hydroisomerization and cracking. The selectivity on the hydroisomerization reaction strongly depended on the B_w/B_s ratio. The *iso*/C values almost had a linear increase with the increase in B_w/B_s ratio. The iso/C ratio increased as follows: Pt/SSZ32-0.6A(3.17) $<$ Pt/SSZ32(3.92) $<$ Pt/SSZ32-0.3AS(3.98) $<$ Pt/SSZ32-0.6AS(4.51) $<$ Pt/SSZ32-0.3A(5.27); the corresponding BW/BS ratio is 1.09, 1.13, 1.18, 1.32, and 1.35, respectively. The synergetic effect of hierarchical structure and acidity regulation promoted the selectivity of monobranched i-C_{16} products. Moreover, the alkali-treated SSZ-32-supported Pt catalysts had no influence on the isomerization mechanism of monobranching via pore mouth or key–lock modes, but the sequential alkali–acid-treated SSZ-32-supported Pt catalysts could enhance the pore mouth mode to form the monobranched i-C_{16} products. This work could provide guidance on designing high-performance bifunctional catalysts and a more comprehensive understanding of the support's roles in the hydroisomerization of long-chain *n*-alkanes. Meanwhile, it also provided an alternative strategy for enhancing the low-temperature performance of diesel or lubricating oils in the petroleum industry.

Author Contributions: X.Y.: investigation, conducting experiments, writing—original draft. W.Z.: investigation, measurement and characterization. L.L.: visualization, methodology, formal analysis. X.N.: formal analysis, data curation. Q.W.: conceptualization, supervision, writing—review and editing, data curation. All authors have read and agreed to the published version of the manuscript.

Funding: This research was funded by National Natural Science Foundation of China [Grant No. 22279088].

Acknowledgments: The financial support by the National Natural Science Foundation of China (Grant No. 22279088) is gratefully acknowledged. The authors would like to thank the workers in Shiyanjia Lab (www.shiyanjia.com (accessed on 31 August 2022)) for the material characterizations.

Conflicts of Interest: The authors declare that they have no known competing financial interests or personal relationships that could have appeared to influence the work reported in this paper.

References

1. Benitez, V.M.; Yori, J.C.; Grau, J.M.; Pieck, C.L.; Vera, C.R. Hydroisomerization and Cracking of n-Octane and n-Hexadecane over Zirconia Catalysts. *Energy Fuels* **2006**, *20*, 422–426. [CrossRef]
2. Corma, A. Transformation of hydrocarbons on zeolite catalysts. *Catal. Lett.* **1993**, *22*, 33–52. [CrossRef]
3. Ward, J.A. Hydrocracking processes and catalysts. *Fuel Process. Technol.* **1993**, *35*, 55–85. [CrossRef]
4. Gao, L.; Shi, Z.Y.; Etim, U.J.; Wu, P.P.; Xing, W.; Zhang, Y.; Bai, P.; Yan, Z.F. Superior catalytic performance of micro-mesoporous Beta-SBA-15 composite with a high indexed isomerization factor in hydroisomerization of n-heptane. *Fuel* **2019**, *252*, 653–665. [CrossRef]
5. Yu, R.; Tan, Y.; Yao, H.; Xu, Y.; Huang, J.; Zhao, B.; Du, Y.; Hua, Z.; Li, J.; Shi, J. Toward n-Alkane Hydroisomerization Reactions: High-Performance Pt–Al_2O_3/SAPO-11 Single-Atom Catalysts with Nanoscale Separated Metal-Acid Centers and Ultralow Platinum Content. *ACS Appl. Mater. Inter.* **2022**, *14*, 44377–44388. [CrossRef]
6. García-Pérez, D.; Blanco-Brieva, G.; Alvarez-Galvan, M.C.; Campos-Martin, J.M. Influence of W loading, support type, and preparation method on the performance of zirconia or alumina-supported Pt catalysts for n-dodecane hydroisomerization. *Fuel* **2022**, *319*, 123704. [CrossRef]
7. Jokar, F.; Alavi, S.M.; Rezaei, M. Investigating the hydroisomerization of n- pentane using Pt supported on ZSM-5, desilicated ZSM-5, and modified ZSM-5/MCM-41. *Fuel* **2022**, *324*, 124511. [CrossRef]
8. Tan, Y.; Hu, W.; Du, Y.; Li, J. Species and impacts of metal sites over bifunctional catalyst on long chain n-alkane hydroisomerization: A review. *Appl. Catal. A Gen.* **2021**, *611*, 117916. [CrossRef]
9. Deldari, H. Suitable catalysts for hydroisomerization of long-chain normal paraffins. *Appl. Catal. A Gen.* **2005**, *293*, 1–10. [CrossRef]
10. Jin, D.; Li, L.; Ye, G.; Ding, H.; Zhao, X.; Zhu, K.; Coppens, M.-O.; Zhou, X. Manipulating the mesostructure of silicoaluminophosphate SAPO-11 via tumbling-assisted, oriented assembly crystallization: A pathway to enhance selectivity in hydroisomerization. *Catal. Sci. Technol.* **2018**, *8*, 5044–5061. [CrossRef]
11. Niu, P.; Xi, H.; Ren, J.; Lin, M.; Wang, Q.; Chen, X.; Wang, P.; Jia, L.; Hou, B.; Li, D. Micropore blocked core–shell ZSM-22 designed via epitaxial growth with enhanced shape selectivity and high n-dodecane hydroisomerization performance. *Catal. Sci. Technol.* **2018**, *8*, 6407–6419. [CrossRef]
12. Liu, S.; He, Y.; Zhang, H.; Chen, Z.; Lv, E.; Ren, J.; Yun, Y.; Wen, X.; Li, Y.-W. Design and synthesis of Ga-doped ZSM-22 zeolites as highly selective and stable catalysts for n-dodecane isomerization. *Catal. Sci. Technol.* **2019**, *9*, 2812–2827. [CrossRef]
13. Li, T.; Wang, W.; Feng, Z.; Bai, X.; Su, X.; Yang, L.; Jia, G.; Guo, C.; Wu, W. The hydroisomerization of n-hexane over highly selective Pd/ZSM-22 bifunctional catalysts: The improvements of metal-acid balance by room temperature electron reduction method. *Fuel* **2020**, *272*, 117717. [CrossRef]
14. Niu, P.; Xi, H.; Ren, J.; Lin, M.; Wang, Q.; Jia, L.; Hou, B.; Li, D. High selectivity for n-dodecane hydroisomerization over highly siliceous ZSM-22 with low Pt loading. *Catal. Sci. Technol.* **2017**, *7*, 5055–5068. [CrossRef]
15. Woo Lee, S.; Ki Ihm, S. Hydroisomerization and hydrocracking over platinum loaded ZSM-23 catalysts in the presence of sulfur and nitrogen compounds for the dewaxing of diesel fuel. *Fuel* **2014**, *134*, 237–243. [CrossRef]
16. Zhang, M.; Chen, Y.; Wang, L.; Zhang, Q.; Tsang, C.-W.; Liang, C. Shape Selectivity in Hydroisomerization of Hexadecane over Pt Supported on 10-Ring Zeolites: ZSM-22, ZSM-23, ZSM-35, and ZSM-48. *Ind. Eng. Chem. Res.* **2016**, *55*, 6069–6078. [CrossRef]
17. Zhang, M.; Li, C.; Chen, Y.; Tsang, C.-W.; Zhang, Q.; Liang, C. Hydroisomerization of hexadecane over platinum supported on EU-1/ZSM-48 intergrowth zeolite catalysts. *Catal. Sci. Technol.* **2016**, *6*, 8016–8023. [CrossRef]
18. Liu, L.; Zhang, M.; Wang, L.; Zhang, X.; Li, G. Construction of ordered mesopores outside MTT zeolite for efficient hydroisomerization. *Appl. Catal. A Gen.* **2020**, *602*, 117664. [CrossRef]
19. Liu, L.; Zhang, M.; Wang, L.; Zhang, X.; Li, G. Modulating acid site distribution in MTT channels for controllable hy-droi-somerization of long-chain n-alkanes. *Fuel Process. Technol.* **2023**, *241*, 107605. [CrossRef]
20. Wang, X.; Zhang, X.; Wang, Q. N-dodecane hydroisomerization over Pt/ZSM-22: Controllable microporous Brönsted acidity distribution and shape-selectivity. *Appl. Catal. A Gen.* **2020**, *590*, 117335. [CrossRef]
21. Maesen, T.; Schenk, M.; Vlugt, T.; Jonge, J.; Smit, B. The Shape Selectivity of Paraffin Hydroconversion on TON-, MTT-, and AEL-Type Sieves. *J. Catal.* **1999**, *188*, 403–412. [CrossRef]
22. Lv, G.; Wang, C.; Wang, P.; Sun, L.; Liu, H.; Qu, W.; Wang, D.; Ma, H.; Tian, Z. Pt/ZSM-22 with Partially Filled Micropore Channels as Excellent Shape-Selective Hydroisomerization Catalyst. *ChemCatChem* **2019**, *11*, 1431–1436. [CrossRef]
23. Ahmed, M.H.M.; Muraza, O.; Al-Amer, A.M.; Yamani, Z.H. Investigation of crucial synthesis parameters of rich Al-MTT framework zeolite: Toward more determination for synthesis zone of SSZ-32. *Microporous Mesoporous Mater.* **2016**, *227*, 48–56. [CrossRef]

24. Ojo, A.F.; Lei, G.; Zhang, Y.; Krishna, K.R. Processes for Producing Diesel from Unconventional Feedstoks. U.S. Patent 2,022,204,867-A1, 30 June 2022.
25. Dai, X.; Cheng, Y.; Si, M.; Wei, Q.; Chen, D.; Huang, W.; Zhou, Y. SAPO-11 molecular sieves synthesized in alcohol-water concentrated gel system with improved acidity, mesoporous volume and hydroisomerization performance. *Fuel* **2022**, *314*, 123131. [CrossRef]
26. Chen, Y.; Li, C.; Chen, X.; Liu, Y.; Liang, C. Synthesis of ZSM-23 zeolite with dual structure directing agents for hydroisomerization of n-hexadecane. *Microporous Mesoporous Mater.* **2018**, *268*, 216–224. [CrossRef]
27. Chen, Y.; Li, C.; Chen, X.; Liu, Y.; Tsang, C.-W.; Liang, C. Synthesis and Characterization of Iron-Substituted ZSM-23 Zeolite Catalysts with Highly Selective Hydroisomerization of n-Hexadecane. *Ind. Eng. Chem. Res.* **2018**, *57*, 13721–13730. [CrossRef]
28. Chen, L.-H.; Sun, M.-H.; Wang, Z.; Yang, W.; Xie, Z.; Su, B.-L. Hierarchically Structured Zeolites: From Design to Application. *Chem. Rev.* **2020**, *120*, 11194–11294. [CrossRef]
29. Serrano, D.P.; Escola, J.M.; Pizarro, P. Synthesis strategies in the search for hierarchical zeolites. *Chem. Soc. Rev.* **2013**, *42*, 4004–4035. [CrossRef]
30. Verboekend, D.; Pérez-Ramírez, J. Design of hierarchical zeolite catalysts by desilication. *Catal. Sci. Technol.* **2011**, *1*, 879–890. [CrossRef]
31. Silaghi, M.-C.; Chizallet, C.; Raybaud, P. Challenges on molecular aspects of dealumination and desilication of zeolites. *Microporous Mesoporous Mater.* **2014**, *191*, 82–96. [CrossRef]
32. Bai, D.; Meng, J.; Li, C.; Zhang, M.; Liang, C. Mesoporosity and Acidity Manipulation in ZSM-23and their n-Hexadecane Hydroisomerization Performance. *Chemistryselect* **2022**, *7*, e202200839. [CrossRef]
33. Guo, K.; Ma, A.; Wang, Z.; Li, J.; Wu, B.; Liu, T.; Li, D. Investigation of n-heptane hydroisomerization over alkali-acid-treated hierarchical Pt/ZSM-22 zeolites. *New J. Chem.* **2022**, *46*, 16752–16763. [CrossRef]
34. Zhou, Q.M.; Wang, S.; Wu, Z.W.; Qin, Z.F.; Dong, M.; Wang, J.G.; Fan, W.B. Aromatization of n-C7–n-C9 alkanes on a Pt/KZSM-5(deAl) catalyst. *Catal. Sci. Technol.* **2023**, *13*, 1009. [CrossRef]
35. Wang, H.Y.; Zhang, X.W.; Li, G.Z. Constructing optimally hierarchical HY zeolite for the synthesis of high-energy-density tricyclic hydrocarbon fuel. *Mol. Catal.* **2023**, *535*, 112871. [CrossRef]
36. Mihályi, R.M.; Kollár, M.; Király, P.; Karolyc, Z.; Mavrodinova, V. Effect of extra-framework Al formed by successive steaming and acid leaching of zeolite MCM-22 on its structure and catalytic performance. *Appl. Catal. A Gen.* **2012**, *417*, 76–86. [CrossRef]
37. Thommes, M.; Kaneko, K.; Neimark, A.V.; Olivier, J.P.; Rodriguez-Reinoso, F.; Rouquerol, J.; Sing, K.S. Physisorption of gases, with special reference to the evaluation of surface area and pore size distribution (IUPAC Technical Report). *Pure Appl. Chem.* **2015**, *87*, 1051–1069. [CrossRef]
38. Groen, J.C.; Moulijn, J.A.; Perez-Ramirez, J. Desilication: On the controlled generation of mesoporosity in MFI zeolites. *J. Mater. Chem.* **2006**, *16*, 2121–2131. [CrossRef]
39. Chen, K.; Horstmeier, S.; Nguyen, V.T.; Wang, B.; Crossley, S.P.; Pham, T.; Gan, Z.; Hung, I.; White, J.L. Structure and Catalytic Characterization of a Second Framework Al(IV) Site in Zeolite Catalysts Revealed by NMR at 35.2 T. *J. Am. Chem. Soc.* **2020**, *142*, 7514–7523. [CrossRef] [PubMed]
40. Yi, X.; Liu, K.; Chen, W.; Li, J.; Xu, S.; Li, C.; Xiao, Y.; Liu, H.; Guo, X.; Liu, S.-B.; et al. Origin and Structural Characteristics of Tri-coordinated Extra-framework Aluminum Species in Dealuminated Zeolites. *J. Am. Chem. Soc.* **2018**, *140*, 10764–10774. [CrossRef] [PubMed]
41. Liu, R.; Fan, B.; Zhi, Y.; Liu, C.; Xu, S.; Yu, Z.; Liu, Z. Dynamic Evolution of Aluminum Coordination Environments in Mordenite Zeolite and Their Role in the Dimethyl Ether (DME) Carbonylation Reaction. *Angew. Chem. Int. Edit.* **2022**, *61*, e202210658.
42. Ravenelle, R.M.; Schüβler, F.; D' Amico, A.; Danilina, N.; van Bokhoven, J.A.; Lercher, J.A.; Jones, C.W.; Sievers, C. Stability of Zeolites in Hot Liquid Water. *J. Phys. Chem. C* **2010**, *114*, 19582–19595. [CrossRef]
43. van Donk, S.; Bitter, J.H.; Verberckmoes, A.; VersluijsHelder, M.; Broersma, A.; de Jong, K.P. Physicochemical characterization of porous materials: Spatially resolved accessibility of zeolite crystals. *Angew. Chem. Int. Edit.* **2005**, *44*, 1360–1363. [CrossRef] [PubMed]
44. Ravi, M.; Sushkevich, V.L.; van Bokhoven, J.A. On the location of Lewis acidic aluminum in zeolite mordenite and the role of framework-associated aluminum in mediating the switch between Brønsted and Lewis acidity. *Chem. Sci.* **2021**, *12*, 4094–4103. [CrossRef]
45. Ravi, M.; Sushkevich, V.L.; van Bokhoven, J.A. Towards a better understanding of Lewis acidic aluminium in zeolites. *Nat. Mater.* **2020**, *19*, 1047–1056. [CrossRef] [PubMed]
46. Palčić, A.; Valtchev, V. Analysis and control of acid sites in zeolites. *Appl. Catal. A Gen.* **2020**, *606*, 117795. [CrossRef]
47. Ono, Y. A survey of the mechanism in catalytic isomerization of alkanes. *Catal. Today* **2003**, *81*, 3–16. [CrossRef]
48. Campelo, J.M.; Lafont, F.; Marinas, J.M. Comparison of the activity and selectivity of Pt/SAPO-5 and Pt/SAPO-11 in n-hexane and n-heptane hydroconversion. *Appl. Catal. A Gen.* **1997**, *152*, 53–62. [CrossRef]
49. Blomsma, E.; Martens, J.A.; Jacobs, P.A. Reaction Mechanisms of Isomerization and Cracking of Heptane on Pd/H-Beta Zeolite. *J. Catal.* **1995**, *155*, 141–147. [CrossRef]
50. Paul, Z.; Zhan, Z.; Manninger, I.; Sachtler, W.M.H. Skeletal Reactions of n-Hexane over Pt-NaY, Pt/SiO2, Hy, and Mixed Pt/Sio_2 + HY Catalysts. *J. Catal.* **1995**, *155*, 43–51. [CrossRef]

51. Martens, J.A.; Jacobs, P.A.; Weitkamp, J. Attempts to rationalize the distribution of hydrocracked products. II. Relative rates of primary hydrocracking modes of long chain paraffins in open zeolites. *Appl. Catal.* **1986**, *20*, 283–303. [CrossRef]
52. Parton, R.; Uytterhoeven, L.; Martens, J.A.; Jacobs, P.A.; Froment, G.F. Synergism of ZSM-22 and Y zeolites in the bifunctional conversion of n-alkanes. *Appl. Catal.* **1991**, *76*, 131–142. [CrossRef]
53. Alvarez, F.; Ribeiro, F.R.; Perot, G.; Thomazeau, C.; Guisnet, M. Hydroisomerization and Hydrocracking of Alkanes: 7. Influence of the Balance between Acid and Hydrogenating Functions on the Transformation ofn-Decane on PtHY Catalysts. *J. Catal.* **1996**, *162*, 179–189. [CrossRef]
54. Claude, M.C.; Martens, J.A. Monomethyl-Branching of Long n-Alkanes in the Range from Decane to Tetracosane on Pt/H-ZSM-22 Bifunctional Catalyst. *J. Catal.* **2000**, *190*, 39–48. [CrossRef]

catalysts

MDPI

Article

Xylene and n-Hexane Adsorption Performance of a Waste Methanol-to-Propylene Catalyst under Acid-Base Treatment

Xiaojing Yong [1,2], Hui Su [2], Nana Zhao [2], Zhengwei Jin [2], Min Yao [1,*] and Yulong Ma [1,*]

1 State Key Laboratory of High-Efficiency Utilization of Coal and Green Chemical Engineering, College of Chemistry and Chemical Engineering, Ningxia University, Yinchuan 750021, China
2 Institute of Coal Chemical Industry Technology, Ningxia Coal Industry Co., Ltd., National Energy Group, Yinchuan 750411, China
* Correspondence: ndglym@163.com (M.Y.); yulongma796@sohu.com (Y.M.)

Abstract: Spent methanol-to-propylene (MTP) catalysts have a large specific surface area and high porosity but are usually directly disposed of in landfills, and recycling is rare. In this study, spent MTP catalyst was moderately dealuminized with acids and etched with an alkali solvent to increase its specific surface area. A novel adsorbent was obtained. XRD, SEM, FT-IR, XRD, XRF, and MAS-NMR characterization shows that the adsorbent maintains a typical ZSM-5 zeolite structure, and the dealumination effect of $H_2C_2O_4$ is better than that of HCl. HCl mainly removes the framework aluminum of the molecular sieve; $H_2C_2O_4$ not only removes the framework aluminum but also dissolves some of the nonframework aluminum, which increases the BET-specific surface area and pore diameter. The spent catalyst maintains an irregular ellipsoidal shape. After alkali treatment, the surface of the spherical particles becomes rough. With increasing alkali concentration, the damage degree increases. After treatment with 4 mol/L $H_2C_2O_4$ and 0.1 mol/L NaOH, the p-xylene and n-hexane adsorption capacities reach the maximum, with values of 141.04 mg/g and 106.87 mg/g, respectively, 20.7% and 16.2% greater than those before treatment. These findings indicate that modified spent MTP catalyst has the potential for application in the removal of VOCs from the air.

Keywords: spent MTP catalyst; ZSM-5; adsorption; p-xylene; n-hexane

Citation: Yong, X.; Su, H.; Zhao, N.; Jin, Z.; Yao, M.; Ma, Y. Xylene and n-Hexane Adsorption Performance of a Waste Methanol-to-Propylene Catalyst under Acid-Base Treatment. *Catalysts* **2022**, *12*, 1028. https://doi.org/10.3390/catal12091028

Academic Editors: Maja Milojević-Rakić and Danica Bajuk-Bogdanović

Received: 6 August 2022
Accepted: 5 September 2022
Published: 10 September 2022

1. Introduction

Methanol-to-propylene (MTP) catalysts are formed of a ZSM-5 molecular sieve. ZSM-5 molecular sieves have a unique pore structure that provides good ion exchange performance and shape selectivity [1–3]. Because of its good thermal stability, hydrothermal stability, wide distribution of Si/Al, and wide range of acid adjustability, ZSM-5 zeolite is a good solid acid catalyst and catalyst carrier [4,5]. It is widely used in catalysis, such as in the petrochemical and coal chemical industries, as well as in the fields of wastewater treatment and selective adsorption [6–9]. Many published studies have reported that MTP catalysts require small grain HZSM-5 molecular sieves with a high Si to Al ratio and a multistage composite pore structure [6,10–12]. Researchers have focused on the design and development of new MTP catalysts [9,13–15], but have rarely studied the comprehensive utilization of functionalized spent MTP catalysts [16,17]. Spent MTP catalysts have a large specific surface area, high porosity, and good adsorption performance [18–20]. In recent years, the massive emission of volatile organic compounds (VOCs) has caused serious harm to the human body and the environment [11,13,21,22]. The porous multistage composite ZSM-5 molecular sieve with an MFI structure can realize the selective adsorption of VOCs on the molecular scale, due to the interconnection of its microporous structure and mesoporous structure [23–27]. This paper systematically investigated the effect of acid-base treatment on the texture properties of waste MTP catalyst, enhanced the adsorption potential of VOCs on a waste molecular sieve, and provided support for the resource utilization of waste catalysts.

2. Results and Discussion

2.1. XRD Characterization of the Adsorbent Products

The XRD spectra of the adsorbent products are shown in Figure 1a,b. The hydrophobic adsorbent products maintain a typical ZSM-5 molecular sieve structure, and the corresponding characteristic peaks of the ZSM-5 (011), (020), (051), and (033), crystal planes appear at 2θ = 7.9°, 8.8°, 23.1°, and 23.8°, respectively, which indicates that the adsorbent channels have long-range order [9–12]. Acid treatment did not change the crystal phase structure of the spent catalyst, but the diffraction peak intensity after sample processing increases with increasing acid concentration. This may be caused by the elimination of impurities over the spent catalyst by acid treatment.

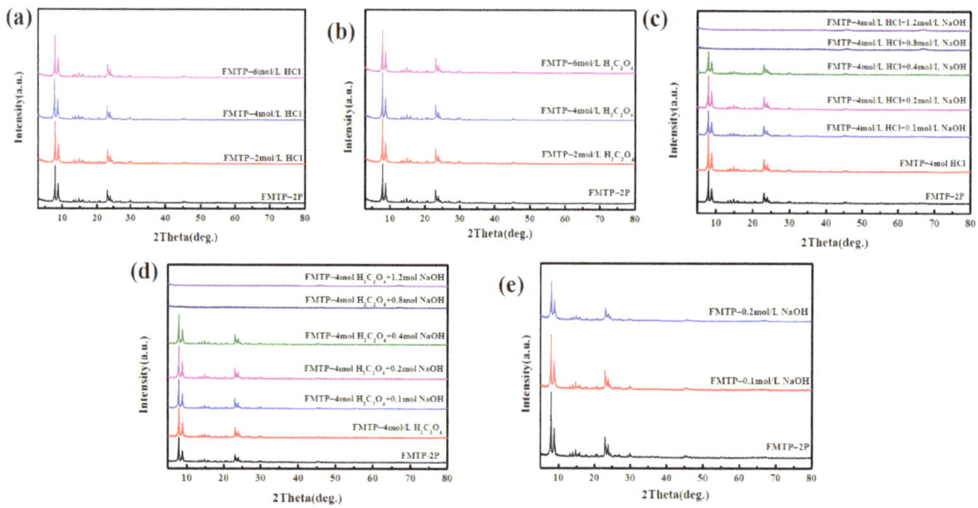

Figure 1. XRD spectra of spent catalyst samples after (**a**) HCl treatment; (**b**) $H_2C_2O_4$ treatment; (**c**) 4 mol/L HCl and alkali combined treatment; (**d**) 4 mol/L $H_2C_2O_4$ and alkali combined treatment; and (**e**) alkali treatment.

Figure 1c,d shows that after treatment with different concentrations of alkali, the crystal phase structure of the spent catalyst is not destroyed, no characteristic peaks disappear, and no new diffraction peaks are observed. However, with increasing alkali concentration, the diffraction peak intensity is reduced, which may be due to framework desilication of the spent catalyst and the generation of local structural defects in the molecular sieve lattice, resulting in a decline in the intensity of diffraction peaks. When the NaOH concentration is higher than 0.8 mol/L, the ZSM-5 molecular sieve framework is seriously damaged and presents an amorphous state. Figure 1e shows that the crystal phase structure of the spent catalyst remains unchanged after alkali treatment alone, but the diffraction peak intensity of the sample decreases, indicating that the lattice structure of the ZSM-5 molecular sieve changes after alkali treatment.

2.2. XRF Analysis

According to Tables 1 and 2, the Si/Al ratio of the waste MTP catalyst before acid treatment is only 6.51, and the Si/Al ratios of the samples after acid treatment are significantly increased. At the same concentration, $H_2C_2O_4$ has a better dealuminization effect than HCl and increases the ratio of silicon to aluminum to a greater degree. The reason may be that $H_2C_2O_4$ is a dihydric acid with a complexation effect. In the dealuminization system, $H_2C_2O_4$ has both acidic and complexation effects on the ZSM-5 molecular sieve framework aluminum. With increasing NaOH concentration, the Si/Al ratio decreases, indicating that the alkali treatment significantly removed Si species on the zeolite. Excessive alkali caused the collapse of the framework structure of ZSM-5 zeolite and caused serious damage to its pore structure.

Table 1. Elemental composition and content of waste catalyst samples after 4 mol/L HCl acid and alkali combined treatment.

Sample	Element Composition (wt%)						SAR
	SiO_2	Al_2O_3	Fe_2O_3	CaO	SO_3	TiO_2	
FMTP-2P	79.02	20.64	0.0291	0.137	0.133	0.047	6.51
FMTP-2 mol/L HCl	85.24	13.9	0.0246	0.324	0.464	0.026	10.41
FMTP-4 mol/L HCl	87.9	11.5	0.019	0.214	0.245	0.034	12.99
FMTP-6 mol/L HCl	88.35	10.8	0.022	0.313	0.479	0.036	13.88
FMTP-4 mol/L HCl + 0.1 mol/L NaOH	87.33	12.2	0.0208	0.16	0.245	0.044	12.17
FMTP-4 mol/L HCl + 0.2 mol/L NaOH	87.85	11.8	0.016	0.15	0.174	0.034	12.66
FMTP-4 mol/L HCl + 0.4 mol/L NaOH	80.87	18.68	0.0277	0.15	0.23	0.051	7.36
FMTP-4 mol/L HCl + 0.8 mol/L NaOH	20.99	78.41	0.085	0.168	0.252	0.102	0.45
FMTP-4 mol/L HCl + 1.2 mol/L NaOH	7.77	91.58	0.0943	0.16	0.275	0.118	0.14

Table 2. Elemental composition and content of samples before and after alkali combined treatment of waste catalyst subjected to 4 mol/L $H_2C_2O_4$ acid treatment.

Sample	Element Composition (wt%)						SAR
	SiO_2	Al_2O_3	Fe_2O_3	CaO	SO_3	TiO_2	
FMTP-2P	79.02	20.64	0.029	0.137	0.133	0.047	6.51
FMTP-2 mol/L $H_2C_2O_4$	91.08	8.03	0.023	0.332	0.503	0.029	19.25
FMTP-4 mol/L $H_2C_2O_4$	93.71	5.88	0.014	0.157	0.201	0.026	27.09
FMTP-6 mol/L $H_2C_2O_4$	87.66	11.5	0.025	0.329	0.476	0.037	12.94
FMTP-4 mol/L $H_2C_2O_4$ + 0.1 mol/L NaOH	92.41	7.26	0.013	0.129	0.156	0.031	21.64
FMTP-4 mol/L $H_2C_2O_4$ + 0.2 mol/L NaOH	93.85	5.81	0.012	0.139	0.164	0.026	27.46
FMTP-4 mol/L $H_2C_2O_4$ + 0.4 mol/L NaOH	90.36	9.31	0.013	0.127	0.156	0.028	16.50
FMTP-4 mol/L $H_2C_2O_4$ + 0.8 mol/L NaOH	43.73	55.1	0.467	0.181	0.338	0.149	1.35
FMTP-4 mol/L $H_2C_2O_4$ + 1.2 mol/L NaOH	12.7	86.27	0.464	0.16	0.295	0.129	0.25

2.3. *Specific Surface Area and Pore Structure Parameters*

Table 3 shows that the specific surface area and pore volume of the samples increase after acid treatment, indicating that acid treatment has a great influence on the specific surface area, pore structure, and properties of the spent catalyst. This is mainly because acid treatment removes the surface of the waste catalyst and the amorphous aluminum accumulated in the channels, which exposes a more effective area and creates a smoother surface and channels. The specific surface area of the sample treated with $H_2C_2O_4$ is larger than that of the sample treated with HCl, which is consistent with the trends in the XRF data. With increasing alkali concentration, the total pore volume, mesoporous specific surface area, and microporous specific surface area of the samples first increase and then decrease. However, when the NaOH concentration is 0.8 mol/L, the total pore volume and mesoporous specific surface area decrease significantly, which may be because the high-intensity desilication process destroyed part of the framework structure of the ZSM-5 molecular sieve.

Table 3. BET-specific surface area and pore structure parameters of samples before and after acid and alkali combined treatment with 4 mol/L HCl and $H_2C_2O_4$.

Sample	SBET (m^2/g)	Smicro (m^2/g)	Smeso (m^2/g)	Vtotal (cm^3/g)	Vmicro (cm^3/g)	Vmeso (cm^3/g)	Pore Width (nm)
FMTP-2P	303.956	105.549	198.407	0.287	0.051	0.236	3.7714
FMTP-4 mol/L HCl	332.671	103.730	228.940	0.321	0.050	0.271	3.8539
FMTP-4 mol/L HCl + 0.1 mol/L NaOH	411.710	144.580	267.131	0.709	0.063	0.646	6.8857
FMTP-4 mol/L HCl + 0.2 mol/L NaOH	357.638	121.853	235.785	0.436	0.064	0.372	4.8711
FMTP-4 mol/L HCl + 0.4 mol/L NaOH	380.460	97.136	283.325	0.799	0.050	0.749	8.4058
FMTP-4 mol/L HCl + 0.8 mol/L NaOH	174.996	52.078	122.918	0.433	0.025	0.408	9.8956
FMTP-4 mol/L $H_2C_2O_4$	355.558	95.656	259.902	0.327	0.045	0.282	3.6792
FMTP-4 mol/L $H_2C_2O_4$ + 0.1 mol/L NaOH	404.337	143.700	260.636	0.302	0.056	0.246	2.9834
FMTP-4 mol/L $H_2C_2O_4$ + 0.2 mol/L NaOH	426.766	174.458	252.309	0.435	0.071	0.364	4.0732
FMTP-4 mol/L $H_2C_2O_4$ + 0.4 mol/L NaOH	424.101	117.667	306.434	0.938	0.059	0.879	8.8502
FMTP-4 mol/L $H_2C_2O_4$ + 0.8 mol/L NaOH	119.542	32.651	86.891	0.474	0.016	0.458	17.7150

2.4. FT-IR Characterization

As seen from Figure 2a,b, the characteristic diffraction peaks of the ZSM-5 molecular sieve appear at wavenumbers of 449 cm^{-1}, 544 cm^{-1}, 798 cm^{-1}, 1090 cm^{-1}, and 1220 cm^{-1}. Among them, the absorption peak at 449 cm^{-1} belongs to the bending vibration of X-O bonds in SiO_4 and AlO_4 tetrahedra (X is Si or Al), and the absorption peaks at 544 cm^{-1} and 1220 cm^{-1} belong to the vibration of the double five-membered ring in the ZSM-5 molecular sieve structure. The absorption peak at 798 cm^{-1} is due to the X-O-X symmetric stretching vibration of external connections, and 1090 cm^{-1} is due to the antisymmetric stretching vibration of X-O-X bonds inside the tetrahedra. With increasing acid concentration, the framework structure of the ZSM-5 molecular sieve essentially does not change [25].

Figure 2c,d shows that the crystallinity of the samples decreases gradually with increasing alkali concentration. When the NaOH concentration reaches 0.8 mol/L, the framework structure collapses, which is consistent with the XRD results in Figure 1. Figure 2e shows that the characteristic diffraction peak intensity of the ZSM-5 molecular sieve framework decreases after alkali treatment alone. From Figure 2f, the pyridine adsorption infrared (Py-IR) spectra indicate that the Brønsted acid (B-acid) and Lewis acid (L-acid) contents of samples treated with HCl and $H_2C_2O_4$ show a decreasing trend. After acid treatment, part of the nonframework aluminum is removed, resulting in a decrease in the acid content of L-acid centers composed of the nonframework aluminum, and part of the framework aluminum is also removed, resulting in the destruction of the Si-Al-O structure of the ZSM-5 molecular sieve. $H_2C_2O_4$ treatment not only removes the framework aluminum of ZSM-5 but also removes part of the nonframework aluminum, while HCl mainly removes the framework aluminum of ZSM-5 but has a weak removal effect on the nonframework aluminum.

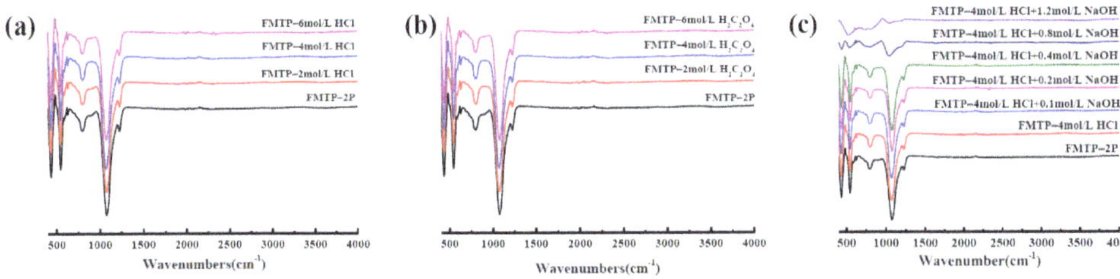

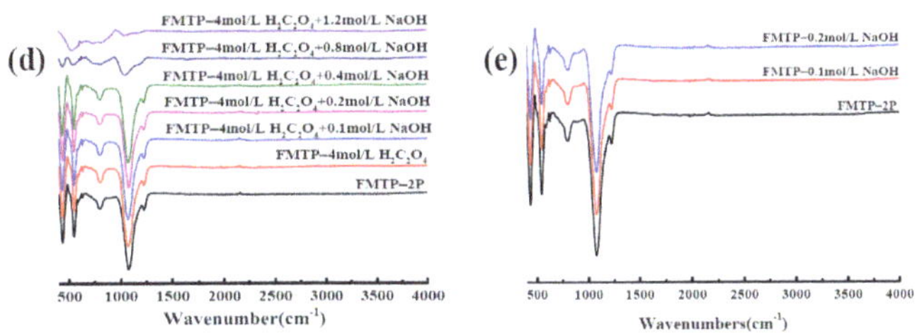

(f)

Sample name	(μmol/g)		B/L
	Bronsted	Lewis	
FMTP-2P	5.2	22.6	0.230
FMTP-4mol/L HCl	3.7	22.1	0.167
FMTP-4mol/L $H_2C_2O_4$	2.4	20.6	0.116

Figure 2. (**a**) FT-IR diagrams of spent catalyst samples before and after HCl treatment; FT-IR diagrams of spent catalyst samples before and after (**b**) $H_2C_2O_4$ treatment; (**c**) alkali combined treatment of 4 mol/L HCl-treated waste catalyst; (**d**) alkali-combined treatment of 4 mol/L $H_2C_2O_4$-treated waste catalyst; (**e**) alkali treatment; (**f**) Brønsted acid (B-acid) and Lewis acid (L-acid) contents of waste catalysts treated with different acids.

2.5. SEM Characterization

As seen from Figure 3, the samples after acid treatment have an irregular ellipsoidal morphology and uniform grain size, with an average grain size of approximately 1.1 μm, which is similar to the properties of the samples before acid treatment. Alkali treatment destroys the framework structure to a certain extent, and the degree of damage increases with increasing concentration of alkali treatment. When the concentration of NaOH is 0.2 mol/L, a honeycomb structure appears on the surface of the sample, indicating that mesopores are formed by alkali treatment. When the concentration of NaOH is increased further, the mesoporous channels become more obvious. When the concentration of NaOH reaches 0.8 mol/L, the particles are dissolved, and the pore structure collapses, which is consistent with the XRD, XRF, FT-IR, and BET data.

2.6. ^{27}Al MAS-NMR Analysis

Figure 4a,b shows that there are two aluminum species on the waste catalyst before acid treatment. With a general chemical shift, the peak at δ = 50~60 ppm belongs to the signal of the framework aluminum in the ZSM-5 molecular sieve, which has tetrahedral coordination, and the peak at δ = 5.0 ppm belongs to the signal of the hexa-coordinated octahedral nonframework aluminum. There is a shoulder peak at δ = 47.8 ppm, which represents the framework aluminum distributed on the outer surface or large pores of the ZSM-5 molecular sieve. After HCl treatment, the chemical shifts of the two aluminum species in the ^{27}Al NMR spectra do not change, but the peak intensity of the framework aluminum signal decreases at δ = 58.0 ppm, the shoulder at δ = 47.8 ppm disappears, the signal peak intensity of the nonframework aluminum at δ = 5.0 ppm remains unchanged, and the ratio of framework aluminum to the nonframework aluminum decreases, indicating that HCl dealumination mainly removes the framework aluminum from the ZSM-5 molecular sieve. After $H_2C_2O_4$ treatment, the ^{27}Al NMR spectrum changes significantly. The peak intensity decreases at δ= 5.0 ppm and δ = 58.0 ppm. The shoulder peak at δ = 47.8 ppm is significantly widened, indicating that $H_2C_2O_4$ not only removes the framework aluminum of the ZSM-5 molecular sieve but also dissolves some of the nonframework aluminum. Therefore, the silicon-to-aluminum ratio after $H_2C_2O_4$ treatment is greater than that after

HCl treatment, which is consistent with the results of Table 1 and the changes in B-acid and L-acid in Figure 2f.

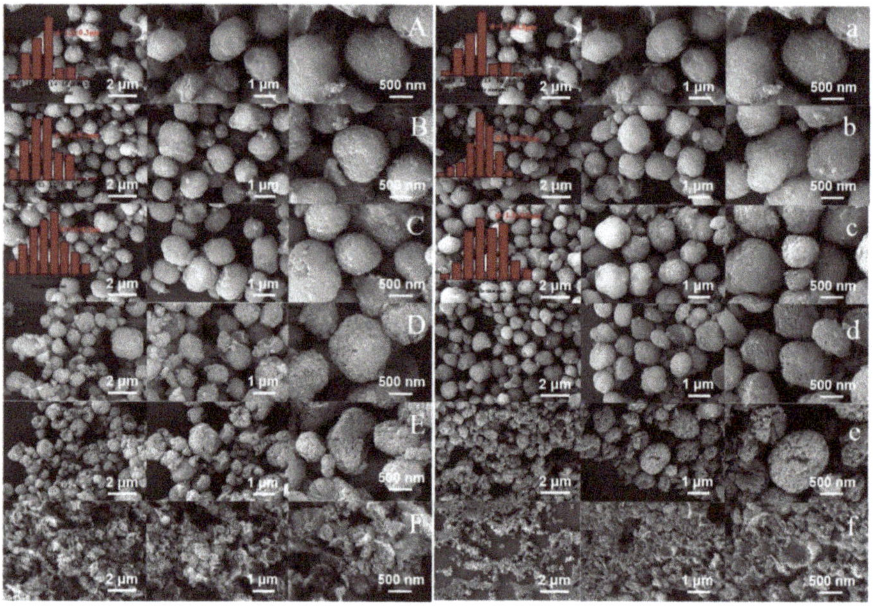

Figure 3. SEM images of (**A**) FMTP-2P; (**B**) 4 mol/L HCl; (**C**) 4 mol/L HCl + 0.1 mol/L NaOH; (**D**) 4 mol/L HCl + 0.2 mol/L NaOH; (**E**) 4 mol/L HCl + 0.4 mol/L NaOH; (**F**) 4 mol/L HCl + 0.8 mol/L NaOH. SEM image of (**a**) FMTP-2P; (**b**) 4 mol/L $H_2C_2O_4$; (**c**) 4 mol/L $H_2C_2O_4$ + 0.1 mol/L NaOH; (**d**) 4 mol/L $H_2C_2O_4$ + 0.2 mol/L NaOH; (**e**) 4 mol/L $H_2C_2O_4$ + 0.4 mol/L NaOH; (**f**) 4 mol/L $H_2C_2O_4$ + 0.8 mol/L NaOH.

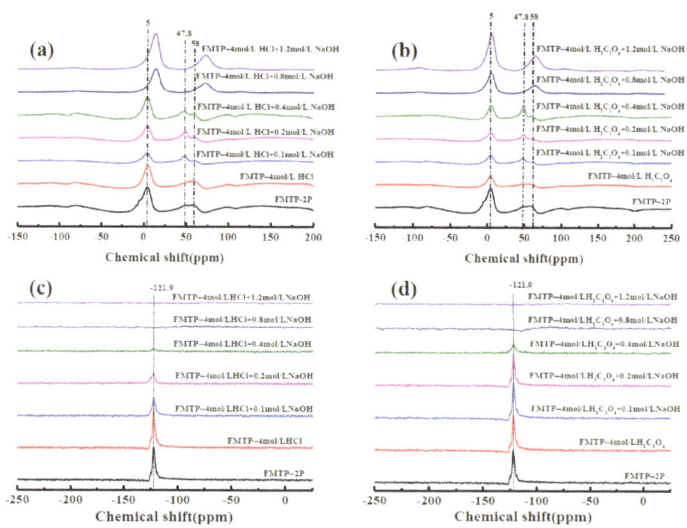

Figure 4. ^{27}Al MAS NMR spectra of waste catalyst samples after (**a**) 4 mol/L HCl treatment and alkali combined treatment; (**b**) 4 mol/L $H_2C_2O_4$ treatment and after alkali combined treatment. ^{29}Si MAS-NMR spectra of waste catalyst samples after (**c**) 4 mol/L HCl and alkali combined treatment; (**d**) 4 mol/L $H_2C_2O_4$ treatment and alkali combined treatment.

Figure 4c,d shows that after acid treatment, the peak intensity of the silicon signal at $\delta = -121.9$ ppm hardly changes, but the chemical shift moves in the direction of high binding energy, and the shift of the sample treated with $H_2C_2O_4$ is greater than that of the sample treated with HCl. This is because the concentration of aluminum species on the surface of the $H_2C_2O_4$-treated sample is low, and the electronegativity of aluminum ($x = 1.61$) is lower than that of silicon ($x = 1.90$). Therefore, the chemical shift is greater. The changes in the spectra of samples treated with different concentrations of alkali are consistent. After treatment with 0.1 mol/L NaOH, the peak intensity of the nonframework aluminum of the ZSM-5 molecular sieve at $\delta = 5.0$ ppm decreases, and the enhanced shoulder signal at $\delta = 47.8$ ppm indicates that alkali treatment further dissolves part of the nonframework aluminum. However, since alkali treatment has the effect of pore expansion, the signal for the framework aluminum in large pores increases. With a further increase in NaOH concentration, the nonframework aluminum signal and shoulder peak aluminum signal gradually increase, the silicon-to-aluminum ratio decreases, the pores collapse, and the aluminum is free in the form of the nonframework aluminum. The peak intensity of the sample after $H_2C_2O_4$ alkali treatment is lower than that of the sample treated with HCl, which is consistent with the XRF data. After acid treatment, the intensity of the silicon peak at $\delta = -121.9$ ppm is almost unchanged. After alkali treatment, the silicon peak decreases, and the chemical shift moves in the direction of high binding energy. When the NaOH concentration reaches 0.8 mol/L, almost no silicon signal can be detected, indicating that the silicon in the sample has been completely dissolved by the high concentration of alkali, and that the framework structure of the ZSM-5 molecular sieve is damaged.

2.7. Adsorption Performance of Waste Catalyst under Acid Treatment

Figures 5 and 6 show that the adsorption capacities of samples treated with HCl and $H_2C_2O_4$ are essentially the same, with values significantly higher than those before treatment. The adsorption capacity increases with increasing acid concentration. When the concentration of acid treatment is 4 mol/L, the adsorption capacity reaches the maximum. When the acid concentration continues to increase to 6 mol/L, the adsorption capacity remains unchanged or even decreases. For a given treatment concentration, the adsorption capacity of samples treated with $H_2C_2O_4$ is greater than that of samples treated with HCl. When $P/P_0 = 0.1$ MPa and the $H_2C_2O_4$ concentration is 4 mol/L, the adsorption capacities of p-xylene and n-hexane reach the maximum, with values of 128.25 mg/g and 103.31 mg/g, respectively, which are 15.4% higher than those of waste catalyst without acid treatment.

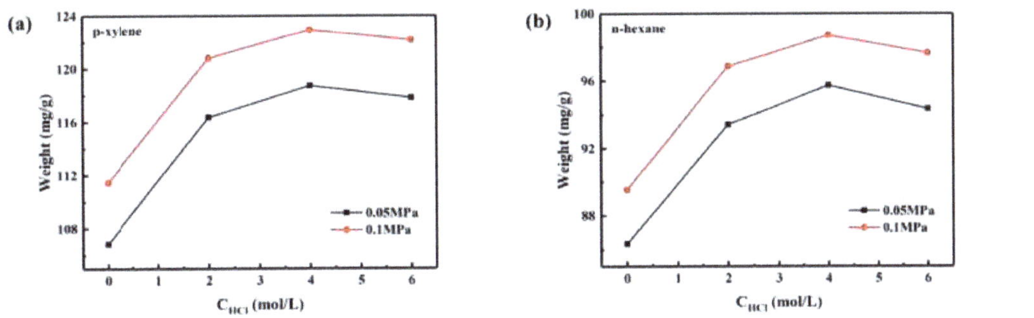

Figure 5. Variation in the adsorption capacities for (**a**) p-xylene and (**b**) n-hexane of samples after HCl treatment.

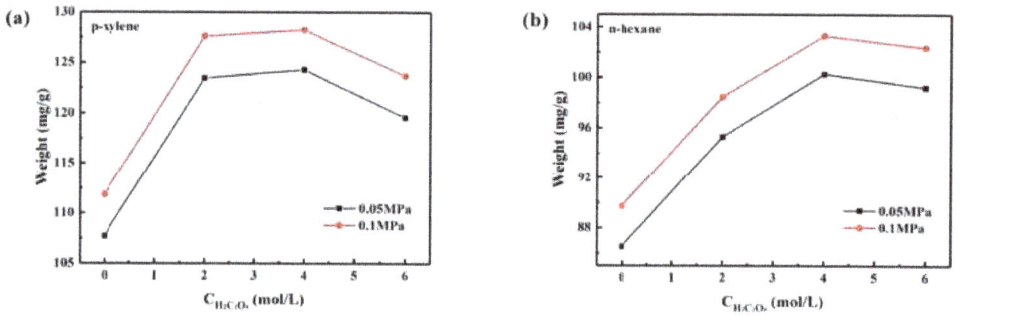

Figure 6. Variation in the adsorption capacities for (**a**) p-xylene and (**b**) n-hexane of samples after $H_2C_2O_4$ treatment.

2.8. Change in the Adsorption Rate of Acid-Treated Spent Catalysts under Different Pressures

Figure 7 shows that at the same pressure of 298 K, the amount of p-xylene and n-hexane adsorbed by waste catalyst samples treated with HCl is higher than that of samples treated with $H_2C_2O_4$. The curve of the amount of p-xylene adsorbed over time has the largest slope at $p = 0.5$ KPa, indicating that the sample has the fastest adsorption rate of p-xylene at this pressure. The slope of the corresponding curve for n-hexane is the largest at $p = 13$ KPa, indicating that the adsorption rate for n-hexane is the fastest at this pressure.

2.9. Adsorption Performance of Waste Catalyst Subjected to Acid-Base Combined Treatment

Figures 8 and 9 show that after treatment with 4 mol/L HCl and $H_2C_2O_4$, the amount of p-xylene and n-hexane adsorbed by the waste catalyst after alkali treatment at different concentrations first increases and then decreases. When $P/P_0 = 0.1$ MPa, the $H_2C_2O_4$ concentration is 4 mol/L and the NaOH concentration is 0.1 mol/l, the maximum adsorption capacity of the sample for p-xylene is 141.04 mg/g, and the maximum adsorption capacity for n-hexane is 106.87 mg/g. Compared with the values obtained after acid treatment, the p-xylene and n-hexane adsorption capacities of the samples subjected to combined treatment increase by 5.94% and 6.89% and are 20.7% and 16.2% higher, respectively, than those of waste catalyst without acid and alkali treatment. This shows that after acid-base treatment, the adsorption performance of waste catalysts for VOCs can be further improved.

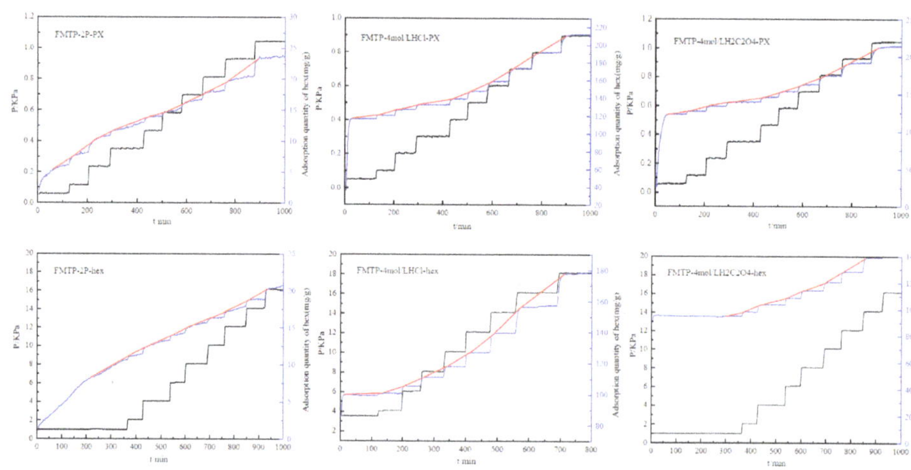

Figure 7. Relationship between sample adsorption capacity and pressure with respect to time before and after HCl and $H_2C_2O_4$ treatment.

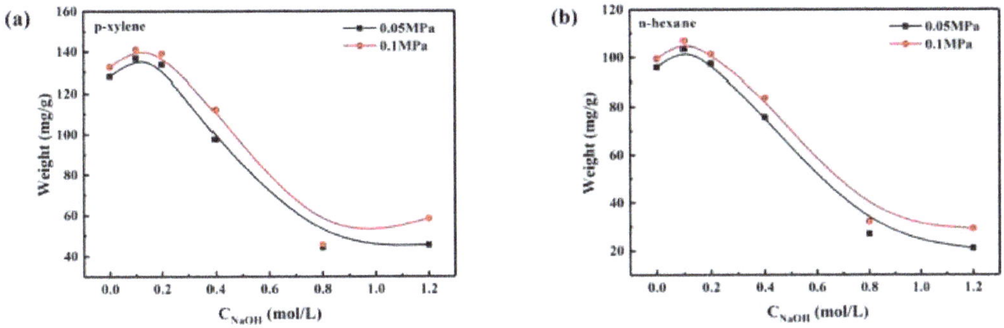

Figure 8. Variation in the adsorption capacities for (**a**) p-xylene and (**b**) n-hexane of samples subjected to alkali after $H_2C_2O_4$ treatment.

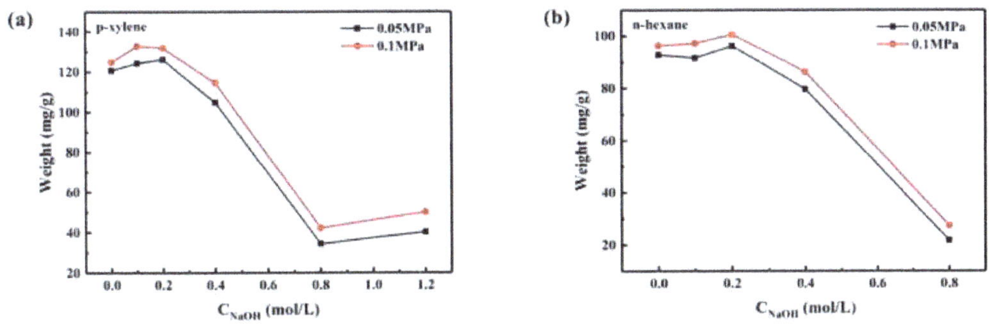

Figure 9. Variation in the adsorption capacities for (**a**) p-xylene and (**b**) n-hexane of samples subjected to alkali after HCl treatment.

3. Preparation of a VOC Adsorbent from Waste MTP Catalyst

3.1. Raw Materials

The raw materials were waste fixed-bed MTP (FMTP) catalyst, hydrochloric acid (HCl), oxalic acid ($H_2C_2O_4$), NaOH, HNO_3 solution (68 wt%), deionized water, p-xylene, and n-hexane.

3.2. Preparation of Adsorbent

First, 10 g of spent MTP catalyst powder (below 200 mesh) was added to a water solution of HCl and $H_2C_2O_4$ at concentrations of 2 mol/L, 4 mol/L and 6 mol/L, stirred continuously, and heated at 95 °C for 4–6 h at constant temperature in a water bath. Next, the solution was centrifuged (8000 r/min, 5 min), washed with deionized water, and dried overnight at 120 °C in a blower drying box (the corresponding samples are denoted FMTP, FMTP-2 mol/L HCl, FMTP-4 mol/L HCl, FMTP-4 mol/L HCl, FMTP-2 mol/L $H_2C_2O_4$, FMTP-4 mol/L $H_2C_2O_4$, and FMTP-6 mol/L $H_2C_2O_4$). Then, the samples treated with 4 mol/L HCl and $H_2C_2O_4$ were added to 0.1 mol/L NaOH, 0.2 mol/L NaOH 0.4 mol/L NaOH, 0.8 mol/L NaOH, and 1.2 mol/L NaOH solutions (liquid-to-solid ratio 5:1). Mechanical stirring treatment was conducted in a 95 °C water bath for 2 h, followed by centrifugation (8000 r/min, 5 min), and the upper waste catalyst mud obtained by centrifugation was then transferred to 250 mL of a 1 mol/L dilute HNO_3 solution under stirring and washing for 30 min to remove the residue after treatment. After washing with deionized water until neutral, the catalyst was put in a blower drying box at 120 °C overnight to dry and calcined at 600 °C for 6 h in a muffle furnace. The powder was pressed into pieces and screened to obtain a 14–30 mesh material, and the adsorbent product after acid-base treatment was obtained (the corresponding samples are denoted FMTP-4 mol/L HCl + 0.1 mol/L

NaOH, FMTP-4 mol/L HCl + 0.2 mol/L NaOH, FMTP-4 mol/L HCl + 0.4 mol/L NaOH, FMTP-4 mol/L HCl + 0.8 mol/L NaOH, FMTP-4 mol/L HCl + 1.2 mol/L NaOH, FMTP-4 mol/L $H_2C_2O_4$ + 0.1 mol/L NaOH, FMTP-4 mol/L $H_2C_2O_4$ + 0.2 mol/L NaOH, FMTP-4 mol/L $H_2C_2O_4$ + 0.4 mol/L NaOH, FMTP-4 mol/L $H_2C_2O_4$ + 0.8 mol/L NaOH, and FMTP-4 mol/L $H_2C_2O_4$ + 1.2 mol/L NaOH).

3.3. *Analysis and Characterization Methods*

The crystal structure of the samples was analyzed by X-ray diffraction (XRD, X-pert3 powder diffractometer, Panaco, The Netherlands). The framework structure of the samples was analyzed by Fourier transform infrared spectrometry (FT-IR, Bruker V70, Germany). The Brunauer–Emmet–Teller (BET)-specific surface area and pore properties of the samples were analyzed by a physical adsorption instrument (ASAP2020 surface analyzer, USA). The composition and content of the samples were analyzed via X-ray fluorescence spectroscopy (XRF, Bruker S8 Tiger, Zeiss Merlin, Germany). The micromorphology of the samples was analyzed using a compact scanning electron microscope (SEM, Carl Zeiss, Germany), and the chemical environment of ^{27}Al and ^{29}Si in the samples was determined by nuclear magnetic resonance (NMR, JNM-ECZ600r NMR, Japan). The adsorption properties of the samples for n-hexane, p-xylene, and water were measured with a Beijing Best 3H-2000PW automatic intelligent gravimetric adsorption (IGA) instrument.

4. Conclusions

After acid-base treatment, spent MTP catalyst maintains the crystal structure of the ZSM-5 molecular sieve. With increasing alkali treatment concentration, framework desilication occurs, local structural defect sites are generated in the molecular sieve lattice, and the silicon-to-aluminum ratio of the sample increases; moreover, the dealumination effect of $H_2C_2O_4$ is better than that of HCl. HCl dealumination mainly removes the framework aluminum of the molecular sieve, while $H_2C_2O_4$ not only removes the framework aluminum of the molecular sieve, but also dissolves some the nonframework aluminum. The BET-specific surface area and pore diameter of the sample are increased, and the spent catalyst maintains an irregular ellipsoidal shape. After alkali treatment, the surface of the spherical particles becomes rough. With increasing alkali concentration, the damage degree increases. The adsorption of p-xylene and n-hexane by the adsorbent product after acid-base treatment is significantly improved. When P/P_0 = 0.1 MPa, the $H_2C_2O_4$ concentration is 4 mol/L and the NaOH concentration is 0.1 mol/L, the adsorption capacity of the sample for p-xylene and n-hexane reaches the maximum, with values of 141.04 and 106.87 mg/g, respectively, which are 20.7% and 16.2% higher than those before treatment, showing good adsorption performance.

Author Contributions: X.Y. and Y.M. designed the experiments and wrote the main manuscript text, H.S. analyzed and prepared Figures 1–5, N.Z. analyzed and prepared Figures 6–9. Z.J. and M.Y. revised the manuscript. All authors have read and agreed to the published version of the manuscript.

Funding: This research was financially supported by the State Key Laboratory of High-Efficiency Utilization of Coal and Green Chemical Engineering (2019-KF-08) and the Ningxia Hui Autonomous Region Top Talent Program Funds.

Conflicts of Interest: The authors declare no competing interests.

References

1. Wei, Y.; Parmentier, T.E.; De Jong, K.P.; Zečević, J. Tailoring and visualizing the pore architecture of hierarchical zeolites. *Chem. Soc. Rev.* **2015**, *44*, 7234–7261. [CrossRef]
2. Betke, U.; Lieb, A. Micro-macroporous composite materials – Preparation techniques and selected applications: A review. *Adv. Eng. Mater.* **2018**, *20*, 1800252. [CrossRef]
3. Srivastava, R. Synthesis and applications of ordered and disordered mesoporous zeolites: Present and future prospective. *Catal. Today* **2018**, *309*, 172–188. [CrossRef]

4. Luo, H.; Law, W.W.; Wu, Y.; Zhu, W.; Yang, E.H. Hydrothermal synthesis of needle-like nanocrystalline zeolites from metakaolin and their applications for efficient removal of organic pollutants and heavy metals. *Microporous Mesoporous Mater.* **2018**, *272*, 8–15. [CrossRef]
5. Cundy, C.S.; Cox, P.A. The hydrothermal synthesis of zeolites: history and development from the earliest days to the present time. *Chem. Rev.* **2003**, *103*, 663–702. [CrossRef] [PubMed]
6. Shi, J.; Wang, Y.; Yang, W.; Tang, Y.; Xie, Z. Recent advances of pore system construction in zeolite-catalyzed chemical industry processes. *Chem. Soc. Rev.* **2015**, *44*, 8877–8903. [CrossRef] [PubMed]
7. Bayat, M.; Javanbakht, V.; Esmaili, J. Synthesis of zeolite/nickel ferrite/sodium alginate bionanocomposite via a co-precipitation technique for efficient removal of water-soluble methylene blue dye. *Int. J. Biol. Macromol.* **2018**, *116*, 607–619. [CrossRef]
8. Burakov, A.E.; Galunin, E.V.; Burakova, I.V.; Kucherova, A.E.; Agarwal, S.; Tkachev, A.G.; Gupta, V.K. Adsorption of heavy metals on conventional and nanostructured materials for wastewater treatment purposes: A review. *Ecotoxicol. Environ. Saf.* **2018**, *148*, 702–712. [CrossRef]
9. Zhang, S.; Gong, Y.; Zhang, L.; Liu, Y.; Dou, T.; Xu, J.; Deng, F. Hydrothermal treatment on ZSM-5 extrudates catalyst for methanol to propylene reaction: Finely tuning the acidic property. *Fuel Process. Technol.* **2015**, *129*, 130–138. [CrossRef]
10. Chen, X.D.; Li, X.G.; Li, H.; Han, J.J.; Xiao, W.D. Interaction between binder and high silica HZSM-5 zeolite for methanol to olefins reactions. *Chem. Eng. Sci.* **2018**, *192*, 1081–1090. [CrossRef]
11. Flores, C.; Batalha, N.; Ordomsky, V.V.; Zholobenko, V.L.; Baaziz, W.; Marcilio, N.R.; Khodakov, A.Y. Direct production of iso-paraffins from syngas over hierarchical cobalt-ZSM-5 nanocomposites synthetized by using carbon nanotubes as sacrificial templates. *ChemCatChem* **2018**, *10*, 2291–2299. [CrossRef]
12. Hu, Z.; Zhang, H.; Wang, L.; Zhang, H.; Zhang, Y.; Xu, H.; Shen, W.; Tang, Y. Highly stable boron-modified hierarchical nanocrystalline ZSM-5 zeolite for the methanol to propylene reaction. *Catal. Sci. Technol.* **2014**, *4*, 2891–2895. [CrossRef]
13. Yao, L.; Zhou, Y.; Guo, S.; Huangfu, C.; Ma, Y.; Liu, Y.; Yu, Z.; Chen, J.; Jin, K.; Jiang, H.; et al. Comparison of VOCs adsorption performance between Y and ZSM-5 zeolite. *Chin. J. Environ. Eng.* **2022**, *16*, 182–189.
14. Wang, Q.; Yang, Y.; Yu, C.; Huang, H.; Kim, M.; Feng, C.; Zhang, Z. Study on a fixed zeolite bioreactor for anaerobic digestion of ammonium-rich swine wastes. *Bioresour. Technol.* **2011**, *102*, 7064–7068. [CrossRef]
15. Gruszecka-Kosowska, A.; Baran, P.; Wdowin, M.; Franus, W. Waste dolomite powder as an adsorbent of Cd, Pb(II), and Zn from aqueous solutions. *Environ. Earth Sci.* **2017**, *76*, 521. [CrossRef]
16. Zhang, Y.; Dong, J.; Guo, F.; Shao, Z.; Wu, J. Zeolite synthesized from coal fly ash produced by a gasification process for Ni^{2+} removal from water. *Minerals* **2018**, *8*, 116. [CrossRef]
17. Jiménez-Castañeda, M.E.; Medina, D.I. Use of surfactant-modified zeolites and clays for the removal of heavy metals from water. *Water* **2017**, *9*, 235. [CrossRef]
18. Chen, Y.; Wang, H.; Zhao, W.; Huang, S. Four different kinds of peels as adsorbents for the removal of Cd (II) from aqueous solution: Kinetics, isotherm and mechanism. *J. Taiwan Inst. Chem. Eng.* **2018**, *88*, 146–151. [CrossRef]
19. Jiang, N.; Shang, R.; Heijman SG, J.; Rietveld, L.C. High-silica zeolites for adsorption of organic micro-pollutants in water treatment: A review. *Water Res.* **2018**, *144*, 145–161. [CrossRef]
20. Mohamadi, M.; Salimi, F.; Sadeghi, S. Reduction of oil, COD and turbidity of Kermanshah oil refinery effluent using modified nano-zeolite by bismuth and iron. *Desalin. Water Treat.* **2017**, *97*, 151–157. [CrossRef]
21. Sui, H.; Liu, J.; He, L.; Li, X.; Jani, A. Adsorption and desorption of binary mixture of acetone and ethyl acetate on silica gel. *J. Chem. Eng. Sci.* **2019**, *197*, 185–194. [CrossRef]
22. Gao, J.; Wang, W.; Zhang, J.; Lei, Z.; Shi, D.; Qu, L. Study on synthesis and adsorption performance of hydrophobic ZSM-5 zeolites for removal of toluene in high-humidity exhaust gas. *CIESC J.* **2020**, *71*, 337–343.
23. Arancibia-Miranda, N.; Baltazar, S.E.; García, A.; Muñoz-Lira, D.; Sepúlveda, P.; Rubio, M.A.; Altbir, D. Nanoscale zero valent supported by zeolite and montmorillonite: Template effect of the removal of lead ion from an aqueous solution. *J. Hazard. Mater.* **2016**, *301*, 371–380. [CrossRef] [PubMed]
24. Carrott, M.R.; Russo, P.A.; Carvalhal, C.; Carrott PJ, M.; Marques, J.P.; Lopes, J.M.; Gerner, I.; Guisnet, M.; Ribeiro, F.R. Adsorption of n-pentane and iso-octane for the evaluation of the porosity of dealuminated BEA zeolites. *Microporous Mesoporous Mater.* **2005**, *81*, 259–267. [CrossRef]
25. Caicedo-Realpe, R.; Pérez-Ramírez, J. Mesoporous ZSM-5 zeolites prepared by a two-step route comprising sodium aluminate and acid treatments. *Microporous Mesoporous Mater.* **2010**, *128*, 91–100. [CrossRef]
26. Wei, Z.; Xia, T.; Liu, M.; Cao, Q.; Xu, Y.; Zhu, K.; Zhu, X. Alkaline modification of ZSM-5 catalysts for methanol aromatization: The effect of the alkaline concentration. *Front. Chem. Sci. Eng.* **2015**, *9*, 450–460. [CrossRef]
27. Groen, J.C.; Zhu, W.; Brouwer, S.; Huynink, S.J.; Kapteijn, F.; Moulijn, J.A.; Pérez-Ramírez, J. Direct demonstration of enhanced diffusion in mesoporous ZSM-5 zeolite obtained via controlled desilication. *J. Am. Chem. Soc.* **2007**, *129*, 355–360. [CrossRef]

catalysts

MDPI

Article

Effects of the Acidic and Textural Properties of Y-Type Zeolites on the Synthesis of Pyridine and 3-Picoline from Acrolein and Ammonia

Israel Pala-Rosas [1], José Luis Contreras [2,*], José Salmones [1], Ricardo López-Medina [2], Deyanira Angeles-Beltrán [3], Beatriz Zeifert [4], Juan Navarrete-Bolaños [5] and Naomi N. González-Hernández [2]

[1] Postgraduate Studies Section, Escuela Superior de Ingeniería Química e Industrias Extractivas, Instituto Politécnico Nacional, Av. Instituto Politécnico Nacional s/n, Col. Nueva Industrial Vallejo, Mexico City 07738, Mexico
[2] Energy Departament, Universidad Autónoma Metropolitana-Azcapotzalco, Av. San Pablo 180, Col. Reynosa, Mexico City 02200, Mexico
[3] Basic Sciences Departament, Universidad Autónoma Metropolitana-Azcapotzalco, Av. San Pablo 180, Col. Reynosa, Mexico City 02200, Mexico
[4] Department of Engineering in Metallurgy and Materials, Escuela Superior de Ingeniería Química e Industrias Extractivas, Instituto Politécnico Nacional, Av. Instituto Politécnico Nacional s/n Col. Nueva Industrial Vallejo, Mexico City 07738, Mexico
[5] Instituto Mexicano del Petróleo, Eje Central Lázaro Cárdenas 152, Col. San Bartolo Atepehuacán, Mexico City 07730, Mexico
* Correspondence: jlcl@azc.uam.mx; Tel.: +52-55-91911047

Citation: Pala-Rosas, I.; Contreras, J.L.; Salmones, J.; López-Medina, R.; Angeles-Beltrán, D.; Zeifert, B.; Navarrete-Bolaños, J.; González-Hernández, N.N. Effects of the Acidic and Textural Properties of Y-Type Zeolites on the Synthesis of Pyridine and 3-Picoline from Acrolein and Ammonia. *Catalysts* **2023**, *13*, 652. https://doi.org/10.3390/catal13040652

Academic Editors: Maja Milojević-Rakić and Danica Bajuk-Bogdanović

Received: 21 January 2023
Revised: 20 March 2023
Accepted: 22 March 2023
Published: 26 March 2023

Abstract: A set of Y-type zeolites with Si/Al atomic ratios between 7–45 were studied as catalysts in the aminocyclization reaction between acrolein and ammonia to produce pyridine and 3-picoline. The catalytic activity tests at 360 °C revealed that the acrolein conversion increased in the order Z45 < ZY34 < ZY7 < ZY17, in agreement with the increase of the total acidity per gram of catalyst. In all cases, pyridine bases and cracking products (acetaldehyde and formaldehyde) were detected in the outflow from the reactor. The total yield of pyridines was inversely proportional to the total acidity for the catalysts, which presented large surface areas and micro- and mesoporosity. The selectivity towards 3-picoline was favored when using catalysts with a Brønsted/Lewis acid sites ratio close to 1. The formation of pyridine occurred more selectively over Lewis acid sites than Brønsted acid sites. The deactivation tests showed that the time on stream of the catalysts depended on the textural properties of zeolites, i.e., large pore volume and large BET area, as evidenced by the deactivation rate constants and the characterization of the spent catalysts. The physicochemical properties of the catalysts were determined by XRD, UV-vis, and Raman spectroscopies, infrared spectroscopy with adsorbed pyridine, N_2 physisorption, and SEM-EDXS. After the reaction, the spent catalysts were characterized by XRD, Raman spectroscopy, TGA, and SEM-EDXS, indicating that the uniform deposition of polyaromatic species on the catalyst surface and within the porous system resulted in the loss of activity.

Keywords: acrolein aminocyclization; pyridine; 3-picoline; Y zeolite; Brønsted/Lewis acid sites; micro/mesoporous zeolites; catalyst deactivation

1. Introduction

Pyridine bases are compounds of interest in several areas of industry and technology, with pyridine (azobenzene) and the series of picolines (methylpyridines) being the most important [1,2]. Due to their cyclic structure which is deficient π-type electrons and contains a basic nitrogen atom in the ring, and the methyl group attached to the ring in the case of picolines, pyridine compounds present unique physical and chemical properties such as weak basicity, dipolar and aprotic character, higher boiling point than benzene, miscibility with water and organic solvents, as well as reactions at the ring atoms and reactions at the substituents attached to the ring. Thus, pyridine bases are widely used as solvents in organic reactions and as precursors of drugs, polymers, insecticides, herbicides, dyes, adhesives, and explosives, or as denaturalizing agents of ethanol and antifreeze [3,4].

The great demand for pyridine compounds caused by the generation of new products has led to the development of different synthesis methods so that, during the last decades, the production of pyridines from several raw materials has been investigated, namely alcohols, aldehydes, ketones, dicarbonyl compounds, dicianoalkanes, alkenes, and alkynes, either in the presence of ammonia (NH_3) or amines. However, the most employed process has been the aminocyclization of aldehydes with NH_3 [2–6]. The reaction was first studied by Tschitschibabin [7], using alumina (Al_2O_3) as a catalyst for the reaction between NH_3 and different aldehydes at temperatures between 300 °C and 400 °C, obtaining pyridine and several mono- and polyalkyl substituted pyridines.

Currently, industrial production of pyridine is performed by the gas-phase reaction between NH_3 and a mixture of formaldehyde and acetaldehyde in the presence of a ZSM-5 catalyst. This process yields 3-picoline (3-methylpyridine) as the main byproduct and depending on the acetaldehyde/formaldehyde molar ratio in the feedstock, variable amounts of 2-picoline (2-methylpyridine) and 4-picoline (4-methylpyridine) are obtained [2]. Nevertheless, the interest in controlling the selectivity of pyridine products has led to research on the use of modified ZSM-5 catalysts and different raw materials.

Sato et al. [8] compared the catalytic activity of a silica-alumina catalyst (SiO_2-Al_2O_3) and different protonated zeolites, namely H-A, H-X, H-Y, H-mordenite, H-ZSM-5, and H-ZSM-11, in the gas-phase aminocyclization reaction between formaldehyde and acetaldehyde at 450 °C and atmospheric pressure. In all cases, pyridine was the main reaction product, while 2-, 3- and 4-picoline were obtained in yields around 6%, 6–14%, and 4–8%, respectively, depending on the catalyst. The H-ZSM-5 zeolite showed the best results with 61% of total pyridine yield, 42% of pyridine, 3% of 2-picoline, 11% of 3-picoline, and 5% of 4-picoline. However, a high amount of acid sites promoted coke formation and the rapid deactivation of the catalysts, so a medium acidity exhibited satisfactory results.

The importance of surface acidity in the reaction between formaldehyde, acetaldehyde, and NH_3 has been explored with a set of H-ZSM-5 zeolites, with Si/Al ratios of 40, 90, 150, and 240 in a continuous flow reactor at atmospheric pressure and 375 °C [9]. It was found that the increase of the Si/Al ratio, from 40 to 240, resulted in a decrease in the amount of acid sites and thus a decrease in the conversion from 98% to 58%. Additionally, the pyridine selectivity increased from 57% to 76%, and the selectivities to 2- and 3-picoline decreased from 15% to 2% and from 24% to 17%, respectively.

Additionally, the use of oxygenated unsaturated compounds, such as allyl alcohol, crotonaldehyde, acrolein, and acrolein dialkyl acetals, to produce pyridine bases has been reported [5,10–12]. Zhang et al. [10] reported the synthesis of pyridine and 3-picoline by the aminocyclization between acrolein and NH_3 in the presence of a ZSM-5 zeolite treated with HF, impregnated with Mg or subjected to both treatments. The best results were obtained using the catalyst sequentially modified (HF/Mg-ZSM-5) in the reaction at 425 °C, reaching pyridine and 3-picoline yields of 27% and 30%, respectively. Compared with the non-treated zeolite, the HF/Mg-ZSM-5 catalyst exhibited less concentration and weaker acid sites. Additionally, the modification of the zeolite resulted in a slight increase in the micro- and mesopore sizes.

The effect of the alkaline- and acid-treatments, and the impregnation with different metals on the catalytic activity of ZSM-5 zeolite in the synthesis of pyridine and 3-picoline from acrolein and acrolein dialkyl acetals, has also been reported [11]. The best results, 61% of the total yield of pyridines, 32% of pyridine yield, and 29% of 3-picoline yield, were obtained with the reaction between acrolein diethyl acetal and ammonia at 450 °C using the zeolite modified with an alkaline-acid treatment and subsequently impregnated with Zn. It was found that the incorporation of Zn changed the acid strength, while the double alkaline-acid treatment originated an increase in the surface area and the mesopore volume.

Similarly, it has been reported that the fluoridation treatment with NH_4-HF of an Mg-ZSM-5 zeolite modifies the porosity and acidity of the catalyst as a result of simultaneous desilication and dealumination. These factors improved the total yield of pyridines in the reaction between acrolein dimethyl acetal and NH_3 to produce pyridine and 3-picoline. In comparison with the use of the parent zeolite, using the modified catalyst enhanced the 3-picoline yield [12].

The advantage of the use of acrolein in the aminocyclization reaction is that it allows for eliminating the production of 2- and 4-picoline, increasing the selectivity towards pyridine and 3-picoline, also resulting in an easier separation of the reaction products, in comparison with the use of the mixture of acetaldehyde and formaldehyde as feedstock [3].

In addition, intensive research has been conducted recently on the catalytic dehydration of glycerol to produce acrolein over solid acid catalysts. This process represents a sustainable alternative for producing acrolein and its derivatives, in contrast with the prevailing industrial process based on the selective oxidation of petroleum-derived propylene [13–15]. This is because glycerol is obtained as a byproduct during the hydrolysis, saponification, and transesterification processing of vegetable oils and fats, resulting in its high availability and low market price [14,16].

On the other hand, using a zeolite catalyst other than ZSM-5 has been barely reported for synthesizing pyridine compounds. The comparison of H-ZSM-5, H-mordenite, and HY zeolites as catalysts in the reaction between ethanol and NH_3 was reported [17]. At 350 °C, the ethanol conversion increased in the order H-ZSM-5 < H-Y < H-mordenite. However, the main difference was reflected in the distribution of reaction products. The main reaction products over the H-ZSM-5 catalyst were ethene (56.9%), pyridine (18%), acetonitrile (4.6%), 2-picoline (3.9%) and CO_2 (10.2%), while H-Y and H-mordenite zeolites produced mainly ethene with pyridine selectivities of 1.6% and <0.1%, respectively, without the presence of any other pyridine base.

Grigor'eva et al. [18] studied the reaction between ethanol, formaldehyde, and NH_3 using H-ZSM-5, H-ZSM-12, and H-β zeolites as catalysts, obtaining pyridine, picolines, and lutidines as the main reaction products. At 300 °C, the conversion and the pyridine selectivity followed the order H-ZSM-12 < H-ZSM-5 < H-β. The higher conversion obtained with the H-β zeolite was related to the higher concentration of acid sites. The improved pyridine selectivity was associated with the structural characteristics of the catalyst. Furthermore, the H-ZSM-12 and H-ZSM-5 zeolites promoted the formation of 3-picoline, reducing the amounts of 2- and 4-picoline.

It is noticeable that a gap exists in the research and development of catalysts with appropriate physicochemical properties for the synthesis of pyridine bases from aldehydes and ammonia. In this sense, the use of mesoporous or hierarchical zeolites represents a strategy to enhance the reaction performance by improving the transport of reactants and products and the accessibility of reactants to the active sites located inside the pore channels [19–22], which is relevant for reactions in which bulky molecules are involved.

Specifically, the Y zeolite has been widely used for the cyclization of different molecules such as phenol [23], citronellal [24], terpenols [25], unsaturated alcohols [26], and aryl methallyl eters [27]. In synthesizing pyridine bases, the use of the Y-zeolite has been barely reported in the reaction between formaldehyde, acetaldehyde, and ammonia [28] and between acrolein diethyl acetal and ammonia [11].

This study presents the influence of the acidic and textural properties of Y-type zeolites, with Si/Al atomic ratios between 7 and 45, on their catalytic activity during the gas-phase reaction between acrolein and NH_3 to produce pyridine and 3-picoline.

2. Results and Discussion

2.1. Characterization of Catalysts

The X-ray diffractograms of the Y-type zeolites are presented in Figure 1. All the samples exhibited the typical reflections of the aluminum silicate hydrated with a faujasite-like structure (JCPDS card 00-043-0168) as the unique crystalline phase. The Y zeolites presented diffraction patterns in agreement with the patterns of ultrastable dealuminated Y zeolites reported in the literature [29].

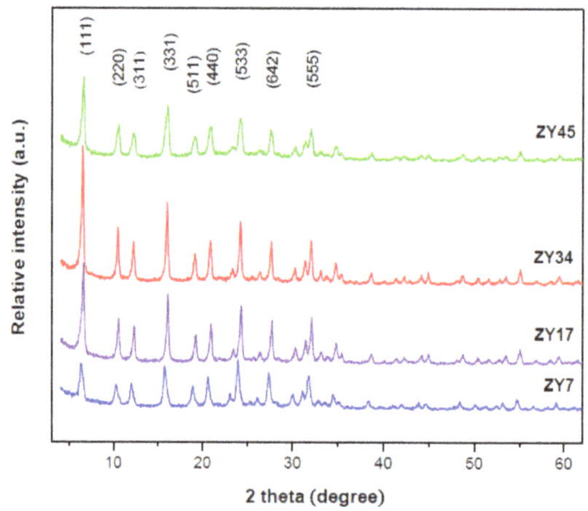

Figure 1. X-ray powder diffractograms of the Y-type zeolites.

The average crystal (L) was estimated from the reflection of the plane (331, considering the Scherrer equation. The average crystal sizes were 19.6 nm, 31.4 nm, 31.2 nm, and 16.1 nm for the ZY7, ZY17, ZY34, and ZY45 zeolites, respectively. Similar results were reported for a series of steam-dealuminated Y-zeolites [30].

The UV-vis spectra of the Y-type zeolite samples (Figure 2a) showed a main band between 280 and 285 nm. The decrease of the Al content in the samples produced a reduction of the absorbance. The band is attributed to the charge transfer of the πp-πd transition between oxygen species (O^{2-}) and tetrahedral aluminum (Al^{3+}) of the zeolite structure [31]. The presence of extra-framework Al cations was evidenced by the extension of the band up to 350 nm, as reported for other zeolite-type solids [31,32].

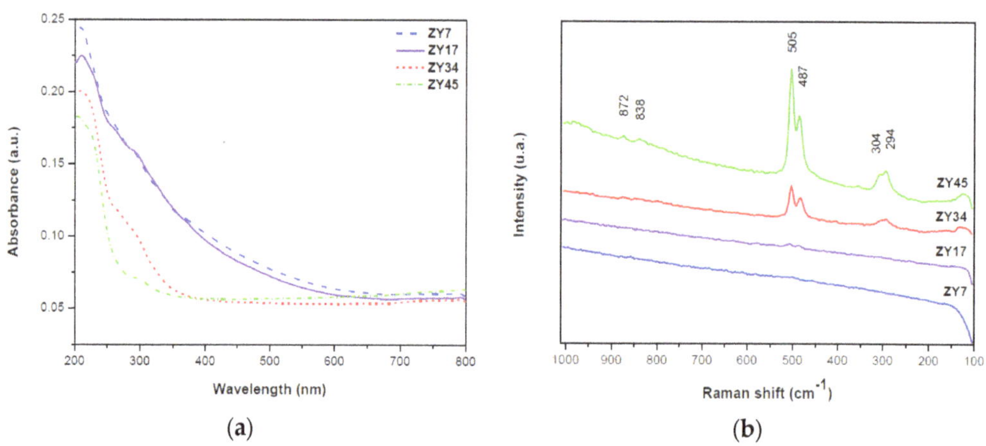

Figure 2. (**a**) UV-vis and (**b**) Raman spectra of the Y-type zeolites.

The Raman spectra of the Y-type zeolites are presented in Figure 2b. The zeolite with a Si/Al atomic ratio of 7 did not show bands, while the sample ZY17 displayed barely noticeable bands at 505 cm^{-1} and 487 cm^{-1}. Both bands were intense for the ZY34 and ZY45 zeolites, and further bands at 872 cm^{-1}, 838 cm^{-1}, 304 cm^{-1}, and 294 cm^{-1} were also displayed. The bands between 200 cm^{-1} and 400 cm^{-1} are attributed to the low-frequency torsional modes and the cation-oxygen modes. The strong bands between 300–600 cm^{-1} are associated with the type of rings in the zeolite structure. Y-type zeolites with high silicon contents show bands at approximately 490 cm^{-1} and 500 cm^{-1}, attributed to the Si-O-Si bond angles and involving the mixed vibration of stretching and bending of the O atom [33–36]. The bands at 873 cm^{-1} and 839 cm^{-1} were assigned to the displacement of Si atoms bonded to four O atoms, besides the stretch vibrations of the Si-O bonds [37,38].

Figure 3a shows the infrared spectra with adsorbed pyridine (IR-Py) of the zeolite samples at 350 °C. For all samples, the spectra displayed bands at 1450 cm^{-1}, 1487 cm^{-1}, 1539 cm^{-1}, and a broad peak between 1576–1661 cm^{-1}. These results agree with the literature, where the bands at 1445–1460 cm^{-1} are reported as a result of the interaction between pyridine and the Lewis acid sites (LAS). The band at 1540 cm^{-1} is related to Brønsted acid sites (BAS) interacting with the pyridine, and the bands between 1485–1489 cm^{-1} are associated with pyridine adsorbed on both Brønsted and Lewis acid sites [39]. The overlapped bands at 1580–1661 cm^{-1} are attributed to ring vibrations of pyridine [40]. Comparing the spectra of the Y zeolites displayed an increment of the absorbance with the rise of the Si/Al ratio from 7 to 17, while a further increase to 34 and 45 resulted in less intense bands, suggesting higher total acidity in the samples with higher aluminum content (Si/Al = 7 and 17) than in the zeolites ZY34 and ZY45.

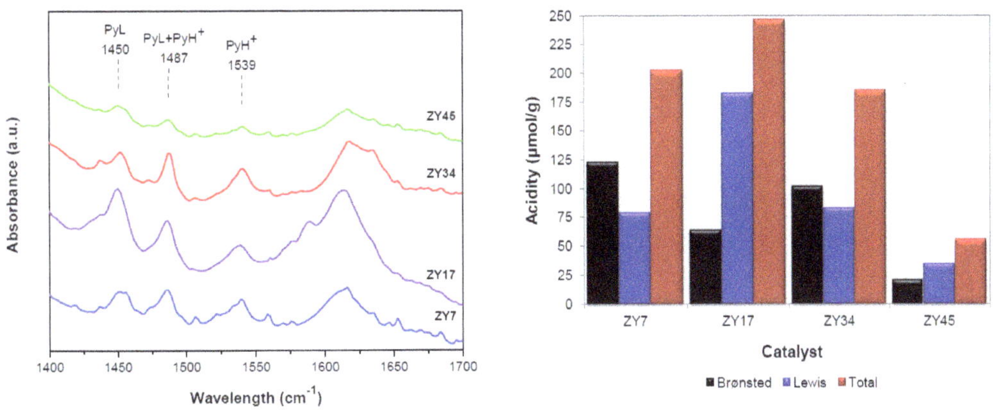

(a) (b)

Figure 3. (**a**) Infrared spectra with adsorbed pyridine, and (**b**) quantification of the Brønsted and Lewis acid sites of the Y-type zeolites at 350 °C.

The quantification of the total, Brønsted and Lewis acidity of the Y-type zeolites at 350 °C is shown in Figure 3b. For the total acidity, the ZY7, ZY17, and ZY34 samples exhibited values of 183–246 µmol Py/g catalyst. The additional increase to an atomic ratio Si/Al = 45 decreased the total acidity, with the lowest value of 56 µmol Py/g catalyst.

The Brønsted acidity exhibited values between 64 and 123 µmol Py/g catalyst for the zeolites with Si/Al atomic ratio between 7 and 34, while the sample ZY45 displayed a value of 21 µmol Py/g catalyst. On the other hand, the Lewis acidity showed values around 79 and 81 µmol Py/g catalyst for the samples ZY7 and ZY34, with maximum and minimum values of 182 and 35 µmol Py/g catalyst for the ZY17 and ZY45 zeolites. As shown by the ratios of the Brønsted and Lewis acidity (BAS/LAS) in Table 1, the Brønsted acidity was predominant in the samples with Si/Al atomic ratios of 7 and 34. The samples ZY17 and ZY45 showed more LAS than BAS, with considerable amounts for the former.

Table 1. Physicochemical properties and composition of the Y-type zeolites.

Catalyst	$L_{(331)}$ [a]	BAS/LAS [b]	Weight Loss [c] (%)	S_{BET} [d] (m^2/g)	V_P [e] (cm^3/g)	Element (Atom %) [f]			Si/Al [g]
						Si	Al	O	
ZY7	19.6	1.56	16.9	a	0.21	29.28	3.73	69.99	7.5
ZY17	31.4	0.35	16.1	736.0	0.25	25.39	1.66	72.95	16.1
ZY34	31.2	1.23	10.8	749.4	0.27	30.19	0.28	69.53	36.4
ZY45	16.1	0.60	12.7	753.8	0.29	27.04	0.23	72.74	47.8

[a] Average crystal size, [b] ratio of Brønsted/Lewis acid sites at 350 °C, [c] determined by TGA at 350 °C, [d] BET surface area, [e] pore volume, [f] determined by EDXS, [g] determined by AAS.

The thermogravimetric analysis (TGA) in Figure 4a showed a pronounced weight loss of the ZY7 sample from room temperature to 260 °C. For the zeolites ZY17, ZY34, and ZY45 the main weight loss occurs up to 145 °C, 110 °C, and 110 °C, respectively. This weight loss was attributed to eliminating water bound to the zeolite surface, as reported for H-ZSM-5 zeolites [41,42]. The change in weight estimated at 350 °C (close to the reaction temperature) is shown in Table 1, and it is observed that the samples that lost the most water are the ZY7 and ZY17 zeolites, while the ZY34 zeolite lost the less water. This behavior suggests that the samples ZY7 and ZY17 could contain more density of hydroxyl groups (OH) by unit weight when the temperature reached 350 °C, which could be related to the Brønsted or Lewis acidity.

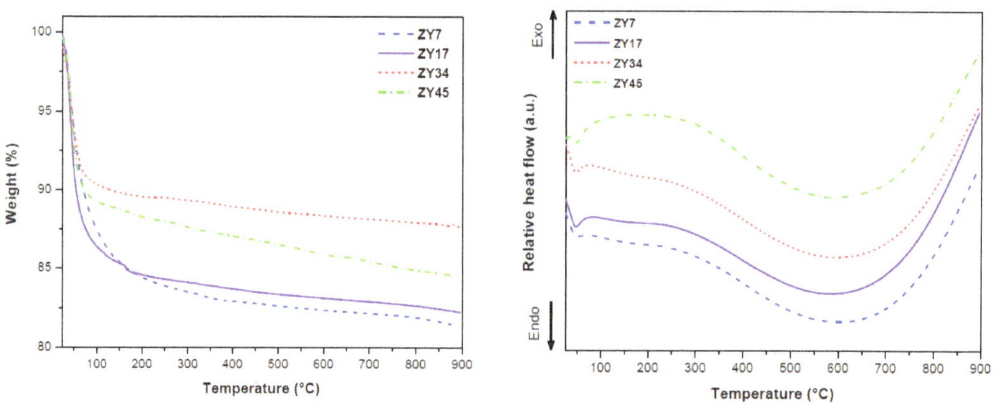

(a) (b)

Figure 4. (**a**) Thermogravimetric analysis and (**b**) differential thermal analysis of the Y-type zeolites.

The differential thermal analysis (DTA), shown in Figure 4b, exhibited endothermic behavior of the samples at temperatures below 150 °C, typical of a dehydration process, and consistent with the loss of physisorbed water. Further temperature increase resulted in the progressive decrease of heat flow, attributed to the release of OH-groups up to 600 °C.

The N_2 isotherms of the Y-type zeolites are presented in Figure 5a. The samples exhibited type IV isotherms and distinctive mesoporous solids. The desorption isotherms presented H4-type hysteresis loops, which are distinguishing agglomerated particles with edges, such as cubes and tetrahedra, resulting in pores with slit shape and uniform size [43,44].

The change of the Si/Al ratio from 7 to 45 increased the specific surface area of the zeolites from 635 to 754 m^2/g, as presented in Table 1. For all samples, the main contribution to the surface area was given by the internal area of the porous system of the solids. However, the external area increased when increasing the Si/Al atomic ratio, representing 21%, 28%, 31%, and 35 % of the total area in the order ZY7, ZY17, ZY34, and ZY45.

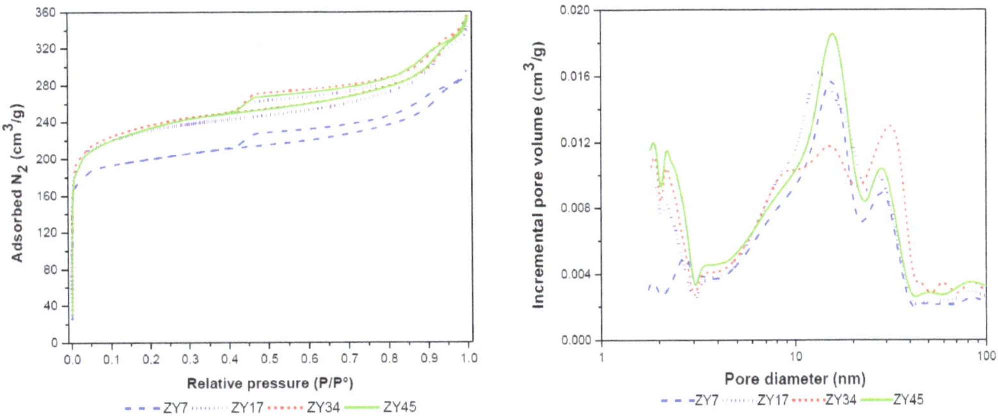

(a) (b)

Figure 5. (**a**) N_2 physisorption isotherms, and (**b**) distribution of pore sizes of the Y-type zeolites.

Additionally, the pore volume increased from 0.21 cm^3/g to 0.287 cm^3/g with the increment in the Si/Al ratio from 7 to 45 (Table 1). The samples showed a proportion of micropores and small mesopores (2–3 nm) besides a mesoporous system with pore sizes between 5 nm and 40 nm.

The scanning electron micrographs (SEM) of the Y-type zeolites are shown in Figure 6. All the samples displayed grains with shapes of the variable and faceted pyramid and hexagonal prisms, which are typical of the Y zeolite [45,46]. Furthermore, the surface of the zeolites was rough, porous, and slightly cracked. The average grain size of the zeolites in this study was 400 nm and did not change significantly with the Si/Al atomic ratio variation. The values correspond approximately with those reported by Taufiqurrahmi et al. [47] of 500 nm for a commercial Y zeolite.

According to the EDXS analysis, the samples are constituted by Si, Al, and O, and the diminishing of the Al content with the increase of the Si/Al atomic ratio was confirmed, as presented in Table 1. The bulk Si/Al atomic ratios of the samples were confirmed by atomic absorption spectroscopy (AAS).

(a) (b)

Figure 6. *Cont.*

(c) (d)

Figure 6. Scanning electron micrographs and particle size distributions of the zeolites (**a**) ZY7, (**b**) ZY17, (**c**) ZY34, and (**d**) ZY45.

2.2. Catalytic Activity

The Y-type zeolites with different Si/Al atomic ratios were active as catalysts in the reaction between acrolein and ammonia at 360 °C. At this point, it is noteworthy to mention that the reaction conditions (T = 360 °C, GHSV = 4994 h^{-1}, molar ratio NH_3/acrolein = 2, and pure acrolein vapor feed) were chosen to evaluate the catalytic performance of the zeolites in terms of their acidic and textural properties, minimizing the effects of the process variables on the reaction.

As presented in Figure 7a, The Y-type zeolites achieved almost total conversion of acrolein at the beginning of the process. The acrolein conversion increased in the order ZY45 < ZY34 < ZY7 < ZY17, similar to the total acidity of the catalysts (Figure 3b).

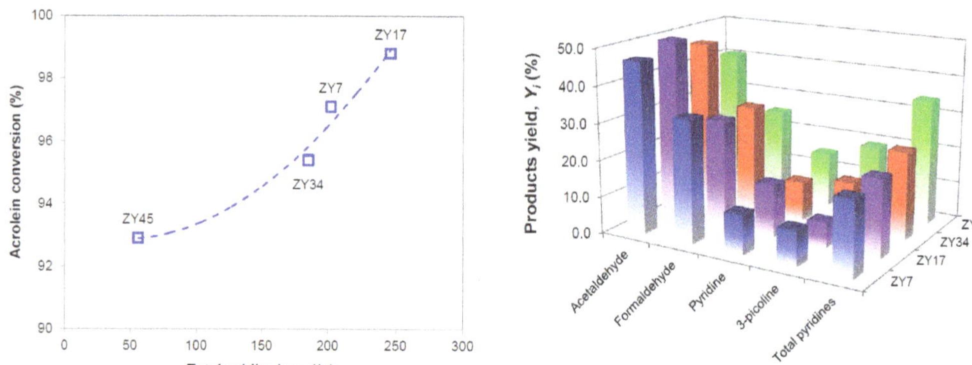

(a) (b)

Figure 7. (**a**) Initial acrolein conversion and, (**b**) initial products yield of the reaction between acrolein and ammonia over the Y-type zeolites. The total yield of pyridine bases = (pyridine + 3-picoline). Reaction conditions: T = 360 °C, molar ratio NH_3/acrolein = 2, GHSV = 4994 h^{-1}, TOS = 5 min.

The compounds identified at the output stream from the reactor were acetaldehyde, formaldehyde, pyridine, and 3-picoline. For all the catalysts, acetaldehyde was the major reaction product (Figure 7b), which is in agreement with the literature where acetaldehyde is reported as a cracking product of acrolein over acid catalysts [48,49]. The total yield of pyridine bases (pyridine + 3-picoline) improved from 20.2% to 34.2% with the change of the

Si/Al atomic ratio from 7 to 45. Pyridine was produced with yields of 10.7%, 14.3%, 13.2%, and 15% with the ZY7, ZY17, ZY34, and ZY45 zeolites, respectively. 3-picoline achieved yields of around 9.5%, 6.7%, 13.2%, and 19.1% in the same order. These low yields can be attributed to the low reaction temperature, and low NH_3/acrolein molar ratio in the feed since it is known [9–12,50–53] that the increase of both variables improves the formation of pyridine compounds and typical reaction conditions for the synthesis of pyridines from aldehydes and ammonia are T = 400–475 °C, NH_3/acrolein molar ratios between 4 and 6, as well as the reactant diluted with water.

It is noticeable that the total yield of pyridine bases followed the order ZY7 < ZY17 < ZY34 << ZY45, following the total acidity of the catalysts, except the ZY45 zeolite that exhibited the highest yield of pyridines but the second highest acidity among the samples, as shown in Figure 8a. The total yield of pyridine bases followed the order ZY7 < ZY17 < ZY34 << ZY45, similar to the total acidity of the catalysts, as shown in Figure 8a. The catalytic activity of zeolites depends on several factors, such as the concentration of sites, their acid strength, and accessibility for reactants. The acid strength increases with the Si/Al atomic ratio. The acid strength of zeolites in this study may be assumed more or less constant. Since the BET areas of the zeolites with Si/Al atomic ratios between 7 and 45 are similar, the accessibility of acrolein to acid sites was also similar among the catalysts. As consequence, the total acidity was the most significant property for the conversion.

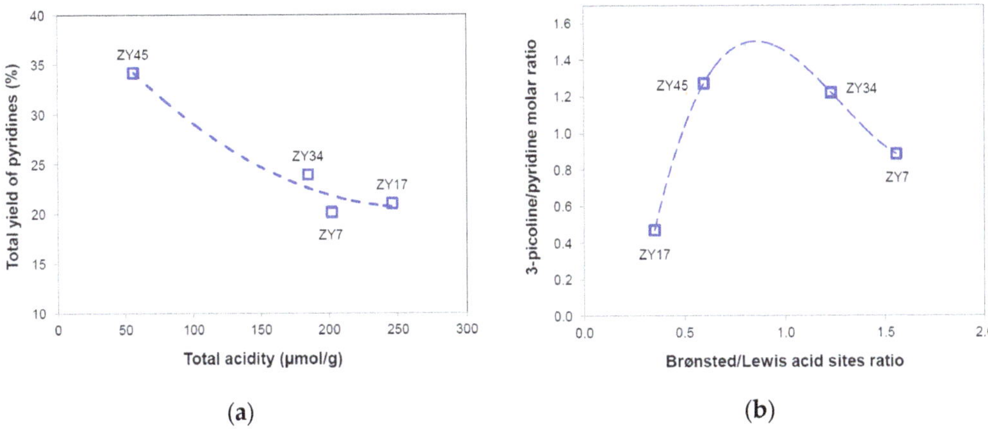

Figure 8. (**a**) Effect of the catalyst total acidity on the total yield of pyridines and (**b**) effect of the Brønsted/Lewis acid sites ratio on the 3-picoline/pyridine molar ratio.

Furthermore, the type and amount of acid sites influenced the selectivity of pyridine products. As shown in Figure 8b, the Brønsted/Lewis acid sites ratio deviated from unity to lower or higher values, 3-picoline was the main pyridine product. However, the 3-P/Py molar ratio approaches 1 (this is similar amounts of 3-picoline and pyridine) in the BAS/LAS range between 0.6 and 1.3, as with catalysts ZY34 and ZY45. The further increase or decrease of the BAS/LAS ratio originated smaller values of the 3-P/Py molar ratio than unity, and the reaction proceeded preferentially towards pyridine. Remarkably, the formation of pyridine occurred more selectively over LAS than BAS; i.e., the ZY17 zeolite, the catalyst with the smallest BAS/LAS acid sites ratio, exhibited a smaller 3-P/Py molar ratio than the catalyst with the highest BAS/LAS ratio, namely the ZY7 zeolite. Similar results were reported for the reaction between acrolein and ammonia over an HF/MgZSM-5 catalyst, where pyridine was found to be produced over strong LAS, and 3-picoline was formed over BAS and weak LAS [54].

2.3. Catalyst Deactivation Behavior

The catalytic activity tests at 360 °C and GHSV = 4994 h^{-1} were monitored with time on stream (TOS). As shown in Figure 9a–c, the activity of the zeolites in the acrolein aminocyclization exhibited differences with TOS. In all cases, the catalysts exhibited a rapid decrease in activity. Both the ZY7 and ZY17 catalysts presented similar behavior, a step decrease in conversion along TOS, reaching 56% and 61% of conversion at 45 min, respectively. Conversely, the decrease of the acrolein conversion for the ZY34 and ZY45 zeolites was less severe during the first 25 min of the process, achieving 65% and 63% of conversion at 45 min, correspondingly. Notably, the catalyst with a Si/Al ratio of 45 showed a steady behavior between 5 and 15 min, with conversions around 92% and 90%. Similarly, a rapid deactivation of ZSM-5 catalysts, due to coke deposited on the catalyst surface, has been reported for the synthesis of pyridines from aldehydes and ammonia [9,10,51,53,55].

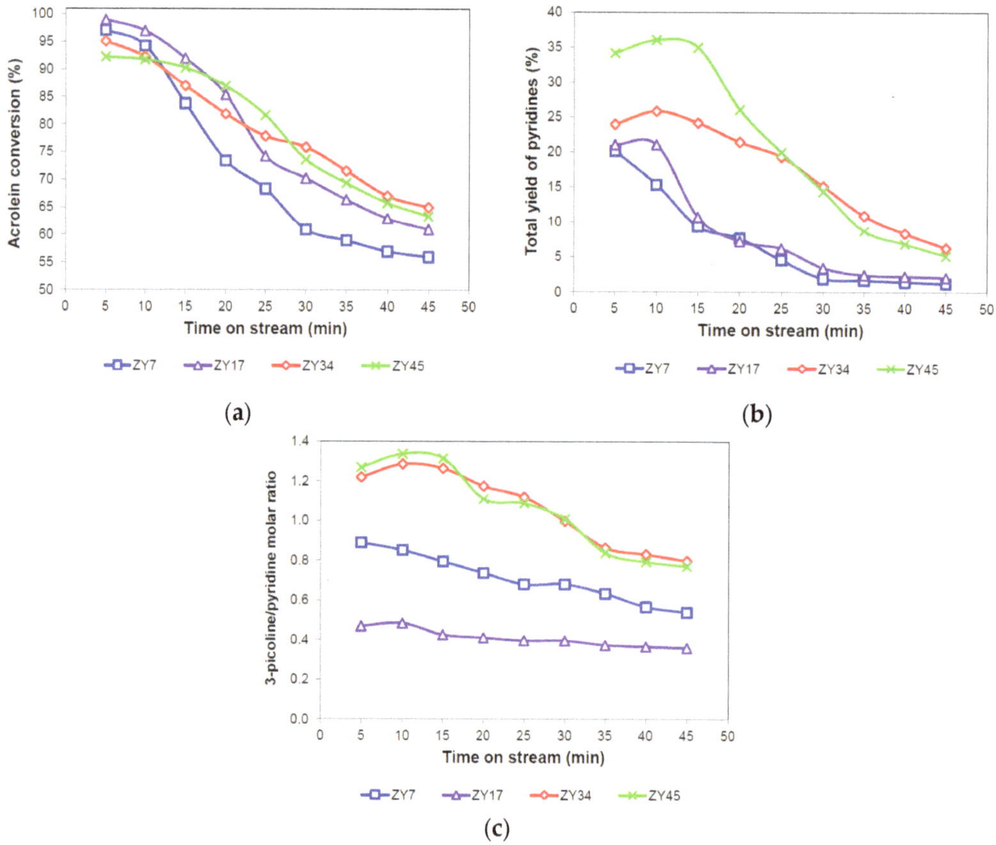

Figure 9. (**a**) Acrolein conversion, (**b**) total yield of pyridines, and (**c**) 3-picoline/pyridine molar ratio over the Y-type zeolites with time on stream.

The deactivation of solid acid catalysts during catalytic reactions of organic reactants is commonly caused by the deposition of heavy secondary products, covering the acid sites and blocking the catalyst pores. These carbonaceous compounds, also known as coke, are produced by polymerization reactions of reactants and products [56,57].

The total yield of pyridines with TOS was also influenced by the catalyst, as presented in Figure 9b. Both the ZY7 and ZY17 zeolites showed a gradual decline in the formation of pyridines with values below 10% after 15 min of TOS and approaching 3% as the minimum value at 45 min. Conversely, the total yield of pyridine bases over the ZY34 catalyst presented a steady behavior between 5 and 20 min, with values around 11%, and smoothly decreased to 7% at 45 min. When using the ZY45 zeolite, pyridines formation maintained higher values than the rest of the catalysts during 25 min, but a significant decrease occurred after 15 min of TOS up to 6% of the total yield of pyridines at 45 min.

Regarding the behavior of pyridine products with TOS, as shown in Figure 9c, the catalysts exhibit a decrease in the 3-P/Py molar ratio in all cases. In the case of the ZY7 and ZY17 catalysts, the change in the pyridine products ratio with TOS was minimal, producing higher amounts of pyridine than 3-picoline throughout the test (3-P/Py < 1). On the other hand, the ZY34 and ZY45 showed a slight increase in the 3-P/Py molar ratio during the first 15 min and a gradual decrease for the rest of the test. Both catalysts maintained higher amounts of 3-picoline than pyridine (3-P/Py > 1) for the initial 30 min. This behavior is consistent with those observed during the reaction with acrolein [10], and formaldehyde with acetaldehyde [50,58] using ZSM-5 zeolites.

The data of the acrolein conversion with TOS were used to establish an equation to model the deactivation of the catalysts during the reaction between acrolein and ammonia. Since the deactivation curves of the Y-type zeolites showed a residual or steady-state activity after initial deactivation the generalized power-law equation (GPLE) deactivation model was considered [59]. The first- and second-order deactivation models are expressed in Equations (1) and (2), respectively:

$$\ln(X - X_s) = \ln(k[W_{cat}/F_{a,0}] - X_s) - k_d t \tag{1}$$

$$\ln\left(\frac{X - X_s}{X + X_s}\right) = \ln\left(\frac{1 - [F_{a,0}X_s/W_{cat}k]}{1 + [F_{a,0}X_s/W_{cat}k]}\right) - [2F_{a,0}X_s k_d/W_{cat}k]t \tag{2}$$

where X is the conversion of acrolein, X_s represents the steady-state conversion, k, and k_d are the reaction and deactivation rate constants, in that order; W_{cat} is the weight of catalyst loaded into the reactor; $F_{a,0}$ is the molar flow rate of acrolein; and t is time. The GPLE deactivation models were fitted to the experimental data by non-linear regression, minimizing the sum of square residuals (SSR) as the objective function and discriminated in terms of their determination coefficients (R^2).

Figure 10 compares the observed activity of each catalyst and the activity estimated with the GPLE models. It is noticeable that the second-order model showed a slightly better fitting than the first-order.

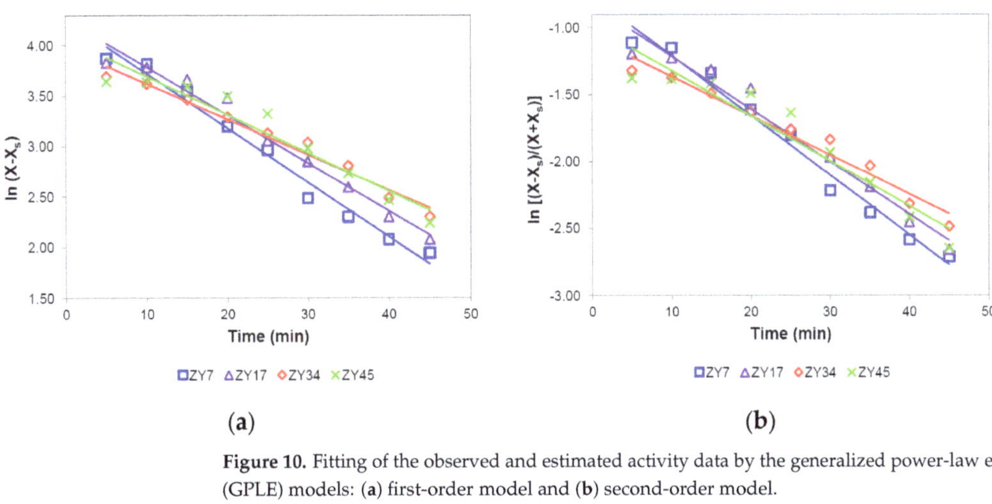

Figure 10. Fitting of the observed and estimated activity data by the generalized power-law equation (GPLE) models: (**a**) first-order model and (**b**) second-order model.

The minimization results presented in Table 2 showed that both models provide reliable results for all the catalysts. However, the second-order model provided a better fit (lower SSR and higher R^2 values) than the first-order model, in agreement with the literature [59–61].

Table 2. Reaction and deactivation rate constants and discrimination parameters for the reaction between acrolein and ammonia over the Y-type zeolites.

Catalyst	$n = 1$				$n = 2$			
	k (mol·g_{cat}^{-1}·min^{-1})	k_d (min^{-1})	SSR	R^2	k (mol·g_{cat}^{-1}·min^{-1})	k_d (min^{-1})	SSR	R^2
ZY7	2.94×10^{-1}	5.37×10^{-2}	0.08	0.98	3.02	6.11×10^{-2}	0.06	1
ZY17	3.04×10^{-1}	4.73×10^{-2}	0.08	0.98	2.98	5.04×10^{-2}	0.07	1
ZY34	2.66×10^{-1}	3.50×10^{-2}	0.05	0.98	3.62	2.98×10^{-2}	0.05	1
ZY45	2.77×10^{-1}	3.77×10^{-2}	0.17	0.93	3.37	3.68×10^{-2}	0.16	0.99

For the first-order model, the reaction rate constants slightly increased with the change of the Si/Al atomic ratio from 7 to 17 and decreased for the ZY34 and ZY45 samples. Conversely, the second-order model gave smaller values of the reaction rate constant for the ZY7 and ZY17 catalysts than those of the zeolites with atomic ratios Si/Al = 34 and 45. Furthermore, for both models, the deactivation rate constants of the ZY34 and ZY45 zeolites displayed the smallest values among the samples, in agreement with the rapid decrease of the acrolein conversion with TOS (Figure 9a).

At this point, it is important to mention that there is a lack of information in the literature on the reaction and deactivation kinetic parameters of the synthesis of pyridines. For this type of reaction, it has only been reported the reaction kinetics and the deactivation rate constant for an H-ZSM-5 catalyst in the synthesis of pyridine bases from acetaldehyde, formaldehyde, and ammonia (k_d = 3.61 × 10^{-4} min^{-1} at 350 ° C, space-time between 147–440 g/h/mol, molar ratio NH_3/aldehyde = 4, and water diluted feed) [62].

2.4. Characterization of Spent Catalysts

Designated as SZY7, SZY17, SZY34, and SZY45, the spent catalysts were characterized to analyze the changes that occurred in the solid samples after the stability test at 360 °C and GHSV = 4994 h^{-1}.

Presented in Figure 11, the X-ray diffractograms of the spent catalysts revealed that after the reaction between acrolein and ammonia, the zeolites maintained their crystalline structure. When comparing with the fresh samples, a loss of crystallinity was evidenced by the decrease of the intensity and a shift towards lower values of 2 thetas of the XRD patterns of spent catalysts, presumably caused by the deposition of carbonaceous compounds in agreement with the literature [63]. The average change in the position (2-theta scale) of representative planes of the spent catalysts regarding the fresh zeolites is indicated in Figure 11. In addition, the reflection plane (331) became the most intense reflection in all cases.

As reported in Table 3 and compared with the (311) plane of the fresh solids, the zeolite samples showed an increase in their average crystallite sizes after the reaction, suggesting the deposition of bulky compounds inside the microporous cages, causing the expansion of the unit cell.

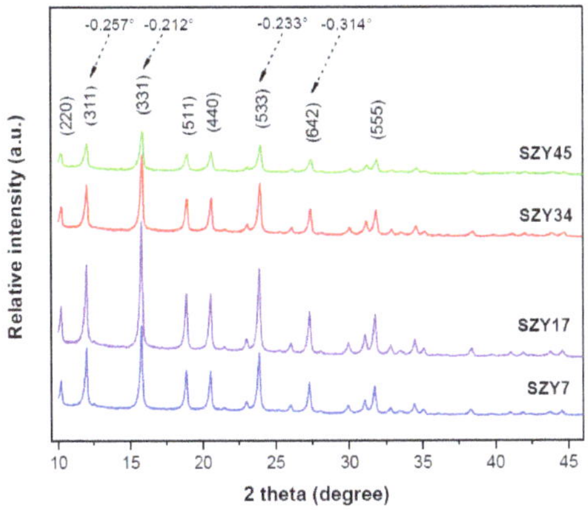

Figure 11. X-ray powder diffractograms of the spent Y-type zeolites.

Table 3. Physicochemical properties and composition of the spent zeolites.

Catalyst	$L_{(331)}$ [a] (nm)	Coke Content [b] (wt %)	Element (atom %) [c]		C/N
			C	N	
SZY7	38.5	25.7	35.77	5.05	7.08
SZY17	37.7	27.1	48.77	4.69	10.39
SZY34	35.2	23.1	38.99	3.97	9.82
SZY45	27.1	22.5	31.99	4.67	6.85

[a] Average crystal size, [b] determined by TGA, [c] determined by EDXS.

The Raman spectra of the spent catalysts are shown in Figure 12. The samples exhibited a noticeable shift at 1611 cm^{-1} and a very weak band at 1372 cm^{-1}. These bands can be related to coke deposited on the catalyst, attributed explicitly to C=C, and to ring stretches of large polyaromatic compounds, presumably with two-dimensional and sheet-like topology [64–66]. This result suggests that coke deposited on the Y-type zeolites could be produced by secondary reactions of acrolein and olefin intermediates, namely condensation and polymerization, in agreement with previous reports [67,68]. Similarly, in the case of the synthesis of pyridine bases from the reaction between ammonia and aldehydes, it has been determined by solid-state ^{13}C NMR and FTIR that the coke formed on ZSM-5 catalysts was constituted by compounds with aliphatic and alkoxyl groups, as well as highly unsaturated polyalkenes and condensed aromatics [69,70].

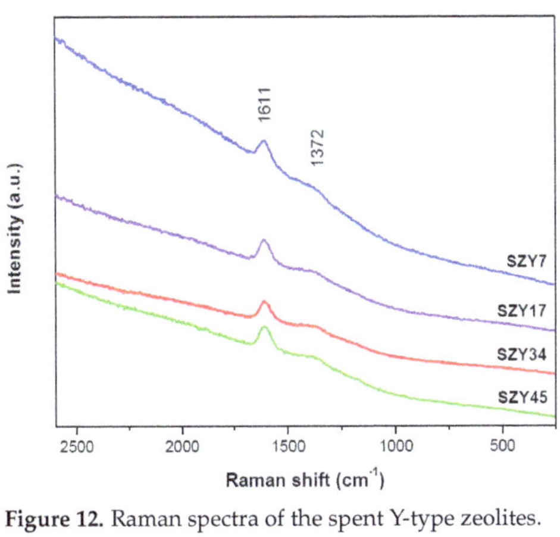

Figure 12. Raman spectra of the spent Y-type zeolites.

Shown in Figure 13a, the thermogravimetric analysis of the spent zeolites exhibited two weight losses: the first one between room temperature and 250 °C attributed to the removal of water adsorbed from the reaction, and a second loss between 300–700 °C ascribed to the combustion of the carbonaceous compounds deposited on the catalyst, which varied from 20% to 27% as presented in Table 3.

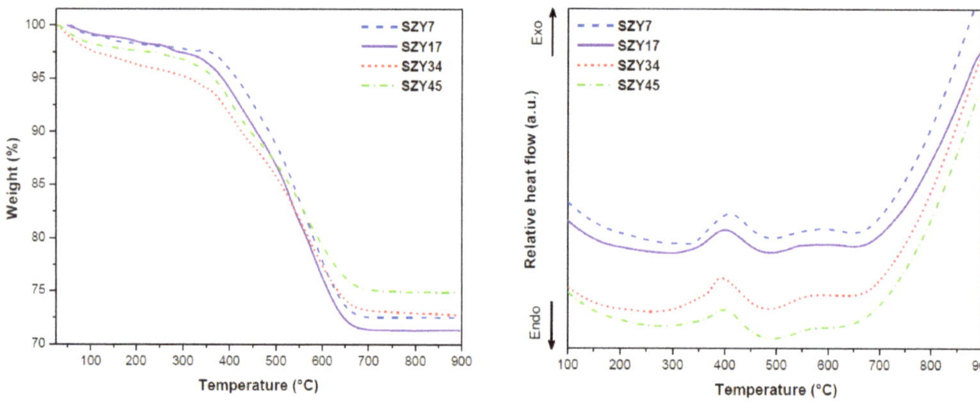

(a) **(b)**

Figure 13. (**a**) Thermogravimetric analysis and (**b**) differential thermal analysis of the spent Y-type zeolites.

The differential thermal analysis (DTA) presented in Figure 13b evidenced that two exothermic events occurred between 315–660 °C, confirming the combustion of coke deposits from the catalyst and suggesting the presence of different carbonaceous species in agreement with the Raman spectroscopy analysis.

From the estimation of the weight loss between 300–700 °C (Table 3), it is noticeable that the coke content increased with the increment of the Si/Al molar ratio from 7 to 17 and subsequently decreased for the samples with Si/Al molar ratios of 34 and 45. As shown in Figure 14, the amount of coke was corresponding to the pore volume of the fresh ZY7, and ZY17 catalysts, while the decrease of the coke content in the spent ZY34 and ZY45 samples can be explained by their larger pore volume and surface areas.

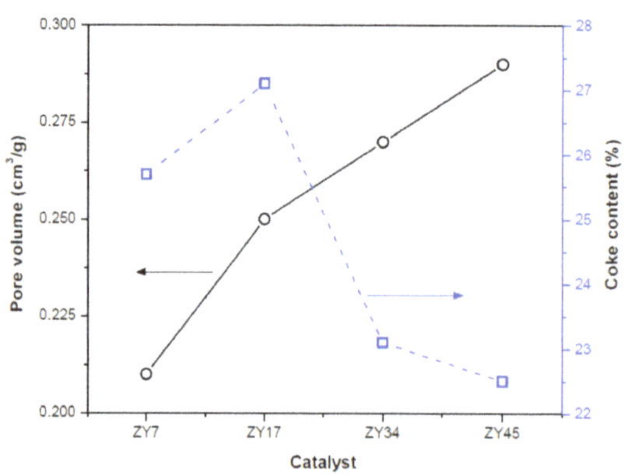

Figure 14. Comparison of the pore volume of the fresh zeolites and the coke content of spent catalysts.

The SEM micrographs of the spent catalysts, presented in Figure 15, showed zeolite grains with the same morphology as the fresh samples. Furthermore, it was revealed that the porosity of the zeolites was preserved. The similarity in micromorphology and surface texture between the fresh and spent catalysts suggested that the carbonaceous compounds could be deposited uniformly inside the porous system and on the outer surface of the zeolites. On the other hand, it was observed an increase in the average grain size of the catalysts after the reaction. The average increment was from 400 nm to 600 nm in the ZY17 and ZY34 catalysts, as shown in the insertions of Figure 15. In the SZY7 and SZY45 samples, an increase was observed in 200 nm and 800 nm particles, respectively.

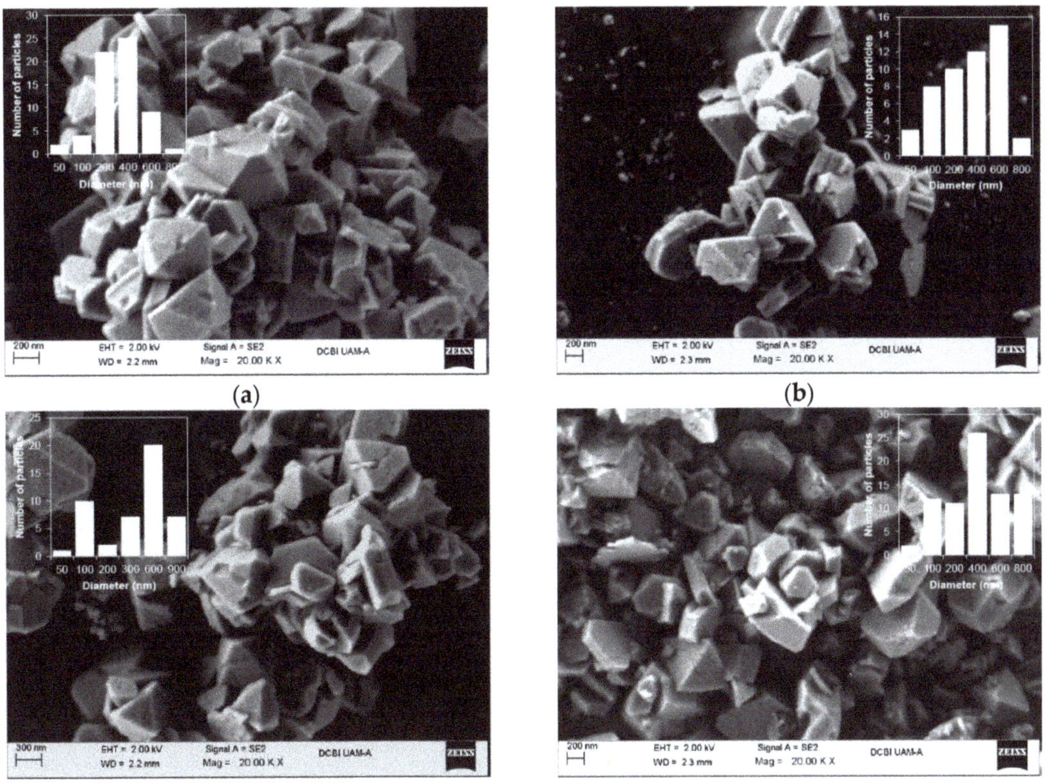

Figure 15. Scanning electron micrographs and particle size distributions of the spent zeolites: (**a**) SZY7, (**b**) SZY17, (**c**) SZY34, and (**d**) SZY45.

The elemental analysis by EDXS confirmed that a large quantity of carbon was deposited on the surface of the zeolite, as reported in Table 3. Additionally, nitrogen was also identified, in proportions between 4 and 5 atom % resulting in C/N atomic ratios between 6.8 and 10.4, in agreement with a previous report [55]. From these compositions, it can be inferred that in addition to heavy compounds derived from aldehydes, the pyridine products and intermediates can also be strongly adsorbed on the acid sites of the catalysts, preventing the reactants from interacting, and being part of the carbonaceous compounds deposited within the porous system, decreasing the activity of the catalyst.

From the above results, it is proposed that the reaction between acrolein and ammonia over Y-type zeolites occurred according to the reaction scheme depicted in Figure 16. Under the reaction conditions of this study (low reaction temperature and NH_3/acrolein molar ratio = 2), acrolein was partially converted to cracking products, namely acetaldehyde, and formaldehyde. These aldehydes could not produce pyridine bases since there was not enough NH_3 to make them react and were easily removed from the reactor due to the high gas hourly space velocity (GHSV = 4994 h^{-1}). On the other hand, acrolein and ammonia react over acid sites to produce propylene imine, as previously reported [54]. Then, this intermediate can undergo a Michael addition reaction with another propylene imine over Brønsted or weak Lewis acid sites, closing the ring structure and producing 3-picoline with the liberation of ammonia. Alternatively, propylene imine can condensate and cyclize with another propylene imine by a Diels-Alder reaction over Lewis acid sites, producing pyridine [71,72]. Due to its basic nature and as suggested by the EDXS results of the spent catalysts, 3-picoline, and pyridine can be strongly absorbed over acid sites, being part of the coke deposited in the catalyst.

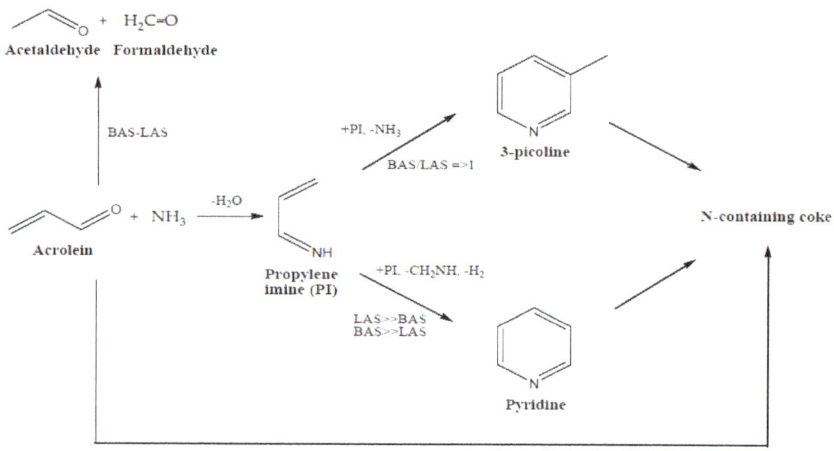

Figure 16. Reaction routes for acrolein and ammonia over the Y-type zeolites.

3. Materials and Methods

3.1. Materials

Y-type zeolites with Si/Al atomic ratios of 7, 17, 34, and 45 were acquired from Zeolyst International (Conshohocken, PA, USA). Previous to the characterization and the catalytic activity experiments, the samples were calcined at 450 °C in an electrical furnace for 4 h, cooled down to room temperature, and passed through 100 mesh. Finally, the samples were identified as ZY7, ZY17, ZY34, and ZY45, where the number indicates the Si/Al atomic ratio. Acrolein was obtained from Sigma-Aldrich (Toluca, México) and used with any further purification.

3.2. Characterization of Catalysts

The structural features of the Y-type zeolites were confirmed by X-ray diffraction, UV-vis, and Raman spectroscopies. A Bruker AXS model D8 diffractometer was used for the XRD analysis, with CuKα radiation (1.5406 Å). The crystalline structure of the zeolites was verified by comparison of the obtained diffractograms with the crystallographic cards of the Joint Committee on Power Diffraction Standards (JCPDS). The average crystal sizes were estimated from the most intense reflection according to the Scherrer equation: $L = 0.9{\cdot}\lambda/\beta{\cdot}\cos\theta$, where L is the crystal size, λ is the X-ray wavelength, β is the line broadening, and θ is the Bragg angle [73].

The UV-vis spectroscopy analysis of the solids was carried out with a spectrophotometer Varian Cary 100 (Agilent Technologies, Mexico city, Mexico) at 25 °C. The powder samples sieved through 100 mesh were analyzed between 200 and 800 nm with a 1 nm sensibility.

The Raman spectra of the zeolites were obtained with a spectrometer Renishaw in-Via (Renishaw Mexico, Apodaca, México) equipped with a cooled CCD detector and a holographic super-Notch filter to remove the elastic scattering, using the green line laser λ = 532 nm, power 9 mW, 1 mW on the sample. The spectral resolution was ca. 3 cm^{-1} and the spectra acquisition consisted of 5 accumulations of 10 s.

The zeolite samples were studied by thermogravimetric analysis in a thermal analyzer model SDT Q600 from TA Instruments, in a temperature range between 25 °C and 900 °C with a heating rate of 10 $°C \cdot min^{-1}$ and an airflow of 10 $mL \cdot min^{-1}$.

The surface acidity of the Y-type zeolites was studied by Fourier transform infrared spectroscopy with adsorbed pyridine (IR-Py) in Nicolet equipment, model 170-SX (Thermo Fisher Scientific Inc., Mexico city, Mexico). In a typical procedure, the powder sample was pressed forming a thin disk, and placed into a glass cell where temperature and vacuum were controlled. Prior to pyridine adsorption, the sample was pretreated with a vacuum at 1×10^{-3} Torr with subsequent heating at 300 °C at a rate of 20 $°C \cdot min^{-1}$, and cooling to room temperature. Then, a mixture of nitrogen and pyridine (4%) was fed to the cell to be in contact with the catalyst sample for 15 min at 25 °C for the chemisorption process.

Physisorbed pyridine was desorbed from the sample by vacuum, and the FTIR spectra were obtained in situ at 350 °C. The Brønsted acid sites were identified by their interaction with pyridine, which exhibits bands at 1540 cm^{-1} and 1640 cm^{-1}, while the coordination of pyridine with Lewis acid sites results in the appearance of typical bands at 1447–1460 cm^{-1}, 1580 cm^{-1} and 1600 cm^{-1} [39]. The amount of Brønsted and Lewis acid sites was determined considering the integrated areas of the bands at 1540 cm^{-1} and 1450 cm^{-1}, correspondingly, and the integrated molar extinction coefficients reported by Emeis [74] of ε_B = 1.67 cm/μmol for Brønsted acid sites and ε_L = 2.22 cm/μmol for Lewis acid sites according to the Beer's law.

For the Al quantification, each sample was subjected to acid digestion with aqua regia (HNO_3 + 3 HCl) at 50 °C for 20 h. Subsequently, the filtered liquid extract was analyzed in a Perkin Elmer Atomic absorption spectrometer model AAnalyst 400 with a calibration curve prepared with the commercial Al standard of the same company.

The N_2 physisorption of the solids was determined at −196 °C using Micromeritics ASAP-2450 equipment. In a typical analysis, the sample was pretreated at 300 °C under vacuum at 1×10^{-3} Torr for 3 h. The adsorption isotherm of each solid was determined by the successive charges of N_2 introduced into the sample container, increasing the pressure and allowing 15 s to pass between each charge to achieve balance. The specific surface area, the pore volume, and the pore size of each sample were estimated considering the Brunauer, Emmet, and Teller model (BET) [75], the t-plot method, and the Barret, Joynet, and Halenda model (BJH) [76], respectively.

Observations of the Y-type zeolites were accomplished in a scanning electron microscope (SEM) Zeiss Supra 55 VP coupled with a probe with energy-dispersive X-ray spectroscopy (EDXS) to obtain the respective spectra, as well as qualitative and quantitative elemental analyses of the samples. For their observation, the fresh and spent catalyst powders were spread on graphite and copper tapes, respectively.

3.3. Catalytic Activity Tests

The gas-phase reactions between acrolein and ammonia were performed at 1 atm and controlled temperature in a system comprised of an acrolein saturator vessel, a stainless steel fixed-bed reactor (32 cm length and 1 cm internal diameter), and an electrical furnace provided with a thermocouple and an electronic temperature controller.

Initially, 0.30 g of catalyst (W_{cat}) was loaded into the reactor over a porous fiberglass bed and activated by heating at 450 °C for 30 min with an N_2 flow at 50 $mL \cdot min^{-1}$. Afterward, the reactor was set to the reaction temperature (360 °C), maintaining the gas flow, and stabilized for 30 min. Subsequently, the acrolein vapor contained in the saturator vessel at 21 °C was carried out with a 50 $mL \cdot min^{-1}$ N_2 flow and fed to the reactor. Due to the feeding conditions of this mixture, the gas hourly space velocity (GHSV) was 4994 h^{-1}. Simultaneously, gaseous ammonia (NH_3) was directly fed at the reactor inlet from a container tank to achieve an NH_3/acrolein molar ratio equal to 2. To mainly exhibit the influence of the catalyst properties on the reaction, no water was added to the feed stream, and the values of reaction temperature and NH_3/acrolein molar ratio were chosen to diminish their positive effect on the formation of pyridine bases.

The output flow from the reactor passed through a gas sampling point, cooled in a condenser at 6 °C, and collected in an ice-cold trap. The acrolein at the input and output of the reactor, and the reaction products, were analyzed in a gas chromatograph GOWMAC 750 equipped with a packed column 15% FFAP Chrom W-A/W 80/100 (2 m × 1/8 inch) and a flame ionization detector. The acrolein conversion (X) and the product yield (Y_i) were calculated according to Equations (1) and (2), respectively:

$$X(\%) = \frac{N_{a,in} - N_{a,out}}{N_{a,in}} \cdot 100 \tag{3}$$

$$Y_i(\%) = \frac{C_i}{C_{a,in}} \cdot 100 \tag{4}$$

where $N_{a,in}$ and $N_{a,out}$ are the moles of acrolein at the inlet and outlet stream of the reactor, correspondingly; C_i and $C_{a,in}$ are the carbon moles in the product i and fed acrolein, respectively.

3.4. Catalyst Deactivation Behavior

The catalytic activity tests at 360 °C and GHSV = 4994 h^{-1} were monitored with time on stream (TOS) until the total yield of pyridine compounds was below 10%. The acrolein conversion data with TOS was used to model the deactivation behavior of each catalyst during the aminocyclization reaction considering the deactivation kinetic model proposed by Fuentes [59],

$$-\frac{da}{dt} = k_d(a^n - a_s^n) \tag{5}$$

where a is the activity, k_d is the deactivation rate constant, a_s is the steady-state activity, and n is the deactivation order.

The kinetic deactivation equations and the experimental results were fitted by non-linear regression, minimizing the sum of square residuals (SSR) as the objective function by use of the Microsoft Excel Solver tool, and discriminated in terms of the determination coefficient (R^2), according to the following equations:

$$SSR = \sum (y_{i,est} - y_{i,obs})^2 \tag{6}$$

$$R^2 = 1 - \frac{SSR}{TSS} \tag{7}$$

$$TSS = \sum (y_{i,obs} - \overline{y}_{obs})^2 \tag{8}$$

where $y_{i,est}$ and $y_{i,obs}$ are the estimated and observed values of the dependent variable, respectively, TSS is the total sum of squares, and $\overline{y}_{obs}$ is the mean of the observed values of the dependent variable.

3.5. Characterization of Deactivated Catalysts

After the deactivation tests, the spent zeolites, denoted as SZY7, SZY17, SZY34, and SZY45, were characterized by XRD, Raman spectroscopy, TGA, and SEM-EDXS under the same conditions as mentioned in Section 3.2.

4. Conclusions

The Y-type zeolites with Si/Al atomic ratios between 7 and 45 were active as catalysts in the reaction between acrolein and ammonia. The acrolein conversion improved with the increase of the total acidity per gram of the catalyst, reaching higher conversion using the zeolite with Si/Al = 17.

The production of pyridine bases was inversely proportional to the total acidity of the catalysts which exhibited similar textural properties (large surface areas, pore volume as well as micro- and mesoporosity) achieving the maximum yield of pyridines (3-picoline + pyridine) using the catalyst with Si/Al = 45. The type and amount of acid sites influenced the selectivity of pyridine products. 3-picoline was produced in higher amounts than pyridine (highest value of the 3-picoline/pyridine molar ratio) when the catalyst presented Brønsted/Lewis acid sites ratios approaching one (in the range of 0.6 and 1.3). The reaction proceeded preferentially towards pyridine with the further decline or increase of the Brønsted/Lewis acid sites ratio. Additionally, the formation of pyridine occurred more selectively over LAS than BAS.

The stability of the catalysts with time on stream was influenced by their textural properties, as indicated by the deactivation tests and the proposed kinetic deactivation models. The loss of activity was caused by the uniform deposition of large amounts of coke on the catalyst surface and within the inner porous network of the zeolites according to the characterization of the spent zeolites by DRX, TGA, and SEM. Additionally, the Raman spectroscopy analysis of the spent catalysts indicated that coke could be composed of large polyaromatic compounds. EDXS results showed that pyridine products and their intermediates can also be part of the coke deposited.

Author Contributions: Conceptualization, I.P.-R., J.L.C. and J.S.; methodology, I.P.-R., J.L.C. and J.S.; software, I.P.-R.; validation, J.L.C. and J.S.; formal analysis, I.P.-R., J.L.C., J.S., R.L.-M., D.A.-B., B.Z., J.N.-B. and N.N.G.-H.; investigation, I.P.-R. and J.L.C.; resources, J.L.C., J.S., R.L.-M., D.A.-B., B.Z., J.N.-B.; data curation, I.P.-R. and J.L.C.; writing—original draft preparation, I.P.-R.; writing—review and editing, I.P.-R., J.L.C., J.S., R.L.-M., D.A.-B., B.Z., J.N.-B. and N.N.G.-H.; visualization, I.P.-R.; supervision, J.L.C. and J.S.; project administration, J.L.C.; funding acquisition, J.L.C. and J.S. All authors have read and agreed to the published version of the manuscript.

Funding: This research received no external funding.

Data Availability Statement: Not applicable.

Acknowledgments: The authors thank the Instituto Politécnico Nacional, the Universidad Autónoma Metropolitana-Azcapotzalco, and the Instituto Mexicano del Petróleo for the support to develop this research. The Scanning Electron Microscopy Laboratory of the Division of Basic Sciences and Engineering of the Universidad Autónoma Metropolitana-Azcapotzalco is acknowledged for the characterization of the catalyst samples by SEM-EDXS. I.P.R. and J.L.C. appreciate the support of the company Síntesis y Aplicaciones Industriales, S.A.

Conflicts of Interest: The authors declare no conflict of interest.

References

1. Weissermel, K.; Arpe, H.J. Chapter 7: Oxydation products of ethylene. In *Industrial Organic Chemistry*, 3rd ed.; VCH Publishers, Inc.: Weinheim, Germany, 1997; pp. 143–190.
2. Shimizu, S.; Abe, N.; Iguchi, A.; Sato, H. Synthesis of pyridine bases: General methods and recent advances in gas phase synthesis over ZSM-5 zeolite. *Catal. Surv. Jpn.* **1998**, *2*, 71–76. [CrossRef]
3. Shimizu, S.; Watanabe, N.; Kataoka, T.; Shoji, T.; Abe, N.; Morishita, S.; Ichimura, H. Pyridine and pyridine derivatives. In *Ullmann's Encyclopedia of Industrial Chemistry*; Wiley-VCH Verlag GmbH & Co. KGaA: Weinheim, Germany, 2012; pp. 557–589.
4. Scriven, E.F.V.; Murugan, R. Pyridine and pyridine derivatives. In *Kirk-Othmer Encyclopedia of Chemical Technology*; John Wiley & Sons, Inc.: Hoboken, NJ, USA, 2005; pp. 1–53.
5. Sagitullin, R.S.; Shkil', G.P.; Nosonova, I.I.; Ferber, A.A. Synthesis of pyridine bases by the Chichibabin method (review). *Chem. Heterocycl. Comp.* **1996**, *32*, 127–140. [CrossRef]
6. Reddy, K.S.K.; Srinivasakannan, C.; Raghavan, K.V. Catalytic vapor phase pyridine synthesis: A process review. *Catal. Surv. Asia* **2012**, *16*, 28–35. [CrossRef]
7. Tschitschibabin, A.E. Über kondensationen der aldehyde mit ammonik zu pyridinbasen. *J. Für Prakt. Chem.* **1924**, *107*, 122–128. [CrossRef]

8. Sato, H.; Shimizu, S.; Abe, N.; Hirose, K. Vapor phase synthesis of pyridine bases from aldehydes and ammonia over pentasil zeolites. Zeolites and related microporous materials: State of the art 1994. In *Studies in Surface Science and Catalysis*; Hölderich, W., Karge, H.G., Weitkamp, J., Pfeifer, H., Eds.; Elsevier Science B.V.: Amsterdam, The Netherlands, 1994; Volume 84, pp. 1951–1958.
9. Reddy, K.S.K.; Sreedhar, I.; Raghavan, K.V. Interrelationship of process parameters in vapor phase pyridine synthesis. *App. Catal. A Gen.* **2008**, *339*, 15–20. [CrossRef]
10. Zhang, X.; Wu, Z.; Liu, W.; Chao, Z.S. Preparation of pyridine and 3-picoline from acrolein and ammonia with HF/MgZSM-5 catalyst. *Catal. Commun.* **2016**, *80*, 10–14. [CrossRef]
11. Luo, C.W.; Chao, Z.S. Unsaturated aldehydes: A novel route for the synthesis of pyridine and 3-picoline. *RSC Adv.* **2015**, *5*, 54090–54101. [CrossRef]
12. Luo, C.W.; Li, A. Synthesis of 3-picoline from acrolein dimethyl acetal and ammonia over NH_4F-HF treated ZSM-5. *React. Kinet. Mech. Catal.* **2018**, *125*, 365–380. [CrossRef]
13. Liu, L.; Ye, X.P.; Bozell, J.J. A comparative review of petroleum-based and bio-based acrolein production. *ChemSusChem* **2012**, *5*, 1162–1180. [CrossRef]
14. Galadima, A.; Muraza, O. A review on glycerol valorization to acrolein over solid acid catalysts. *J. Taiwan Inst. Chem. Eng.* **2016**, *67*, 29–44. [CrossRef]
15. Pala Rosas, I.; Contreras Larios, J.L.; Zeifert, B.; Salmones Blásquez, J. Catalytic dehydration of glycerine to acrolein. In *Glycerine Production and Transformation—An Innovative Platform for Sustainable Biorefinery and Energy*; Frediani, M., Bartoli, M., Rosi, L., Eds.; IntechOpen: London, UK, 2019; pp. 9–27.
16. Kong, P.S.; Aroua, M.K.; Daud, W.M.A.W. Conversion of crude and pure glycerol into derivatives: A feasibility evaluation. *Renew. Sustain. Energy Rev.* **2016**, *63*, 533–555. [CrossRef]
17. Van Der Gaag, F.J.; Louter, G.; Oudejans, J.C.; Van Bekkum, H. Reaction of ethanol and ammonia to pyridines over zeolite ZSM-5. *Appl. Catal.* **1986**, *26*, 191–201. [CrossRef]
18. Grigor'eva, N.G.; Filippova, N.A.; Tselyutina, M.I.; Kutepov, B.I. Synthesis of pyridine and methylpyridines over zeolite catalysts. *Appl. Petrochem. Res.* **2015**, *5*, 99–104. [CrossRef]
19. Van Donk, S.; Janssen, A.H.; Bitter, J.J.; de Jong, K.P. Generation, characterization, and impact of mesopores in zeolite catalysts. *Catal. Rev.* **2003**, *45*, 297–319. [CrossRef]
20. Perego, C.; Bosetti, A. Biomass to fuels: The role of zeolite and mesoporous materials. *Microporous Mesoporous Mater.* **2011**, *144*, 28–39. [CrossRef]
21. Bu, L.; Nimlos, M.R.; Robichaud, D.J.; Kim, S. Diffusion of aromatic hydrocarbons in hierarchical mesoporous H-ZSM-5 zeolite. *Catal. Today* **2018**, *312*, 73–81. [CrossRef]
22. Zhang, H.; Samsudin, I.B.; Jaenicke, S.; Chuah, G.K. Zeolites in catalysis: Sustainable synthesis and its impact on properties and applications. *Catal. Sci. Technol.* **2022**, *12*, 6024. [CrossRef]
23. Subba Rao, Y.V.; Kulkarni, S.J.; Subrahmanyam, M.; Rama Rao, A.V. A novel acylative cyclization reaction of phenol over modified Y zeolites. *J. Chem. Soc. Chem. Commun.* **1993**, *3212*, 1456–1457.
24. Vajglová, Z.; Kumar, N.; Mäki-Arvela, P.; Eranen, K.; Peurla, M.; Hupa, L.; Nurmi, M.; Toivakka, M.; Murzin, D.Y. Synthesis and physicochemical characterization of shaped catalysts of β and Y zeolites for cyclization of citronellal. *Ind. Eng. Chem. Res.* **2019**, *58*, 18084–18096. [CrossRef]
25. Yu, W.; Bian, F.; Gao, Y.; Yang, L.; Liu, Z.L. Y-zeolite-catalyzed cyclizations of terpenols. *Adv. Synth. Catal.* **2006**, *348*, 59–62. [CrossRef]
26. Pérez-Mayoral, E.; Matos, I.; Fonseca, I.; Cejka, J. Zeolites efficiently promote the cyclization of nonactivated unsaturated alcohols. *Chem. Eur. J.* **2010**, *16*, 12079–12082. [CrossRef]
27. Kim, H.J.; Seo, G.; Kim, J.N.; Choi, K.H. HY zeolite catalyzed one-pot synhtesis of 2,3-dihydro-2,2-dimethylbenzofurans from aryl methallyl ethers. *Bull. Korean Chem. Soc.* **2004**, *25*, 1726–1728.
28. Shimizu, S.; Abe, N.; Iguchi, A.; Dohba, M.; Sato, H.; Hirose, K. Synthesis of pyridine bases on zeolite catalysts. *Microporous Mesoporous Mater.* **1998**, *21*, 447–451. [CrossRef]
29. Treacy, M.M.J.; Higgins, J.B. *Collection of Simulated XRD Patterns for Zeolites*; Elsevier B.V.: Amsterdam, The Netherlands, 2001; pp. 146–153.
30. Lutz, W.; Rüscher, C.H.; Gesing, T.M. Investigations of the mechanism of dealumination of zeolite Y by steam: Tuned mesopore formation versus the Si/Al ratio. *Stud. Surf. Sci. Catal.* **2004**, *154*, 1411–1417.
31. Li, C.; Wu, Z. Chapter 11 Microporous materials characterized by vibrational spectroscopies. In *Handbook of Zeolite Science and Technology*; Auerbach, S.M., Carrado, K.A., Dutta, P.K., Eds.; CRC Press: Boca Raton, FL, USA, 2003; pp. 423–514.
32. Meng, Y.; Genuino, H.C.; Kuo, C.H.; Huang, H.; Chen, S.Y.; Zhang, L.; Rossi, A.; Suib, S.L. One-step hydrothermal synthesis of Manganese-Containing MFI-type zeolite, Mn-ZSM-5, characterization, and catalytic oxydation of hydrocarbons. *J. Am. Chem. Soc.* **2013**, *135*, 8594–8605. [CrossRef]
33. Dutta, P.K.; Twu, J. Influence of framework Si/Al ratio on the Raman spectra of faujasitic zeolites. *J. Phys. Chem.* **1991**, *95*, 2498–2501. [CrossRef]
34. Dutta, P.K.; Rao, K.M.; Park, J.Y. Correlation of Raman spectra of zeolites with framework architecture. *J. Phys. Chem.* **1991**, *95*, 6654–6656. [CrossRef]

35. Brémard, C.; Le Maire, M. Low-frequency Raman spectra of dehydrated faujasitic zeolites. *J. Phys. Chem.* **1993**, *97*, 9695–9702. [CrossRef]
36. Yu, Y.; Xiong, G.; Li, C.; Xiao, F.S. Characterization of aluminosilicate zeolites by UV Raman spectroscopy. *Microporous Mesoporous Mater.* **2001**, *46*, 23–34. [CrossRef]
37. Laughlin, R.B.; Joannopoulous, J.D. Phonons in amorphous silica. *Phys. Rev. B* **1977**, *16*, 2942–2952. [CrossRef]
38. De Man, A.J.M.; van Beest, B.H.W.; Leslie, M.; van Santen, R.A. Lattice dynamics of zeolitic silica polymorphs. *J. Phys. Chem.* **1990**, *94*, 2524–2534. [CrossRef]
39. Parry, E.P. An infrared study of pyridine adsorbed on acidic solids. Characterization of surface acidity. *J. Catal.* **1963**, *2*, 371–379. [CrossRef]
40. Sandoval-Díaz, L.E.; González-Amaya, J.A.; Trujillo, C.A. General aspects of zeolite acidity characterization. *Microporous Mesoporous Mater.* **2015**, *215*, 229–243. [CrossRef]
41. de Castro, P.R.D.S.; Barreto Maia, A.A.; Simões Angélica, R. Study of the thermal stability of faujasite zeolite synthesized from kaolin waste from the Amazon. *Mater. Res.* **2019**, *22*, e20190321. [CrossRef]
42. Ayad, Z.; Hussein, H.Q.; Al-Tabbakh, B.A. Synthesis and characterization of high silica HY zeolite by basicity reduction. *AIP Conf. Proc.* **2020**, *2213*, 020168.
43. Leofanti, G.; Padovan, M.; Tozzola, G.; Venturelli, B. Surface area and pore texture of catalysts. *Catal. Today* **1998**, *41*, 207–219. [CrossRef]
44. Thommes, M.; Kaneko, K.; Neimark, A. Physisorption of gases, with special reference to the evaluation of surface area and pore size distribution (IUPAC Technical Report). *Pure Appl. Chem.* **2015**, *87*, 1051–1069. [CrossRef]
45. Huang, Q.; Wang, J.; Sun, Y.; Li, X.; Wang, X.; Zhao, Z. Gas-sensing properties of composites of Y-zeolite and SnO_2. *J. Mater. Sci.* **2018**, *53*, 6729–6740. [CrossRef]
46. Liu, M.; Ren, Y.; Wu, J.; Wang, Y.; Chen, J.; Lei, X.; Zhu, X. Effect of cations on the structure, physico-chemical properties and photocatalytic behaviors of silver-doped zeolite Y. *Microporous Mesoporous Mater.* **2020**, *293*, 109800. [CrossRef]
47. Taufiqurrahmi, N.; Mohamed, A.R.; Bhatia, S. Nanocrystalline zeolite Y: Synthesis and sharacterization. *IOP Conf. Ser. Mater. Sci. Eng.* **2011**, *17*, 012030. [CrossRef]
48. Li, A.; Luo, C.; Liu, Y.; Li, L.; Lin, Y.; Liu, K.; Zhou, C. Zn-promoted Hβ zeolite for gas-phase catalyzed aza-heterocyclic-aromatization of acrolein dimethyl acetal and aniline to quinolines. *Mol. Catal.* **2020**, *486*, 110833. [CrossRef]
49. Martinuzzi, I.; Azizi, Y.; Devaux, J.F.; Tretjak, S.; Zahraa, O.; Leclerc, J.P. Reaction mechanism for glycerol dehydration in the gas phase over a solid acid catalyst determined with on-line gas chromatography. *Chem. Eng. Sci.* **2014**, *116*, 118–127. [CrossRef]
50. Kuppi Reddy, S.K.R.; Sreedhar, I.; Raghavan, K.V.; Kulkarni, S.J.; Ramakrishna, M. Optimization of vapor phase pyridine synthesis hindered by rapid catalyst deactivation. *Int. J. Chem. React. Eng.* **2008**, *6*, 1–20. [CrossRef]
51. Liu, Y.; Yang, H.; Jin, F.; Zhang, Y.; Li, Y. Synthesis of pyridine and picolines over Co-modified HZSM-5 catalyst. *Chem. Eng. J.* **2008**, *136*, 282–287. [CrossRef]
52. Luo, C.W.; Li, A.; An, J.F.; Feng, X.Y.; Zhang, X.; Feng, D.D.; Chao, Z.S. The synthesis of pyridine and 3-picoline from gas-phase acrolein diethyl acetal with ammonia over ZnO/HZSM-5. *Chem. Eng. J.* **2015**, *273*, 7–18. [CrossRef]
53. Zhang, W.; Duan, S.; Zhang, Y. Enhanced selectivity in the conversion of acrolein to 3-picoline over bimetallic catalyst 4.6%Cu–1.0%Ru HZSM-5 (38) with hydrogen as carrier gas. *React. Kinet. Mech. Catal.* **2019**, *127*, 391–411. [CrossRef]
54. Zhang, X.; Wu, Z.; Chao, Z.S. Mechanism of pyridine bases prepared from acrolein and ammonia by in situ infrared spectroscopy. *J. Mol. Catal. A Chem.* **2016**, *411*, 19–26. [CrossRef]
55. Kumar Reddy, K.S.; Sreedhar, I.; Venkateshwar, S.; Vijay Raghavan, K. Behavior of a deactivating HZSM-5 with varying Al^{3+} content in vapor phase aminocyclization. *Catal. Lett.* **2008**, *125*, 110–115. [CrossRef]
56. Guisnet, M.; Magnoux, P. Deactivation by coking of zeolite catalysts. Prevention of deactivation. Optimal conditions for regeneration. *Catal. Today* **1997**, *36*, 477–483. [CrossRef]
57. Guisnet, M.; Magnoux, P. Organic chemistry of coke formation. *Appl. Catal. A Gen.* **2001**, *212*, 83–96. [CrossRef]
58. Jin, F.; Cui, Y.; Li, Y. Effect of alkaline and atom-planting treatment on the catalytic performance of ZSM-5 catalyst in pyridine and picolines synthesis. *Appl. Catal. A Gen.* **2008**, *350*, 71–78. [CrossRef]
59. Fuentes, G.A. Catalyst deactivation and steady-state activity: A generalized power-law equation model. *Appl. Catal.* **1985**, *15*, 33–40. [CrossRef]
60. Argyle, M.D.; Frost, T.S.; Bartholomew, C.H. Cobalt Fischer–Tropsch catalyst deactivation modeled using generalized power law expressions. *Top. Catal.* **2014**, *57*, 415–429. [CrossRef]
61. Fadaeerayeni, S.; Shan, J.; Sarnello, E.; Xu, H.; Wang, H.; Cheng, J.; Li, T.; Toghiani, H.; Xiang, Y. Nickel/gallium modified HZSM-5 for ethane aromatization: Influence of metal function on reactivity and stability. *Appl. Catal. A Gen.* **2020**, *601*, 117629. [CrossRef]
62. Kuppi Reddy, S.K.R.; Sreedhar, I.; Raghavan, K.V. Kinetic studies on vapour phase pyridine synthesis and catalyst regeneration studies. *Can. J. Chem. Eng.* **2011**, *89*, 854–863. [CrossRef]
63. Bauer, F.; Karge, H.G. Characterization of coke on zeolites. In *Characterization II*; Molecular Sieves. Science and Technology; Karge, H.G., Weitkamp, J., Eds.; Springer: Berlin/Heidelberg, Germany, 2007; Volume 5, pp. 249–364.
64. Li, C.; Stair, P.C. Utraviolet Raman spectroscopy characterization of coke formation inzeolites. *Catal. Today* **1997**, *33*, 353–360. [CrossRef]

65. Chua, Y.T.; Stair, P.C. An ultraviolet Raman spectroscopic study of coke formation in methanol to hydrocarbons conversion over zeolite H-MFI. *J. Catal.* **2003**, *213*, 39–46. [CrossRef]
66. An, H.; Zhang, F.; Guan, Z.; Liu, X.; Fan, F.; Li, C. Investigating the coke formation mechanism of H-ZSM-5 during methanol dehydration using operando UV-Raman spectroscopy. *ACS Catal.* **2018**, *8*, 9207–9215. [CrossRef]
67. Pala-Rosas, I.; Contreras, J.L.; Salmones, J.; Zeifert, B.; López-Medina, R.; Navarrete-Bolaños, J.; Hernández-Ramírez, S.; Pérez-Cabrera, J.; Fragoso-Montes de Oca, A.A. Catalytic deactivation of HY zeolites in the dehydration of glycerol to acrolein. *Catalysts* **2021**, *11*, 360. [CrossRef]
68. Jiang, X.C.; Zhou, C.H.; Tesser, R.; Di Serio, M.; Tong, D.S.; Zhang, J.R. Coking of catalysts in catalytic glycerol dehydration to acrolein. *Ind. Eng. Chem. Res.* **2018**, *57*, 10736–10753. [CrossRef]
69. Jin, F.; Li, Y. The effect of H_2 on Chichibabin condensation catalyzed by pure ZSM-5 and Pt/ZSM-5 for pyridine and 3-picoline synthesis. *Catal. Lett.* **2009**, *131*, 545–551. [CrossRef]
70. Luo, C.W.; Feng, X.Y.; Liu, W.; Lia, X.Y.; Chao, Z.S. Deactivation and regeneration on the ZSM-5-based catalyst for the synthesis of pyridine and 3-picoline. *Microporous Mesoporous Mater.* **2016**, *235*, 261–269. [CrossRef]
71. Jiang, D.; Wang, S.; Li, W.; Xu, L.; Hu, X.; Barati, B.; Zheng, A. Insight into the mechanism of glycerol dehydration and subsequent pyridine synthesis. *ACS Sustain. Chem. Eng.* **2021**, *9*, 3095–3103. [CrossRef]
72. Calvin, J.R.; Davis, R.D.; McAteer, C.H. Mechanistic investigation of the catalyzed vapor-phase formation of pyridine and quinoline bases using $^{13}CH_2O$, $^{13}CH_3OH$, and deuterium-labeled aldehydes. *Appl. Catal. A Gen.* **2005**, *285*, 1–23. [CrossRef]
73. Azaroff, L.V.; Buerger, M.J. *The Powder Method in X-ray Crystallography*; McGraw-Hill: New York, NY, USA, 1958.
74. Emeis, C.A. Determination of Integrated molar extinction coefficients for infrared absorption bands of pyridine adsorbed on solid acid catalysts. *J. Catal.* **1993**, *141*, 347–354. [CrossRef]
75. Brunauer, S.; Emmett, P.H.; Teller, E. Adsorption of gases in multimolecular layers. *J. Am. Chem. Soc.* **1938**, *60*, 309–319. [CrossRef]
76. Barret, E.P.; Joyner, L.G.; Halenda, P.P. The determination of pore volume and area distributions in porous substances. I. Computations from nitrogen isotherms. *J. Am. Chem. Soc.* **1951**, *73*, 373–380. [CrossRef]

catalysts

Article

Thermogravimetry Applied for Investigation of Coke Formation in Ethanol Conversion over Heteropoly Tungstate Catalysts

Orsina Verdeş, Alexandru Popa *, Silvana Borcănescu, Mariana Suba and Viorel Sasca

"Coriolan Drăgulescu" Institute of Chemistry, Bl. Mihai Viteazul 24, 300223 Timisoara, Romania
* Correspondence: alpopa_tim2003@yahoo.com; Tel.: +40-256-491818

Abstract: Thermogravimetric analysis (TGA) was used to evaluate the thermal stability and the amount of coke deposition resulting from the deactivation of catalysts during ethanol dehydration reaction in a fixed bed continuous flow reactor. In this study, a series of catalysts containing 30% of Pd doped and pure 12-tungstophosphoric acid and its insoluble $Cs_{2.5}H_{0.5}PW_{12}O_{40}$ salt supported on SBA-15 were prepared. The catalytic efficiency of ethanol dehydration reaction was also evaluated. Two types of coke are identified from the TPO (Temperature programmed oxidation) profiles and assigned to the coke precursor and hard coke, respectively. The results indicate that cesium salts reduced the formation of hard coke. The amount of total coke formed was significantly reduced by supporting the catalysts on mesoporous SBA-15 molecular sieves.

Keywords: heteropoly acid; thermogravimetric analysis; ethanol conversion; coke formation; catalyst stability; cesium salts

Citation: Verdeş, O.; Popa, A.; Borcănescu, S.; Suba, M.; Sasca, V. Thermogravimetry Applied for Investigation of Coke Formation in Ethanol Conversion over Heteropoly Tungstate Catalysts. *Catalysts* **2022**, *12*, 1059. https://doi.org/10.3390/catal12091059

Academic Editor: Enrique Sastre

Received: 19 August 2022
Accepted: 14 September 2022
Published: 16 September 2022

Publisher's Note: MDPI stays neutral with regard to jurisdictional claims in published maps and institutional affiliations.

1. Introduction

The catalytic conversion of ethanol over bulk and supported Brönsted solid acid catalysts based on tungsten Keggin heteropoly acids (HPAs) is an alternative route for the production of ethylene (ET) and diethyl ether (DEE) [1–5]. From an economic point of view, ethylene is an interesting starting material in the chemical industry. Currently, special attention is directed toward green alternatives for the manufacture of ethylene, namely, the catalytic dehydration of bioethanol would lead to the reduction of greenhouse gas emissions and a decrease in dependence on fossil fuels. Diethyl ether (DEE) is a valuable renewable fuel that can be used as a fuel additive for diesel engines. Dehydration of ethanol on supported and unsupported heteropoly compounds (HPCs) indicates that at temperatures lower than 180 °C, the main product formed is diethyl ether. At high temperatures, ethylene is obtained in large quantities and is the main reaction product [3–5].

At the same time, heteropolyacids (HPA) as a green solid catalyst have become attractive due to their low toxicity, being environmentally friendly, and being easy to recover and recycle [6–8]. According to literature data [9,10], the most active heteropoly acids (HPAs) catalysts are $H_3PW_{12}O_{40}$ (HPW) and its bulk heteropoly salt $Cs_{2.5}H_{0.5}PW_{12}O_{40}$ (Cs2.5PW). Cs2.5PW catalyst was shown by Misono and co-workers [9] to be more efficient than its parent acid due to high surface acidity. Until now, numerous studies have been conducted on the Cs2.5PW salt due to its remarkable activity in a large number of reactions. In comparison with basic heteropolyacids, which have a low specific surface area below 10 m^2/g, Cs2.5PW salt has a higher specific surface area (130 m^2/g) due to its micro- and mesoporous structure that allows very efficient use of protons. In order to increase the specific surface area of heteropoly acids, it, therefore, becomes necessary to disperse HPAs on supports that possess a large surface area. Highly ordered mesoporous materials with a two-dimensional porous structure such as SBA-15 can be incorporated as a support in the catalyst leading to good dispersion of the active phase during the synthesis of the catalysts. SBA-15 is a mesoporous silica molecular sieve with uniform tubular channels whose pore diameter is variable from 50 to 300 Å. Compared with other supports, for instance, MCM-41, SBA-15 has a larger pore diameter, thicker pore wall, and higher hydrothermal stability [11–13].

During organic reactions catalyzed by heteropoly compounds (HPCs), a deactivation process occurs due to the formation of carbonaceous deposits (coke), which cause pore blockage and/or poisoning of active sites [14–17]. Generally, the formation and deposition of coke take place on the external surface of the catalysts. Coke consists of aliphatic molecules (precursor coke) as well as polyaromatic molecules (hard coke).

Ultrastable Y (US-Y) zeolite was studied quantitatively by thermogravimetric method during catalytic cracking of normal n-hexane and 1-hexene in order to identify the coke formation and the role of coke precursors in catalytic cracking reaction. The amount of coke was estimated by weight loss during coke burning and composition by coke extraction after the dissolution of the US-Y in HF [18,19].

The ^{13}C MAS NMR and TGA/TPO measurements are shown that modification of heteropoly acid catalysts, e.g., $H_3PW_{12}O_{40}$, by doping with Pt or Pd, significantly decreases the temperature of coke gasification with oxygen. In the case of propene oligomerization reaction on PW/SiO_2 catalyst by adding Pd, only the soft coke is observed [16]. The undoped catalyst forms two types of coke: soft coke (high molecular weight aliphatic oligomers) and hard coke (polynuclear aromatics). At the same time, the addition of palladium increases the regeneration of catalysts without loss of catalytic activity [16,17].

The aim of this paper is to examine the thermal stability and quantity of coke formed during ethanol conversion at different temperatures using the thermal analysis method. Thermogravimetry (TG/DTG) is regarded as an essential method to determine the mass loss of coke precursor and hard coke.

2. Results and Discussion

2.1. Characterization of the Catalyst

2.1.1. Structure and Thermal Stability of Heteropoly Compounds

The high angles XRD patterns (2θ = 5–60°) of pure and Pd doped HPW and Cs2.5PW salt exhibit all reflections corresponding to a cubic crystalline structure (*Pn3m* phase), similar to other literature data [20]. In the case of silica-supported catalysts, the characteristic peaks for HPCs are not detected (not shown). This indicated that HPW, PdPW, Cs2.5PW, and PdCs2.5PW were finely dispersed on the surface, incorporated in the pore walls, or inside the hexagonal channels of SBA-15 [2,20]. The highly ordered hexagonal SBA-15 structure was confirmed by the presence of three main diffraction peaks below 2.0° (2θ) corresponding to the Miller indices (100), (110), and (200) [20].

Figure 1 showed the low angle XRD patterns of SBA-15, HPW/SBA-15, PdHPW/SBA-15, Cs2.5PW/SBA-15 and PdCs2.5PW/SBA-15 catalysts. All supported samples exhibited one intense diffraction peak at 0.92 indexed to (100) and two peaks at 1.59 and 1.83 indexed to (110) and (200) that almost disappeared in the case of HPAs supported SBA-15. The results indicated that the mesostructured ordering is slightly modified by the impregnation of the catalysts into SBA-15.

The presence of Keggin anion in organized mesoporous silica-HPCs composites was identified by Fourier transform infrared spectroscopy (FTIR) in the range of 400–1200 cm^{-1} (Figure 2).

FTIR spectra represent a real fingerprint of the Keggin unit for these compounds, and the results were briefly reported before [2]. The four principal bands of Keggin anion vibration appear at 1080 cm^{-1} due to $\nu_{as}(PO_i)$ vibration, 980 cm^{-1} due to the terminal $\nu_{as}(W = O_t)$ vibration, 890 and 789 cm^{-1} assigned to corner-sharing $\nu_{as}(W\text{-}O_c\text{-}W)$ and edge-sharing $\nu_{as}(W\text{-}O_e\text{-}W)$, respectively. In addition, a weaker adsorption band appeared at 520 cm^{-1}, which was attributed to $\nu_s(W\text{-}O\text{-}W)$. All characteristic infrared adsorption bands are similar to the literature data [21–23]. The IR spectrum of Cs2.5PW salt was identical to that of the acid, but it has narrower vibration bands due to a weaker interaction between cesium and the Keggin anion. No significant difference between pure and Pd-doped HPW and Cs2.5 PW was observed. It could be concluded that by adding cesium and palladium to the heteropoly compounds, the structure of the primary structure is not changed. In the case of supported catalysts, all bands are preserved, but they are broadened and partially overlapped because of the strong absorption bands of mesoporous silica from SBA-15 (1100, 950, 807, and 450 cm^{-1}). When comparing the IR spectra of catalysts within the structural vibration region, it was observed that the primary Keggin structure of HPW, Cs2.5PW, PdPW, and PdCs2.5PW were preserved after impregnation on the SBA-15 molecular mesoporous sieve.

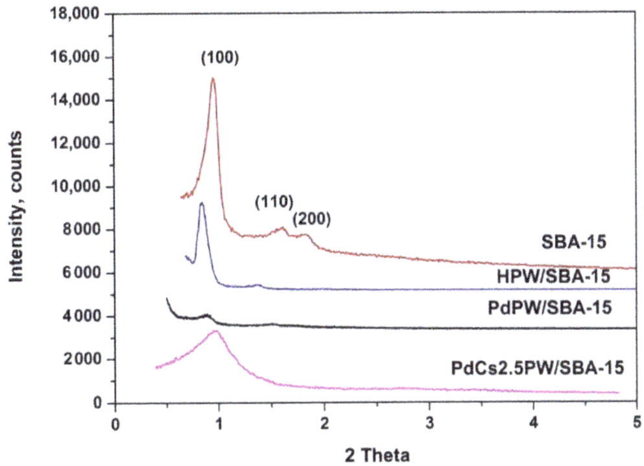

Figure 1. Low-angle XRD patterns of the catalysts.

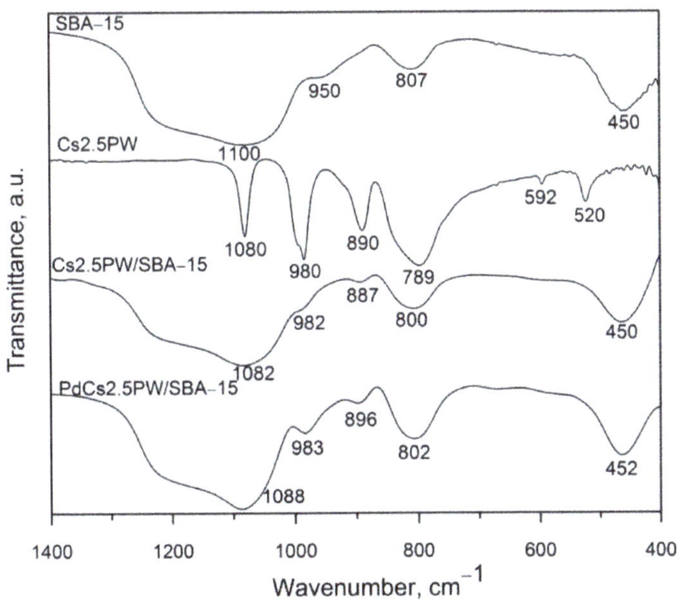

Figure 2. FTIR spectra of synthesized catalysts.

The thermal decomposition of catalysts was determined by thermal analysis and studied in detail in our previous papers between 25 and 650 °C [2,23]. Some considerations from the previous papers are given below. According to TG curves, the stage until 110 °C indicated the removal of the physically adsorbed water. The stage between 110−300 °C corresponds to the loss of water molecules, in which the waters are hydrogen-bonded to the acidic protons to form the [H_2O ... H+ ... OH_2]. At the last stage, over 300 °C corresponds to the loss of constitutional water and decomposition of Keggin structure with the formation of corresponding oxides (WO_3 and P_2O_5). The oxides crystallize between 570 and 590 °C with an exothermic effect on the DTA curves.

On the other hand, in the case of supported catalysts, the removed hydrated water is higher than for unsupported ones due to the thermal overlap effects of the desorbed water from the silica surface and the loss of HPCs physisorbed water. The departure of water molecules which are linked by hydrogen bonds with acidic protons due to removing some traces of the surfactants used for the synthesis of mesoporous sieve SBA-15, takes place in the second step. An exothermic peak that appears on DTA curves around 590 °C could be assigned to the decomposition of the active phase.

It can be noted that the HPW/SBA-15, PdPW/SBA-15, Cs2.5PW/SBA-15, and PdCs2.5PW/SBA-15 catalysts may be considered slightly more stable than pure catalysts, probably due to the interaction of catalyst with silanols on the surface.

2.1.2. Textural Properties of Heteropoly Compounds

The results of textural characteristics analysis (specific surface area, pore volume, and average pore diameter for catalysts are summarized in Table 1. In our previous articles [2,23] was reported that upon Cs incorporation, the surface area increases significantly, from 3 $m^2\ g^{-1}$ for HPW to 130 $m^2\ g^{-1}$ for Cs2.5PW. By adding palladium, the surface area of Cs2.5PW decreases from 130 $m^2\ g^{-1}$ to 98 $m^2\ g^{-1}$, which explains the microstructural model [23] according to the Cs3PW crystallites are formed at first, and then HPW is adsorbed epitaxially on the surface of Cs3PW during the synthesis.

Table 1. Specific surface area, pore volume and average pore diameter of catalysts.

Catalyst	Surface Area (m^2/g)	Pore Volume BJH_{Des} (cm^3/g)	Average Pore Diameter BJH_{Des} (nm)
SBA-15	725	1.100	6.2
HPW/SBA-15	194	0.191	6.0
PdPW/SBA-15	137	0.211	5.8
Cs2.5PW/SBA-15	303	0.830	6.1
PdCs2.5PW/SBA-15	298	0.790	5.9

The significant decrease in surface area from 725 m^2/g (SBA-15) to 194 m^2/g (HPW/SBA-15) is due to the dispersion of the acid on the large surface of the SBA-15 molecular sieve. For supported cesium compounds, a slight decrease is taking place from 725 m^2/g (SBA-15) to 303 m^2/g (Cs2.5PW/SBA-15) and 298 m^2/g (PdCs2.5PW/SBA-15) respectively which indicates a partial blockage of particles on the external surface of the mesoporous SBA-15 sieve. For all catalysts, the average pore diameter tends to increase with the increased surface area.

The N_2 adsorption–desorption isotherms of HPW/SBA-15, PdPW/SBA-15, Cs2.5PW/SBA-15 and PdCs2.5PW/SBA-15 catalysts were depicted in Figure 3. All catalysts exhibited a Type IV (a–e) isotherm according to the IUPAC classification [24] with a hysteresis loop H1 type and a sharp increase in volume adsorbed at a relative pressure of p/p0 = 0.6–0.9, characteristic of highly ordered mesoporous silica.

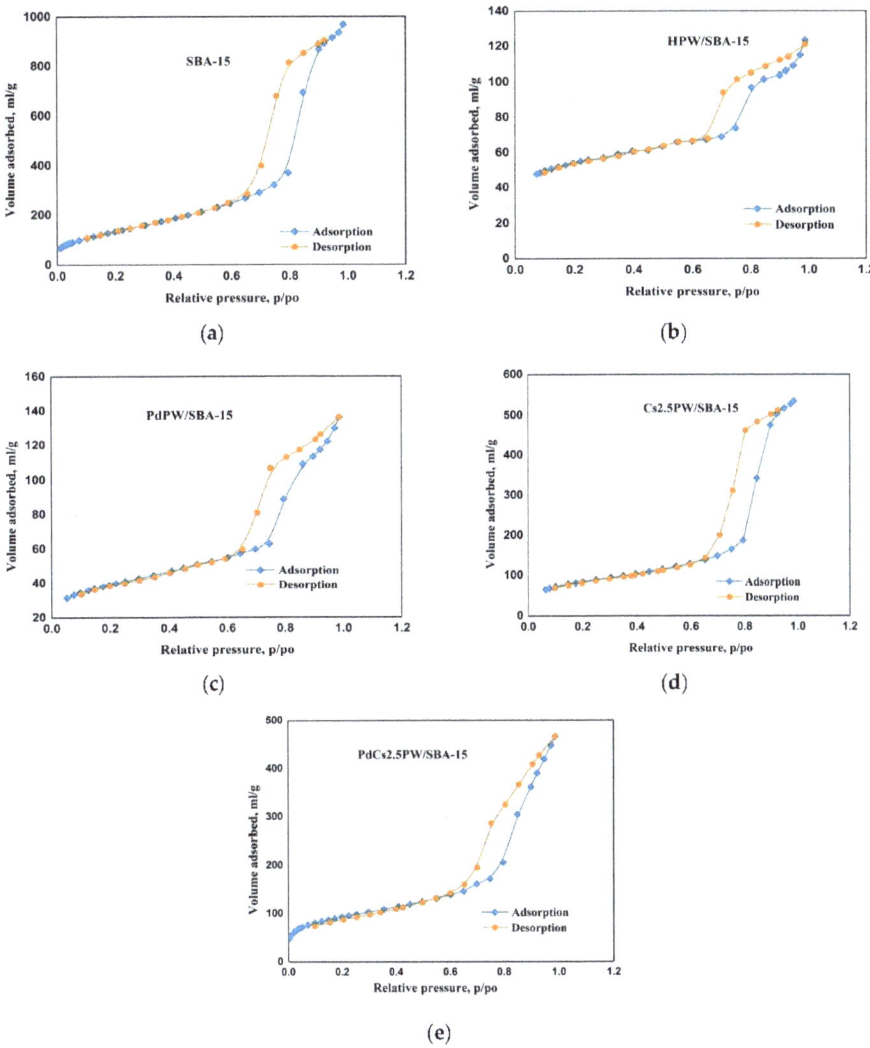

Figure 3. Nitrogen adsorption–desorption isotherms of SBA-15 (**a**), HPW/SBA-15 (**b**), PdPW/SBA-15 (**c**), Cs2.5PW/SBA-15 (**d**) and PdCs2.5PW/SBA-15 (**e**).

Comparing adsorption–desorption isotherms, it was observed that all catalysts have a hysteresis loop similar to the parent support. This indicates that the 2D hexagonal structure of SBA-15 is preserved after Pd doping and Cs2.5 impregnation.

2.2. Catalytic Ethanol Transformation

Ethanol conversion over catalysts was found to produce a range of products, with ethylene and diethyl ether being the main products. It should be noted that, during ethanol dehydration, by products such as small amounts of hydrocarbons were also obtained [1,2,4,5].

In a previous work, we show that the conversion and selectivity of synthesized catalysts in ethanol conversion, are strongly influenced by temperature. Catalyst activity increases with temperature and the optimum reaction temperature is between 275 and 300 °C. The catalysts begin to deactivate above 300 °C [2].

For pure compounds, significant quantities of DEE (around 20%) were produced via dehydration of ethanol at 200 °C. For silica supported catalysts DEE was obtained in very small amount due to its decomposition to ethylene (Table 2). A comparison of ethanol conversion of catalysts shows that the supported catalysts have significantly higher conversion than the unsupported ones.

Table 2. Ethanol conversion, ethylene and diethyl ether selectivity for pure and supported catalysts as a function of TOS at 200 °C.

Sample	Ethanol Conversion (%)		Ethylene Selectivity (%)		Diethyl Ether Selectivity (%)	
	Time on Stream (TOS)					
	35 min	*350 min*	*35 min*	*350 min*	*35 min*	*350 min*
HPW	51.2	45.5	43.2	39.0	18.1	19.2
PdPW	43.2	39.0	38.8	36.0	19.2	19.9
Cs2.5PW	78.6	76.1	28.4	25.2	26.3	23.2
PdCs2.5PW	84.2	79.2	50.9	42.6	18.3	19.7
HPW/SBA-15	93.4	89.6	90.1	80.1	0.3	0.5
PdPW/SBA-15	94.7	84.7	88.4	80.6	0.2	0.1
Cs2.5PW/SBA-15	98.9	90.1	93.2	89.4	0.3	0.4
PdCs2.5PW/SBA-15	94.9	87.2	91.2	86.4	0.1	0.3

Deactivation of the catalysts has been examined for ethanol conversion at 300 °C under atmospheric pressure. The results obtained for unsupported catalysts are shown in Figure 4. As can be seen a temperature increase causes an increase in ethanol conversion especially for Cs2.5PW which exhibiting the highest activity. Cs2.5 PW is well known to be an excellent acid catalyst in various gas-phase reactions exhibiting the highest activity in ethanol conversion due to its larger surface area (130 m^2/g). It is regarded as the most active catalyst in the gaseous phase reaction. The other catalysts begin to deactivate after 1 h of TOS due to the cokes formation which is leading to blocking the acidic centers responsible for ethylene formation. The lower catalytic activity of HPW and PdPW in ethanol conversion reaction is due to a very low specific area, 3 and 5 m^2/g, respectively which are in agreements with literature date [9,10]. In the same conditions the highest ET selectivity values were obtained for Cs2.5PW salt catalysts (Figure 5).

A significant increase in catalytic activity on the same catalysts but supported on an SBA-15 molecular sieve as a function of TOS at 300 °C was observed (Figure 6). This could be explained by the fact that a large specific area and a good dispersion of active sites on support leads to high catalytic activities.

The highest ethanol conversion was obtained for Cs2.5PW/SBA-15 salt. By adding palladium, the catalytic activity slowly decreases for both pure and supported HPW/SBA-15 and Cs2.5/SBA-15 catalysts. I.V. Kozhevnikov et al. studied the coking during propene oligomerization over silica-supported HPAs, namely, HPW and its Pd-doped form (1.6–2.5 wt.% Pd), and subsequent catalyst regeneration. It has revealed that the Keggin structure of the catalyst was unaffected by coke deposition in both unmodified PW/SiO2 and Pd-modified form, but it should be noted that under the conditions studied, the Pd-modified HPW deactivated a bit faster than the unmodified HPW [16,17].

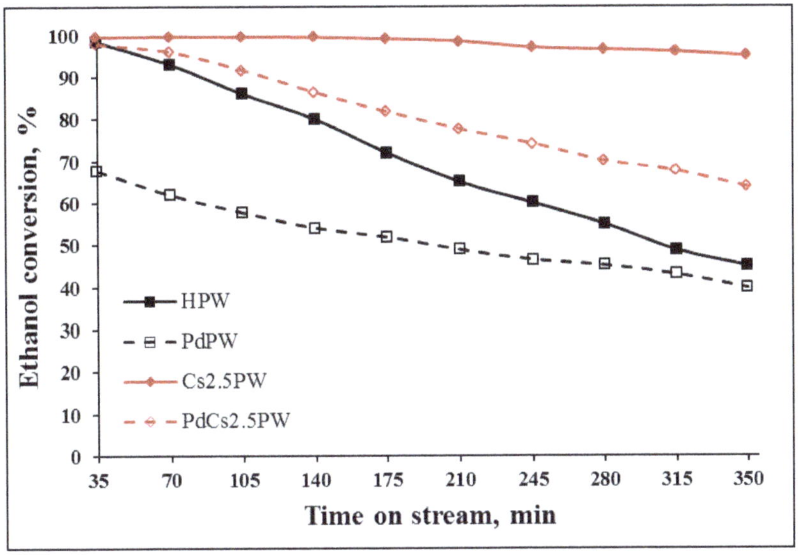

Figure 4. Conversion of ethanol over pure and Pd doped HPW and Cs2.5 PW as a function of TOS at 300 °C.

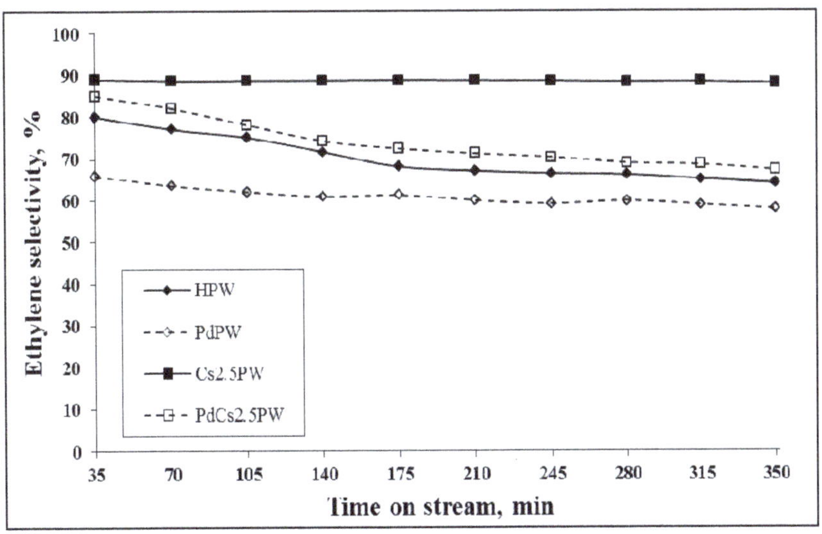

Figure 5. Ethylene selectivity over pure and Pd doped HPW and Cs2.5 PW as a function of TOS at 300 °C.

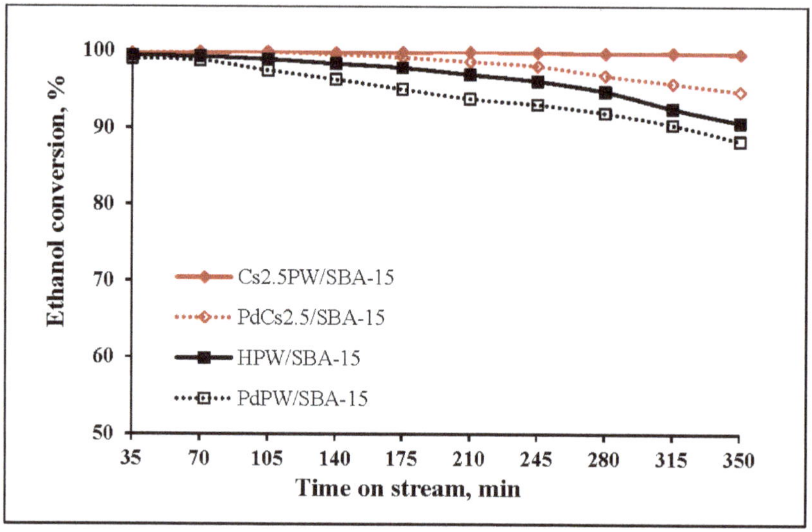

Figure 6. Conversion of ethanol over pure and Pd doped HPW/SBA-15 and Cs2.5 PW/SBA-15 as a function of TOS at 300 °C.

For all catalysts, the ethanol conversion has increasing values during the temperature increase from 200 to 300 °C, but the conversion values are higher for the supported ones. The ethylene selectivity rises with increasing temperature from 200 °C to 300 °C for both pure and supported catalysts. Anyway, as in the case of ethanol conversion, the values of ethylene selectivity decrease slowly with time on stream (Figure 7). Generally, the catalytic activity of supported mesoporous SBA-15 catalysts was higher than that of bulk catalysts.

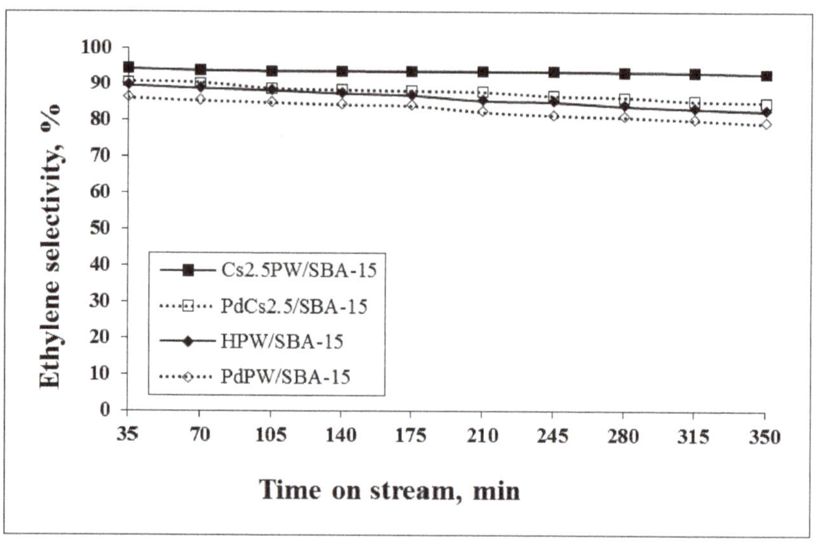

Figure 7. Ethylene selectivity over pure and Pd doped HPW/SBA-15 and Cs2.5 PW/SBA-15 as a function of TOS at 300 °C.

2.3. *Temperature Programmed Oxidation (TPO) of Spent Catalysts*

As we mentioned before, heteropoly compounds have a high catalytic activity for ethanol conversion but undergo a deactivation process due to the formation of coke [15–18]. Coke deposition of spent catalysts was studied by the temperature-programmed oxidation (TGA/TPO) method. The quantities of coke precursor and hard coke were determined from TG curves. The measurements were conducted in the range of 25–650 °C in a nitrogen atmosphere with an isothermal step at 300 °C. Finally, the catalysts were oxidized under air flow. TGA/TPO data for coke precursors and hard coke of spent catalysis are presented in Figures 8–11.

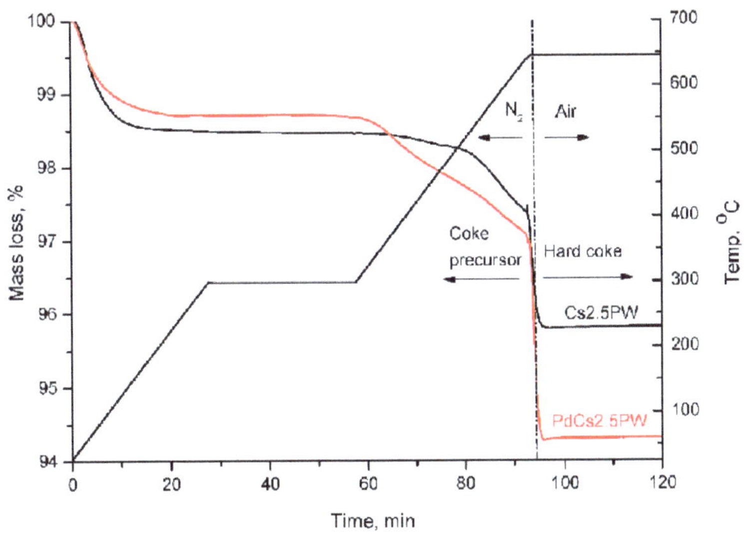

Figure 8. Coke precursors and hard coke of Cs2.5PW and PdCs2.5PW at 300 °C.

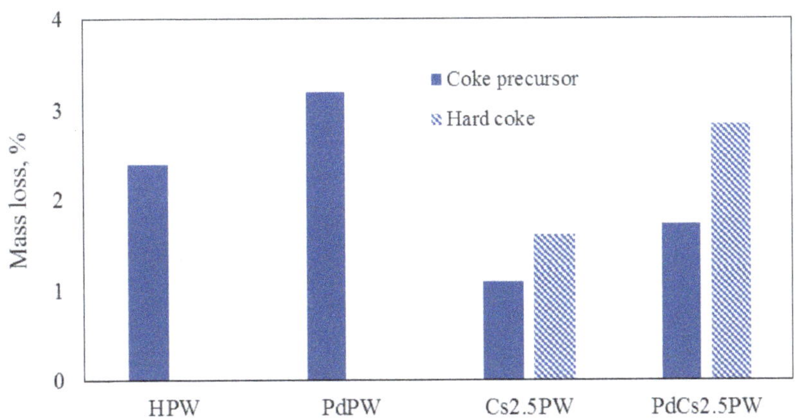

Figure 9. Quantities of coke precursor and hard coke for unsupported catalysts formed by ethanol conversion at 300 °C.

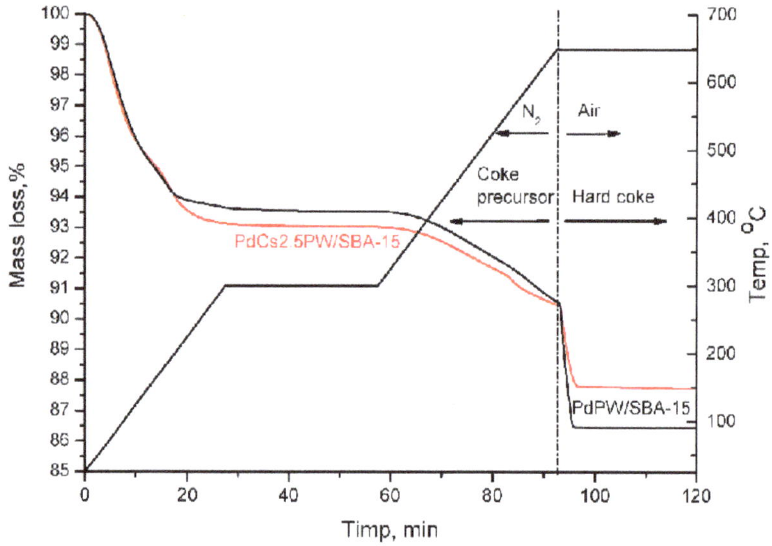

Figure 10. Coke precursors and hard coke of PdCs2.5PW/SBA-15 and PdPW/SBA-15 at 300 °C.

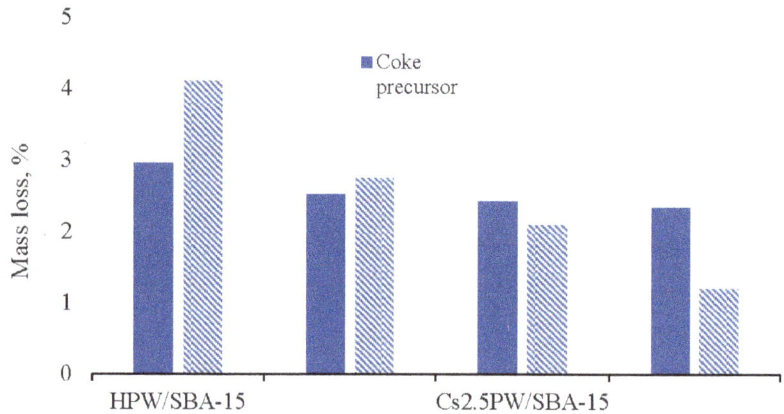

Figure 11. Quantities of coke precursor and hard coke for supported catalysts formed by ethanol conversion at 300 °C.

An increase in the total amount of coke can be clearly observed in the case of unsupported PdCs2.5PW compared to the pure Cs2.5PW catalyst (Figure 8). However, the quantities of coke decrease significantly when PdCs2.5PW is supported on a mesoporous SBA-15 sieve (Figure 10).

Also, there is a significant decrease in hard coke formation for pure and Pd-doped Cs2.5PW (Figure 9). By adding cesium, the Brönsted acidity increases (the number of acid centers increases), which explains the decrease in coke formation. In previous work, the Brönsted acidity of these compounds was determined by TG curves [2].

The number of total coke decreases for Cs2.5PW/SBA-15 and PdCs2.5PW/SBA-15, especially hard coke (Figure 11). As the supported Cs2.5PW and PdCs2.5PW salts contain a mesoporous sieve with an SBA 15 two-dimensional structure, deactivation is slower, resulting in a reduction in the quantity of coke. In the case of supported compounds, the interaction between the heteropoly acid and hydroxyl groups (-OH) present in the mesoporous silica SBA-15 is important; the degree of dispersion influences the acid concentration of these sites [11,13,25].

The IR frequencies of catalysts were studied after the formation of coke during ethanol conversion in a fixed-bed flow reactor at 300 °C. All characteristic bands of the Keggin unit (primary structure) are preserved after depositing the coke on the catalysts. Deactivation of these compounds does not affect the primary structure, as is evident in Figure 12.

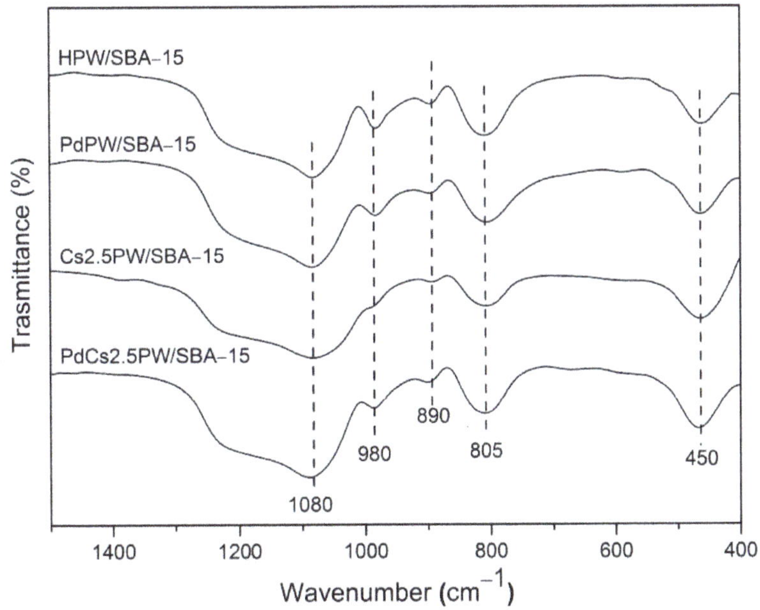

Figure 12. Infrared spectra for spent supported catalysts after ethanol conversion in a fixed-bed flow reactor at 300 °C.

3. Experimental

3.1. Chemicals and Materials

Tungstophosphoric acid ($H_3PW_{12}O_{40}xH_2O$), abbreviated as HPW, was prepared from sodium phosphate dibasic ($NaHPO_4 \cdot H_2O$, ACS reagent, ≥99%, Merck, Darmstadt, Germany) and sodium tungstate dehydrates ($Na_2WO_4 \cdot H_2O$, ACS reagent, ≥99%, Merck, Darmstadt, Germany). The $CsNO_3$ and $Pd(NO_3)_2 \cdot 2H_2O$ salts were purchased from Sigma-Aldrich (St. Louis, MI, USA). Tetraethyl orthosilicate (TEOS) reagent grade, 98% and *Pluronic P123* (EO20PO70EO20) Poly (ethylene glycol)-block-poly(propylene glycol)-block-poly(ethylene glycol)-PEG-PPG-PEG used for preparation of SBA-15 mesoporous silica were purchased from Merck (Darmstadt, Germany). All chemicals were used directly without further purification.

3.2. Measurement

The thermal analyses (TGA and DTA) of both bulk and supported catalysts were carried out by using a thermoanalyzer system Mettler TGA/SDTA 851/LF/1100 (Columbus, OH, USA). The measurements were conducted in a dynamic atmosphere of air (50 mL/min) in alumina crucibles of 150 μL. The mass samples were about 20–30 mg. The heating rate was about 10 °C/min in the temperature range of 25–650 °C.

Fourier transform infrared spectroscopy (FT-IR) was performed on a Jasco 430 spectrometer (Tokio, Japan) for all prepared compounds using KBr pellets at ambient temperature. The IR spectra were obtained in the region from 4000 to 400 cm^{-1} range with 256 scans at 2 cm^{-1} resolution.

Powder X-ray diffraction data were collected using an XD 8 Advanced Bruker automated diffractometer (Karlsruhe, Germany) using the Cu Kα radiation at small angles (2θ = 0.5–5°) and wide angles (2θ = 5–60°), respectively.

A Quantachrome instrument Nova 2000 series (Boynton Beach, FL, USA) was applied to measure the nitrogen adsorption–desorption isotherms. The samples were preheated and degassed at 250 °C for 2 h under vacuum. The specific surface areas of samples were calculated based on Brunauer–Emmet–Teller (BET) equation according to a linear part of the adsorption curves. The pore diameters were calculated by Barrett–Joyner–Halenda (BJH) method applied to the desorption branches of the isotherms.

3.3. *Catalyst Preparation*

Tungstophosphoric acid $H_3PW_{12}O_{40}$ (HPW) was synthesized according to the J. C. Bailar method [25,26]. The $Cs_{2.5}$ $H_{0.5}PW_{12}O_{40}$ (Cs2.5PW) salt was prepared by precipitation from the aqueous solution of acid after adding the stoichiometric amount of $CsNO_3$ under continuous stirring according to our previous report [2]. The HPW doped with 0.25 Pd/Keggin unit (KU), namely $Pd_{0.25}H_{2.5}PW$ (PdPW), was prepared through the reaction between $Pd(NO_3)_2$ and HPW 0.1 M aqueous solutions in the proper ratio. The Cs2.5H0.5PW salt doped with 0.25 at Pd/KU (denoted as PdCs2.5PW) was prepared by adding an amount of $Pd(NO_3)_2$ as aqueous solution 0.1 M into the HPW aqueous solution 0.1 M and after, the required stoichiometric quantity of $CsNO_3$ (aqueous solution) was poured drop by drop under continues stirring. For all catalysts, the pH was lower than 1.5 during all steps of the synthesis. All samples were dried at 50–60 °C under stirring and heated up to 250 °C in air for nitrate anion total decomposition. *SBA-15 mesoporous* silica was prepared by the hydrolysis of tetraethyl orthosilicate (TEOS) using Pluronic P123 block copolymer as surfactant, according to Zhao et al. [27]. The HPW, PdPW, and PdCs2.5 were impregnated by aqueous incipient wetness onto SBA-15 mesoporous molecular sieve as loading of 30 wt.% concentration to obtain a monolayer array of HPC on SBA-15. The final products obtained were denoted as HPW/SBA-15, PdPW/SBA-15 and PdCs2.5/SBA-15.

3.4. *Catalytic Activity Measurement*

The catalytic activities of the as-prepared catalysts for ethanol conversion were carried out in a homemade flow microreactor (10 mm inner diameter) placed into an electric furnace. The experimental conditions are presented in Table 3. The catalyst sample was placed in the middle of the reactor and supported by quartz packing at both ends. The reactor temperature was adjusted by a temperature controller within ± 1% in the temperature range of 200−350 °C.

Table 3. Summary of experimental conditions.

Variable	Values
Amount catalyst unsupported (mg)	100
Amount catalysts supported (mg)	330
Catalyst loading (wt.%)	30
Operating temperature (°C)	200, 250, 300, 350
Ethanol flow rate (mL/h)	1.2
Nitrogen flow rate (mL/min)	30
TOS (min)	350

Liquid ethanol was introduced by a Hamilton syringe pump into an evaporator heated at 150 °C. The composition of the reactor effluent stream was analyzed by a gas chromatograph equipped with two detectors: thermal conductivity detector (TCD) and flame ionization detector (FID). In order to separate the products, GC-packed columns filled with Porapak QS 80–100 mesh were used. A temperature program (keep the temperature at 50 °C for 5 min, increase the temperature up to 200 °C with a heating rate of 20 °C/min and maintain the temperature for 12 min at 200 °C) was used in GC technique. The main reaction products have the following retention times: ethylene at 5.6, ethanol at 14.8, and diethyl ether at 18.8 min. Secondary products such as methane, C2 (ethane, ethylene), C3 (propane, propene), C4 (butane, butane), C5 (pentane, pentene), C6 (hexane, hexane), and diethyl ether (DEE) have been detected. Prior to the reaction, catalysts were pre-treated "in situ" under the nitrogen flow at 250 °C for 1 h.

The ethanol conversion and ethylene selectivity were calculated based on the carbon balance as follows:

$$\text{EtOH conversion\%} = (\text{moles of EtOH reacted/moles of EtOH total}) \times 100 \quad (1)$$

$$\text{Product selectivity\%} = (\text{moles of product/moles of reacted EtOH}) \times 100 \quad (2)$$

3.5. Thermal Analysis of Coke

The amount of coke was analyzed by temperature-programmed oxidation (TGA/TPO) method on the same thermo analyzer system, Mettler TGA/SDTA 851/LF/1100. Experiments were carried out using about 20 mg samples into alumina crucibles of 150 µL. The heating rate was 10 °C/min in the range of temperature 25–650 °C with an isothermal step at 300 °C for 30 min. The measurements were conducted in dynamic atmosphere of nitrogen (50 mL/min) and finally under the flow of air for 30 min to oxidize the catalysts. By switching from N_2 to air at the same flow rate, the remaining coke deposited on the catalysts was burnt out, as shown in Figure 13.

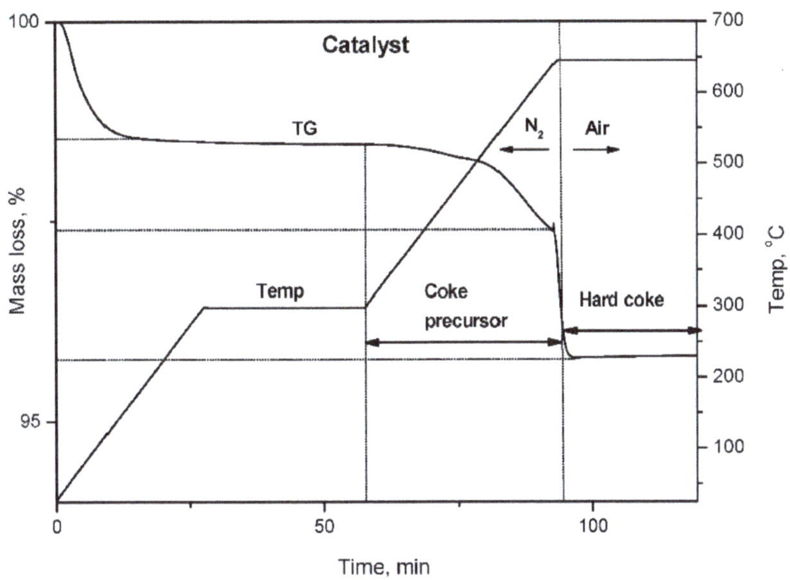

Figure 13. Coke precursor and hard coke catalyst during TGA-TPO method.

The quantities of coke precursor present in the catalysts were calculated as difference between the initial mass of spent catalyst sample after isothermal heating at 300 °C and the sample mass heated in nitrogen at 650 °C. Soft coke is removed from the samples through volatilization in inert nitrogen and refers to high molecular weight aliphatic oligomers. The amount of hard coke present in the catalysts was calculated as difference between samples mass heated in nitrogen at 650 °C and mass loss of sample at 650 °C in air when the coke was burnt out. Hard coke refers to heavy polynuclear aromatic.

4. Conclusions

A series of pure and Pd-doped HPW/SBA-15 and Cs2.5PW/SBA-15 catalysts have been prepared, and their catalytic performance and coke formation in ethanol conversion have been investigated.

The XRD results indicated that the mesostructured ordering is slightly modified by the impregnation of the catalysts into SBA-15.

By comparing adsorption–desorption isotherms, it was observed that all catalysts have a hysteresis loop similar to the parent support, which indicates that the hexagonal structure of SBA-15 is preserved after Pd doping and Cs2.5PW impregnation.

In the case of supported samples, the effect of the support leads to higher values of ethanol conversion and ethylene selectivity. As a result, during the time on stream (TOS), the amount of coke decreases significantly. The effect of palladium doping on the coke formation was studied; thus, it can be concluded that Pd increases the coke formation for pure catalysts while Cs decrease it. However, in the case of catalysts supported on SBA-15, Pd decreases significantly the quantities of coke, especially hard coke.

FTIR studies show that the formation of coke deposits during the ethanol conversion, although the deactivation occurs, it does not affect the Keggin structure of catalysts supported on SBA-15.

Author Contributions: Conceptualization, O.V., V.S. and A.P.; methodology, O.V. and A.P.; formal analysis, S.B., M.S. and V.S.; investigation, S.B., M.S. and O.V.; resources, V.S. and A.P; data curation, S.B. and M.S.; writing—original draft preparation, O.V., S.B, M.S. and A.P.; writing—review and editing O.V. and A.P.; visualization, S.B., M.S. and O.V.; supervision, V.S. and A.P. All authors have read and agreed to the published version of the manuscript.

Funding: This research was funded by Romanian Academy Project No. 4.3.

Conflicts of Interest: The authors declare no conflict of interest. The funders had no role in the design of the study; in the collection, analyses, or interpretation of data; in the writing of the manuscript; or in the decision to publish the results.

References

1. Mohsenzadeh, A.; Zamani, A.; Taherzadeh, M.J. Bioethanol production from ethanol: A review and techno-economical evaluation. *Chem. Bio. Eng. Rev.* **2017**, *4*, 1–18.
2. Verdeş, O.; Sasca, V.; Popa, A.; Suba, M.; Borcanescu, S. Catalytic activity of heteropoly tungstate catalysts for ethanol dehydration reaction: Deactivation and regeneration. *Catal. Today* **2021**, *366*, 123–132. [CrossRef]
3. Zhang, M.; Yu, Y. Dehydration of Ethanol to Ethylene. *Ind. Eng. Chem. Res.* **2013**, *52*, 9505–9514. [CrossRef]
4. Varisli, D.; Dogu, T.; Dogu, G. Ethylene and diethyl-ether production by dehydration reaction of ethanol over different heteropolyacid catalysts. *Chem. Eng. Sci.* **2007**, *62*, 5349–5352. [CrossRef]
5. Rawan, A.; Kozhevnikova, E.F.; Kozhevnikov, I.V. Diethyl Ether Conversion to Ethene and Ethanol Catalyzed by Heteropoly Acids. *ACS Omega* **2021**, *6*, 9310–9318.
6. Esmi, F.; Masoumi, S.; Dalai, A.K. Comparative Catalytic Performance Study of 12-Tungstophosphoric Heteropoly Acid Supported on Mesoporous Support for Biodiesel Production from Unrefined Green Canola Oil. *Catalysts* **2022**, *12*, 658. [CrossRef]
7. Heravi, M.M.; Fard, M.V.; Faghihi, Z. Heteropoly acids-catalyzed organic reactions in water: Doubly green reactions. *Green Chem. Lett. Rev.* **2013**, *6*, 282–300. [CrossRef]
8. Diamantopoulos, N.; Panagiotaras, D.; Nikolopoulos, D. Comprehensive Review on the Biodiesel Production using Solid Acid Heterogeneous Catalysts. *J. Thermodyn. Catal.* **2015**, *6*, 1–8. [CrossRef]
9. Misono, M. Unique acid catalysis of heteropoly compounds (heteropolyoxometalates) in the solid state. *Chem. Commun.* **2001**, 1141–1152. [CrossRef]
10. Okuhara, T.; Watanabe, H.; Nishimura, T.; Inumaru, K.; Misono, M. Microstructure of Cesium Hydrogen Salts of 12-Tungstophosphoric Acid Relevant to Novel Acid. *Chem. Mater.* **2000**, *12*, 2230–2238. [CrossRef]
11. Kokunesoskia, M.; Gulicovskia, J.; Matovica, B.; Logarb, M.; Milonjica, S.K.; Babic, B. Synthesis and surface characterization of ordered mesoporous silica SBA-15 Mater. *Chem. Phys.* **2010**, *124*, 1248–1252.
12. Brahmkhatri, V.; Patel, A. 12-Tungstophosphoric acid anchored to SBA-15: An efficient, environmentally benign reusable catalysts for biodiesel production by esterification of free fatty acids. *Appl. Catal. A.* **2011**, *403*, 161–172. [CrossRef]
13. Popa, A.; Sasca, V.; Verdes, O.; Barvinschi, P.; Holclajtner-Antunovica, I. Acidic and neutral caesium salts of 12-molybdophosphoric acid supported on SBA-15 mesoporous silica. The influence of Cs concentration and surface coverage on textural and structural properties. *Mater. Res. Bull.* **2014**, *50*, 312–322. [CrossRef]

14. Wang, B.; Manos, G. A novel thermogravimetric method for coke precursor characterisation. *J. Catal.* **2007**, *250*, 121–127. [CrossRef]
15. Zhou, J.; Zhao, J.; Zhang, J.; Zhang, T.; Ye, M.; Liu, Z. Regeneration of catalysts deactivated by coke deposition: A Review Chinese. *J. Catal.* **2020**, *41*, 1048–1406. [CrossRef]
16. Siddiqui, M.R.H.; Holmes, S.; Smith, W.; He, H.; Coker, E.N.; Atkins, M.P.; Kozhevnikov, I.V. Coking and regeneration of palladium-doped $H_3PW_{12}O_{40}/SiO_2$ catalysts. *Catal. Lett.* **2000**, *66*, 53–57. [CrossRef]
17. Kozhevnikov, I.V.; Holmes, S.; Siddiqui, M.R.H. Coking and regeneration of $H_3PW_{12}O_{40}/SiO_2$ catalysts. *Appl. Catal. A Gen.* **2001**, *214*, 47–58. [CrossRef]
18. Chen, S.; Manos, G. Study of coke and coke precursors during catalytic cracking of n-hexane and 1-hexene over ultrastable Y zeolite. *Catal. Lett.* **2004**, *96*, 195–200. [CrossRef]
19. Chen, S.; Manos, G. In situ thermogravimetric study of coke formation during catalytic cracking of normal hexane and 1-hexene over ultrastable Y zeolite. *J. Catal.* **2004**, *226*, 343–350. [CrossRef]
20. Jalil, P.A.; Faiz, M.; Tabet, N.; Hamdan, N.M.; Hussain, Z. A study of the stability of tungstophosphoric acid, $H_3PW_{12}O_{40}$, using synchrotron XPS, XANES, hexane cracking, XRD, and IR spectroscopy. *J. Catal.* **2003**, *217*, 292–297. [CrossRef]
21. Essayem, N.; Holmqvist, A.; Gayraud, P.Y.; Vedrine, J.C.; Taarit, Y.B. In Situ FTIR Studies of the Protonic Sites of $H_3PW_{12}O_{40}$ and Its Acidic Cesium Salts $M_xH_{3x}PW_{12}O_{40}$. *J. Catal.* **2001**, *197*, 273–280. [CrossRef]
22. Bielański, A.; Lubańska, A.J. FTIR investigation on Wells–Dawson and Keggin type heteropolyacids: Dehydration and ethanol sorption. *Mol. Catal. A Chem.* **2004**, *224*, 179–187. [CrossRef]
23. Sasca, V.; Verdeş, O.; Avram, L.; Popa, A.; Erdöhelyi, A.; Oszko, A. The $Cs_xH_{3-x}PW_{12}O_{40}$ catalysts microstructure model. *Appl. Catal. A Gen.* **2013**, *451*, 50–57. [CrossRef]
24. Thommes, M.; Kaneko, K.; Neimark, A.V.; Olivier, J.P.; Rodriguez-Reinoso, F.; Rouquerol, J.; Sing, K.S.W. Physisorption of gases, with special reference to the evaluation of surface area and pore size distribution (IUPAC Technical Report). *Pure Appl. Chem.* **2015**, *87*, 1051–1069. [CrossRef]
25. Bailar, J.C. Phosphotungstic acid. *Inorg. Synth.* **1939**, *1*, 132–133.
26. Misono, M.; Mizuno, N.; Katamura, K.; Kasai, A.; Konishi, Y.; Sakata, K.; Okuhara, T.; Yoneda, Y. Catalysis by heteropoly compounds III. The structure and properties of 12-heteropolyacids of molybdenum and tungsten ($H_3PMo_{12-x}W_xO_{40}$) and their salts pertinent to heterogeneous catalysis. *Bull. Chem. Soc. Jpn.* **1982**, *55*, 400–406. [CrossRef]
27. Zhao, D.; Feng, J.; Huo, Q.; Melosh, N.; Fredrickson, G.H.; Chmelka, B.F.; Stucky, G.D. Triblock Copolymer Syntheses of Mesoporous Silica with Periodic 50 to 300 Angstrom Pores. *Science* **1998**, *279*, 548–552. [CrossRef]

catalysts

MDPI

Article

Insights into Synergy of Copper and Acid Sites for Selective Catalytic Reduction of NO with Ammonia over Zeolite Catalysts

Wenyi Zhao [1,2], Menglin Shen [1,2], Yueran Zhu [2], Xudong Ren [2] and Xingang Li [1,3,*]

1 Collaborative Innovation Center of Chemical Science and Engineering (Tianjin), Tianjin Key Laboratory of Applied Catalysis Science & Technology, School of Chemical Engineering & Technology, Tianjin University, Tianjin 300350, China
2 National Key Laboratory for Research and Comprehensive Utilization of Rare Earth Resources in Baiyun Obo, Baotou Research Institute of Rare Earths, Baotou 014030, China
3 Institute of Shaoxing, Tianjin University, Zhejiang 312300, China
* Correspondence: xingang_li@tju.edu.cn; Tel.:+86-22-27403389

Abstract: Herein, we report the function of copper sites in Cu-SSZ-13, Cu-ZSM-5 and Cu-Beta catalysts with the same Si/Al ratio (14) and Cu/Al ratio (0.4) on selective catalytic reduction of NO with NH_3 (NH_3-SCR) and reveal the relationship between active sites (Cu sites, acid sites) and catalytic activity. The results show that the amount of isolated Cu^{2+} ions in the catalysts directly determines the formation of strong Lewis acid sites and reaction intermediate NO_3^- ions, thus affecting the low-temperature SCR performance, while the amount of highly stable Cu^+ ions and Brønsted acid sites is related to the high-temperature SCR performance of the catalysts. Consequently, it contains enough isolated Cu^{2+} ions, highly stable Cu^+ ions and Brønsted acid sites, which endows Cu-SSZ-13 with excellent NH_3-SCR activity.

Keywords: NH_3-SCR; zeolite; copper cations; acid sites; NO_3^- species

Citation: Zhao, W.; Shen, M.; Zhu, Y.; Ren, X.; Li, X. Insights into Synergy of Copper and Acid Sites for Selective Catalytic Reduction of NO with Ammonia over Zeolite Catalysts. *Catalysts* **2023**, *13*, 301. https://doi.org/10.3390/catal13020301

Academic Editors: Maja Milojević-Rakić and Danica Bajuk-Bogdanović

Received: 4 January 2023
Revised: 20 January 2023
Accepted: 26 January 2023
Published: 28 January 2023

1. Introduction

In 2020, the China national automobile nitrogen oxide (NO_x) emissions was 6.137 million tons, and the diesel vehicle NO_x emissions had reached 88.8% of the total automobile emissions. Diesel engines are oxygen enriched combustion and the oxygen content in the exhaust is high. It is very difficult to directly carry out a selective catalytic reduction of NO_x, which is a worldwide problem. At present, ammonia selective catalytic reduction technology (NH_3-SCR) is mainly used to treat the diesel engine NO_x [1]. As NH_3-SCR is a dynamic process, reactive gases such as NO, NO_2, O_2 and NH_3 will be adsorbed and activated on the active site to form an active species. The active species and reaction gas are constantly consumed, thus generate important reaction intermediates. Consequently, developing an efficient NH_3-SCR catalyst is the key to solving the above issues [2].

With the increasingly strict environmental emission regulations, the NH_3-SCR catalyst has been developed from the vanadium tungsten titanium (VWTi) catalyst, which can meet the national IV and V emission standards to copper-based zeolite catalysts that can meet the national VI. In the 1980s, Liu et al. [3,4] first reported that the Cu-ZSM-5 zeolite had good SCR activity for NO decomposition. Consequently, with the gradual deepening of research, more Cu-containing zeolites (such as BEA, MFI, FAU, MOR and CHA etc.) have attracted extensive attention due to their efficient removal of NO [5,6]. However, the Beta zeolite has relatively large 12-rings channels compared with ZSM-5. The research shows that the hydrothermal stability of Cu-Beta zeolite is higher than that of Cu-ZSM-5 with relatively small channel structure [7,8]. The SSZ-13 zeolite has a CHA structure and is a three-dimensional microporous zeolite composed of D6R and CHA composite units alternately. The copper exchanged SSZ-13 catalyst has become a commercial NH_3-SCR catalyst due to it having higher activity and durability than any earlier SCR catalyst [9]. However, with the follow-up implementation of the "China national VI b" and above emission standards, the higher index requirements have been proposed for the low-temperature SCR performance, hydrothermal stability and exhaust by-products (N_2O, NH_3, etc) of the NH_3-SCR catalysts, which has become a current research hotspot. According to the literature [10–12], the nature of active sites such as metal sites and acid sites on the surface of metal-exchanged zeolite catalysts have an important influence on the NH_3-SCR reaction. Generally, this can be achieved by tuning the type, number and location of active sites to ameliorate the catalytic activity of the NH_3-SCR catalyst. Chen et al. [13] found that the low-temperature SCR activity of Cu-Ce-La-SSZ-13 was enhanced by controlling the position of Cu^{2+} ions, but did not mention that regulating the position of Cu^{2+} ions caused the change of acid sites on the catalyst' surface. Luo et al. [14] indicated through DRIFTS research that NH_3 was adsorbed on Lewis acid sites and Brønsted acid sites on the Cu-SSZ-13' surface, but did not point out the relationship between acid sites and catalytic activity in the article. Hence, the roles of copper and acid sites on the surface of copper-based zeolites for the NH_3-SCR reaction were compared and analyzed, and the correlation between them and catalytic activity is critical to continue to improve the catalytic performance of the Cu-SSZ-13 catalyst.

In this paper, the intrinsic physicochemical properties of their active sites in the Cu-SSZ-13, Cu-ZSM-5 and Cu-Beta NH_3-SCR catalysts with the same Si/Al ratio (14) and Cu/Al ratio (0.4) were comparatively studied, which is crucial to establish the relationship between active site and catalytic performance. We analyzed the reaction processes of NH_3 adsorption, NO adsorption and hourly adsorption of NO + O_2 and NO + O_2 + NH_3 in an attempt to understand the effect of active sites on the reaction intermediates formed on the surface of copper-based zeolites in the Standard SCR reaction. The comprehensive characterization results allow us to explain the important reasons for the differences in the NH_3-SCR performance among Cu-SSZ-13, Cu-ZSM-5 and Cu-Beta catalysts.

2. Results and Discussion

2.1. Catalytic Comparison

Figure 1a exhibits the NO conversion as a function of temperature in the standard SCR reaction over Cu-SSZ-13, Cu-ZSM-5 and Cu-Beta catalysts. As shown in Figure 1a, the reaction temperature window (200–500 °C) of the Cu-SSZ-13 catalyst is significantly wider than Cu-ZSM-5 (200–400 °C) and Cu-Beta (300–600 °C), indicating that the Cu-SSZ-13 has the excellent $deNO_x$ performance in the standard SCR reaction. If the reaction temperature is lower than 350 °C, the order of the NO conversion of the Cu based zeolites is Cu-ZSM-5 > Cu-SSZ-13 > Cu-Beta, which suggests that Cu-SSZ-13 and Cu-ZSM-5 catalysts have the superior low-temperature $deNO_x$ activity under the standard SCR conditions. If the reaction temperature is higher than 350 °C, the order of the NO conversion of the Cu based zeolites is Cu-Beta > Cu-SSZ-13 > Cu-ZSM-5. Especially at temperatures higher than 500 °C, the high-temperature SCR performance advantage of the Cu-Beta catalyst is more obvious.

The SCR performance of the Cu-SSZ-13 catalyst before and after hydrothermal aging is shown in Figure 1b. The durability and stability of the catalysts are expressed by the deterioration rate, and this is shown in Figure 1c. In Figure 1c, if the aging temperature is less than 950 °C, the degradation rate of the aged samples does not change significantly with the increase of the aging temperature. After aging at 650 °C, 700 °C and 800 °C, the degradation rates of the samples are 0%, 2% and 1%, respectively. This shows that the Cu-SSZ-13 has a superior hydrothermal stability and durability. If the aging temperature continues to rise to 950 °C, the degradation rate of the aged sample can reach 23.5%. Especially at the reaction temperature higher than 400 °C, the reaction activity of the aged sample is obviously reduced, and the NO conversion at the reaction temperature of 550 °C and above is reduced to zero. This clearly indicates that the Cu-SSZ-13 catalyst still has good catalytic activity and retains the integrity of CHA skeleton structure after high-temperature hydrothermal aging treatment, but too high aging temperature ($\geq$950°C) will lead to the CHA structure collapse and complete deactivation of the Cu-SSZ-13 catalyst.

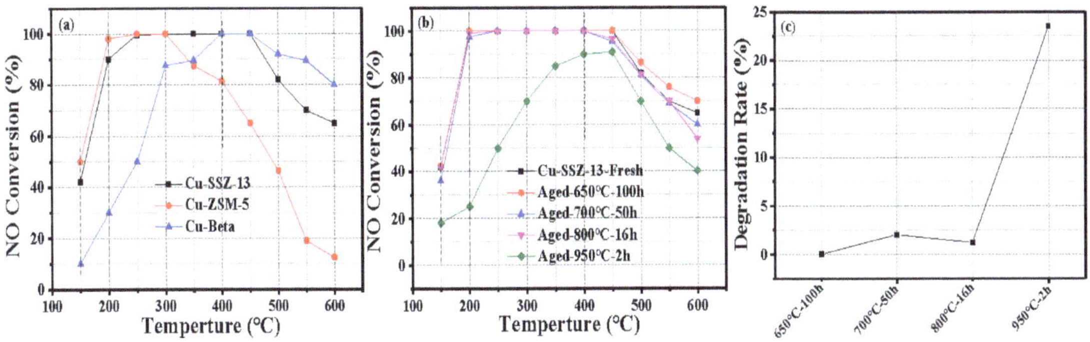

Figure 1. NO conversion as a function of temperature in the standard SCR reaction over Cu-SSZ-13, Cu-ZSM-5 and Cu-Beta catalysts: (**a**) fresh activity of the Cu based zeolites; (**b**) NO conversion of the Cu-SSZ-13 before and after hydrothermal aging; and (**c**) deterioration rate of the Cu-SSZ-13.

2.2. Composition and Textural Properties of the Catalysts

The XRD patterns of Cu-SSZ-13, Cu-ZSM-5 and Cu-Beta catalysts before and after copper ion exchange are presented in Figure 2. The chemical composition of the copper-based zeolites is listed in Table 1. As shown in Table 1, the Cu/Al ratios of the Cu-SSZ-13, Cu-ZSM-5 and Cu-Beta catalysts are 0.40, 0.38 and 0.40, and their corresponding copper ion exchange degrees are 79%, 75% and 80%, respectively. This indicates that the copper ion exchange degree of Cu-SSZ-13, Cu-ZSM-5 and Cu-Beta catalysts prepared by the high-temperature one-time ion exchange method is basically the same, and the Cu loading amounts are 2.4%, 2.3% and 2.5%, respectively. It can be seen from Figure 2a,b that the powder XRD patterns of these catalysts with different zeolite support structures are in excellent agreement with those of zeolite supports (i.e., H-SSZ-13, H-ZSM-5, H-Beta) and no diffraction peak of crystalline phase CuO is detected in all samples (2θ = 35.6°and 38.7°), except for minor differences in the relative X-ray peak intensity. The results show that they maintain structural integrity during the copper ion exchange. The Cu species can stably exist in the pore structure system of zeolite supports, and the crystal size of Cu-SSZ-13, Cu-ZSM-5 and Cu-Beta catalysts decreased gradually. The XRD patterns of the copper-based zeolites after the NH_3-SCR reaction are presented in Figure 2c. Compared with fresh samples (Figure 2b), after the NH_3-SCR reaction, the CHA and BEA structures in Cu-SSZ-13 and Cu-Beta catalysts still exist stably and the peak intensity is obviously enhanced, while the peak intensity of Cu-ZSM-5 is evidently reduced, and its crystal form has a tendency to develop into amorphous state, which may be caused by the collapse of the MFI structure of Cu-ZSM-5, and thus affect its high-temperature SCR performance.

Table 1. Chemical composition of the copper-based zeolites.

Catalyst ID	Cu [1] (wt %)	Si/Al [1]	Cu/Al [1]	Copper Ion Exchange Degree [1] (%)	BET Surface Area ($m^2 \cdot g^{-1}$)
Cu-SSZ-13	2.4	14	0.4	79%	767.3
Cu-ZSM-5	2.3	14	0.4	75%	438.2
Cu-Beta	2.5	14	0.4	80%	702.7

[1] Determined by elemental analysis.

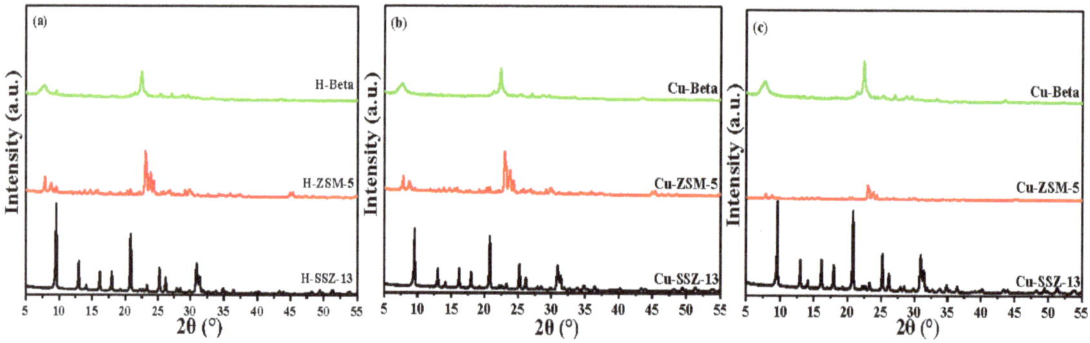

Figure 2. XRD patterns of zeolite supports (**a**), copper-based catalysts (**b**) and copper-based zeolites after the NH_3-SCR reaction (**c**).

The specific surface area and pore structure parameters of the catalysts with different zeolite support structures are shown in Figure 3a. It can be seen from Figure 4a that the specific surface area and the total pore volume of Cu-SSZ-13, Cu-ZSM-5 and Cu-Beta catalysts are 767.30 m^2/g, 0.2835 cm^3/g, 438.15 m^2/g, 0.1901 cm^3/g and 702.68 m^2/g, 0.4176 cm^3/g, respectively. It is clear that the Cu-SSZ-13 catalyst has the largest specific surface area, while the total pore volume of the Cu-Beta catalyst is relatively larger than Cu-SSZ-13 and Cu-ZSM-5. To some extent, this shows that the larger the specific surface area and the pore volume, the more favorable the contact between the active center and the reactant molecules, which results in an improved catalytic activity [15]. Figure 3b shows the proportion of micropores and mesopores in the Cu based zeolites. As shown in Figure 3b, the order of proportion of microporous volume to total pore volume of the catalysts is Cu-SSZ-13 (94%) > Cu-ZSM-5 (66%) > Cu Beta (49%), indicating that the Cu-SSZ-13 catalyst is a typical microporous material, while Cu-Beta and Cu-ZSM-5 are micro-mesoporous composite materials [16,17], which is consistent with the results shown in Figure S1.

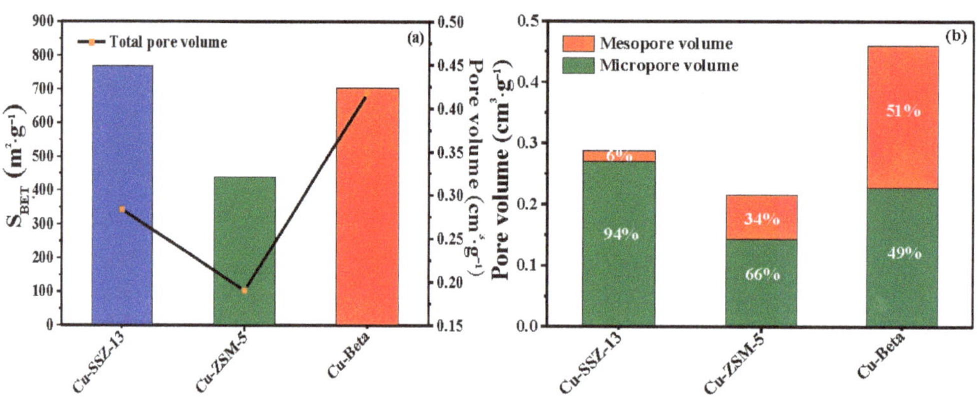

Figure 3. Specific surface area and pore structure parameters (**a**) and proportion of micropores and mesopores (**b**) of the copper-based zeolites.

The SEM images of the catalysts are exhibited in Figure 4. As shown in Figure 4a–c, the crystal morphology of the Cu-SSZ-13 catalyst is a regular cubic structure, and the crystal surface is flat and smooth, while Cu-ZSM-5 and Cu-Beta exhibits irregular aggregates and particles adhere to each other. Through a comparison of appearance and particle size test of these catalysts, it is found that the Cu-SSZ-13 sample has a large particle size and excellent mobility, while the Cu-Beta has a relatively small particle size and poor mobility, and the powder particles are relatively loose and have large electrostatic attraction, which is consistent with the XRD results. In addition, in the ^{27}Al-NMR signal, the signal peak of non-skeleton aluminum species is not found in Cu-SSZ-13, Cu-ZSM-5 and Cu-Beta catalysts, indicating that all Al atoms in these catalysts exist in the form of skeleton aluminum and the silicon (aluminum) oxygen skeleton structure is relatively stable (Figure S2).

Figure 4. SEM images of Cu-SSZ-13 (**a**), Cu-ZSM-5 (**b**) and Cu-Beta (**c**).

2.3. *Nature of Active Sites*

EPR is an effective method of characterization for identifying copper species in the copper-based zeolites. Cu^{2+} ions with one unpaired electron have a strong EPR response, while it should be noted that the Cu species such as Cu^{+}, $[Cu(OH)^{+}]$ and CuO_X clusters are all EPR-silent [13]. Consequently, the EPR signal of the Cu based zeolites can be attributed to the isolated Cu^{2+}. The isolated Cu^{2+} is usually considered as the active site of a NH_3-SCR reaction, and its content often determines the activity of the catalysts [18]. To compare the difference in the amount of isolated Cu^{2+} ions in Cu-SSZ-13, Cu-ZSM-5 and Cu-Beta catalysts, all samples were characterized by EPR. Figure 5a presents the EPR spectra for the Cu based zeolites recorded at 110 K. As shown in Figure 5a, Cu-SSZ-13, Cu-ZSM-5and Cu-Beta catalysts all have signal peaks of isolated Cu^{2+} ions at $g_{\perp} = 2.06$ and $g_{||} = 2.39$ (or $g_{||} = 2.38$), while the negative peaks at high field ($g_{\perp} = 2.06$) reflect the axial symmetry of the divalent copper ions [19]. The relative intensity of isolated Cu^{2+}ions is obtained by the double integration of the EPR spectra of all samples, and the results are shown in Table 2. The order of the amount of Cu^{2+} ions in the Cu based zeolites is Cu-ZSM-5 $\approx$ Cu-SSZ-13 > Cu-Beta. It can be seen that the amount of isolated Cu^{2+} in Cu-SSZ-13 and Cu-ZSM-5 catalysts is basically the same, but they are obviously higher than Cu-Beta. The lack of enough isolated Cu^{2+} may also be an important reason why the Cu-Beta catalyst exhibits poor NH_3-SCR performance in the low temperature range (150~250 °C). Figure 5b displays the hyperfine structures of copper ions for all samples at low field. The EPR characteristic signal peaks for the Cu based zeolites at $g_{||} = 2.39$ (or $g_{||} = 2.38$) should be attributed to Cu^{2+} located in the cationic sites around 6-rings, which is consistent with the reports in the literature [20].

Table 2. EPR signal relative intensity for the Cu based zeolites.

Sample	Relative Intensity (%)
Cu-SSZ-13	98
Cu-ZSM-5	100 [1]
Cu-Beta	43

[1] The intensity of the Cu-ZSM-5 sample was defined to be 100%.

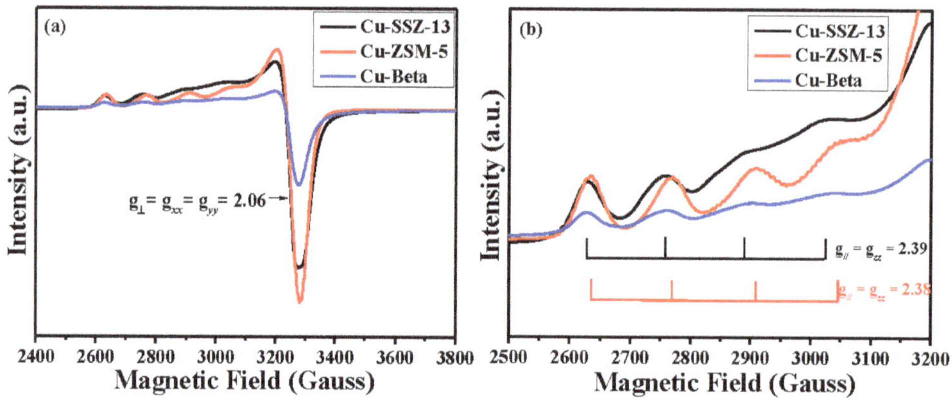

Figure 5. EPR spectra recorded at 110 K (**a**) and magnification of the low-field hyperfine structure (**b**) for the copper-based zeolites.

To further understand the copper species status on the surface of the copper-based zeolites, the catalysts' surface were analyzed by X-ray photoelectron spectroscopy (XPS). The XPS spectra reveal that different Cu species existed on the surface of the copper-based zeolites associated with Cu contents. The XPS spectra for Cu 2p are presented in Figure 6. In Figure 6, the XPS spectra of Cu 2p in Cu-SSZ-13, Cu-ZSM-5 and Cu-Beta catalysts have two obvious peaks concentrated at ~952.8 eV and ~933.2 eV, which belong to Cu $2p_{1/2}$ and Cu $2p_{3/2}$, respectively. Through Gaussian fitting, the XPS spectra of Cu 2p of all samples can be fitted with four peaks. The peaks at 953.8~954.4 and 933.9~934.1 eV are attributed to Cu^{2+}, the peaks at 952.6~952.9 and 932.9~933.2 eV are assigned to Cu^{+} [21,22]. Table 3 reveals the relative content of Cu^{+} and Cu^{2+} ions on the catalysts' surface. The ratio of $Cu^{2+}/(Cu^{+} + Cu^{2+})$ was calculated according to the deconvoluted peak areas and used to measure the concentration of the Cu^{2+} species on the surface of the catalyst [23]. According to the area ratio of the Cu peaks, the surface Cu^{2+} concentrations of Cu-SSZ-13, Cu-ZSM-5 and Cu-Beta catalysts were 30%, 29% and 23%, respectively. The indicates that the concentration of Cu^{2+} ions on the surface of Cu-SSZ-13 and Cu-ZSM-5 is significantly higher than Cu-Beta, which may also be an important reason for the better low-temperature SCR activity of Cu-SSZ-13 and Cu-ZSM-5 catalysts. In addition, the ratio of $Cu^{+}/(Cu^{+} + Cu^{2+})$ on the surface of Cu-SSZ-13, Cu-ZSM-5 and Cu-Beta catalysts can all reach more than 70%, so the copper-based zeolites' surface is mainly Cu^{+} ions. Luo et al. [24,25] believed that more Cu^{+} ions may increase the amount of oxygen vacancies on the catalysts' surface, which is conducive to the adsorption, activation and migration of oxygen in the gas phase.

Table 3. The integral areas of Cu 2p determined from XPS spectra over Cu-SSZ-13, Cu-ZSM-5 and Cu-Beta catalysts.

Sample	Peak Area (a.u.)		$Cu^{2+}/Cu^{+} + Cu^{2+}$ (%)	$Cu^{+}/Cu^{+} + Cu^{2+}$ (%)
	Cu^{2+}	Cu^{+}		
Cu-SSZ-13	10,355	24,187	30	70
Cu-ZSM-5	8010	19,652	29	71
Cu-Beta	6710	22,718	23	77

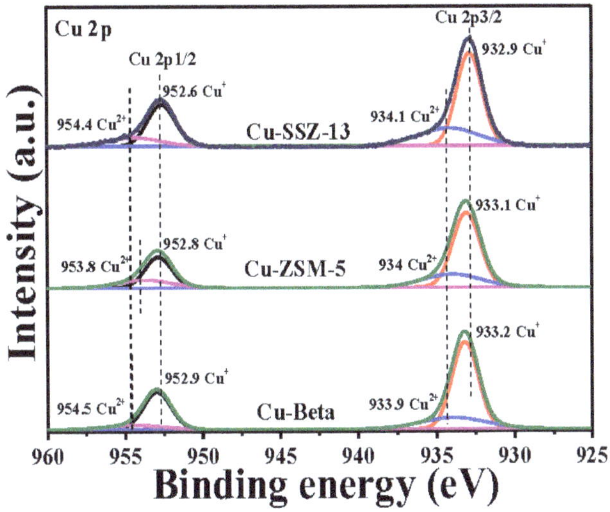

Figure 6. XPS spectra of Cu 2p for the copper-based zeolites.

It is generally accepted that the redox properties are an important factor affecting the NH_3-SCR reaction. The H_2-TPR patterns of the copper-based zeolites are shown in Figure 7a. As shown in Figure 7a, the H_2 consumption profiles of Cu-SSZ-13 and Cu-Beta catalysts can be divided into four reduction peaks by Gaussian fitting, while only three reduction peaks can be fitted for the Cu-ZSM-5 catalyst. According to the literature [26,27], the reduction of isolated Cu^{2+} ions located at the ion exchange sites often need to go through two steps. First, the isolated Cu^{2+} ions are reduced to Cu^{+} ions at a lower temperature, and then the Cu^{+} ions are completely reduced to Cu^{0} at a higher temperature. These Cu^{+} ions include the intermediate products of the two-step reduction of isolated Cu^{2+} ions and the Cu^{+} ions existing in the catalyst. Therefore, the low-temperature reduction peak in the temperature region 255~333°C can be due to the reduction of the isolated Cu^{2+} to Cu^{+} [28,29]. The reduction peak of 356~489 °C is attributed to the reduction of the unstable Cu^{+} to Cu^{0}. The higher-temperature reduction peak at 549~665 °C indicates the reduction of the highly stable Cu^{+} to Cu^{0}. It can be seen from the above analysis that the reduction temperature of the isolated Cu^{2+} ions in Cu-SSZ-13 catalyst (255 °C) is significantly lower than that of Cu-ZSM-5 (275 °C) and Cu-Beta (333 °C) catalysts, indicating that the reduction of Cu^{2+} to Cu^{+} in Cu-SSZ-13 is easier. According to Figure 7b, the total H_2 consumption of the Cu based zeolites is Cu-SSZ-13 > Cu-ZSM-5 > Cu-Beta, which shows that the Cu-SSZ-13 catalyst has the strongest redox properties among them, especially its low-temperature reduction capacity.

Figure 7c illustrates the proportion of different copper species in the copper-based zeolites. The ratio of various Cu ions to total Cu was calculated according to the H_2 reduction peak areas and used to measure the concentration of Cu ions in the catalysts. It can be seen from the Figure 7c that the proportion of the stable Cu^{+} to Cu in the catalysts is Cu-SSZ-13 (65%) > Cu-Beta (42%) > Cu-ZSM-5 (8%), and the reduction temperature of the stable Cu^{+} in Cu-SSZ-13 (573~665 °C) and Cu-Beta (614 °C) catalysts is significantly higher than Cu-ZSM-5 (549 °C). This indicates that the high-temperature stability of Cu species in Cu-SSZ-13 and Cu-Beta zeolite frameworks is significantly better than Cu-ZSM-5. It is worth noting that the reduction temperature of the highly stable Cu^{+} ions is generally considered as the temperature at which the framework structure of zeolite begins to collapse [30,31]. Consequently, this may also be the cause of the poor high-temperature SCR performance of the Cu-ZSM-5 catalyst.

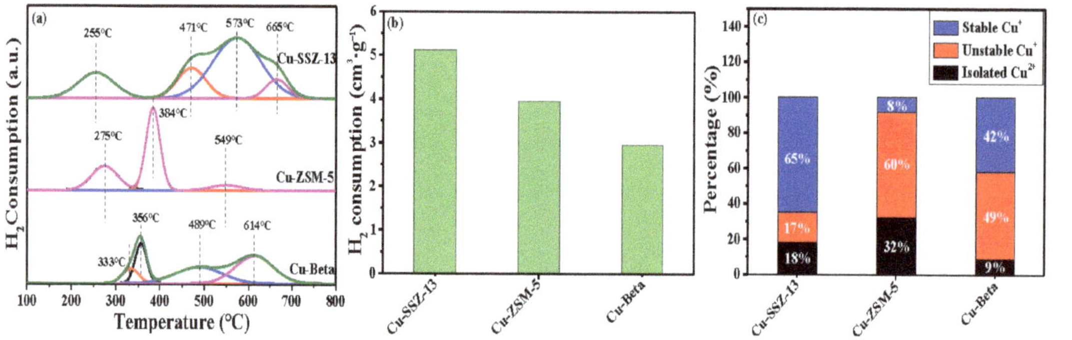

Figure 7. H_2-TPR patterns (**a**), H_2 consumption (**b**) and proportion of different Cu species (**c**) for the copper-based zeolites.

To further understand the local environment of the Cu^{2+} species in the copper-based zeolites, the DRIFTS-NO adsorption experiments were used to explore the nature of the Cu species in the catalysts [32]. The Cu-SSZ-13, Cu-ZSM-5 and Cu-Beta catalysts were exposed to 500 ppm NO in N_2 at room temperature for 60 min followed by purging with N_2 for 30 min. The NO adsorption properties were obtained by recording the DRIFTS-NO spectra of the catalysts. The DRIFTS spectra of the adsorbed NO on the Cu based zeolites are shown in Figure 8. The absorption peaks at 1950, 1930, 1895 and 1826 cm^{-1} for the catalysts are seen in the spectra. The absorption peak characteristics of Cu-SSZ-13 (1950, 1930 and 1895 cm^{-1}), Cu-ZSM-5 (1950 and 1826 cm^{-1}) and Cu-Beta (1930 cm^{-1}) catalysts are slightly distinct, expressing that there are differences in the location and distribution of copper ions among the three catalytic systems. For the Cu-SSZ-13 catalyst, the peaks at 1950, 1930 and 1895 cm^{-1} correspond to the NO species adsorbed on the two different types of Cu^{2+} ion active sites [33]. The peak at 1950 cm^{-1} is due to the adsorption of NO on the isolated Cu^{2+} ions in the D6R, while the peaks at 1930 and 1895 cm^{-1} are attributed to the NO species adsorbed on the isolated Cu^{2+} ions in the 8-rings window of the CHA cage [34]. Compared with Cu-SSZ-13 catalyst, the Cu-ZSM-5 catalyst appears to show a new absorption peak at 1826 cm^{-1} in addition to a peak at 1930 cm^{-1}, which is attributed to the NO species adsorbed on Cu^+ ions [35], while the peak of Cu-ZSM-5 and Cu-Beta at 1930 cm^{-1} represents the interaction between NO and isolated Cu^{2+} [36]. Therefore, it can be seen from the above analysis that NO is mainly adsorbed on isolated Cu^{2+}, thus forming the Cu^{2+}-NO active species. In addition, the order of the NO absorption peak intensity in the 1800–2000 cm^{-1} region is Cu-ZSM-5 $\approx$ Cu-SSZ-13 > Cu-Beta, proclaiming that the adsorption capacity of Cu-ZSM-5 and Cu-SSZ-13 catalysts for NO is significantly better than Cu-Beta. According to the EPR analysis results, the amount of isolated Cu^{2+}ions in Cu-ZSM-5 and Cu-SSZ-13 catalysts is more than Cu-Beta, which is also an important reason for their strong NO adsorption capacity.

In situ DRIFTS-NH_3 adsorption experiments were carried out to examine the NH_3 adsorption properties of the catalysts [37]. The Cu-SSZ-13, Cu-ZSM-5 and Cu-Beta catalysts were exposed to 500 ppm NH_3 in N_2 at room temperature for 60 min followed by purging with N_2 for 30 min. The NH_3 adsorption properties were obtained by recording the DRIFTS-NH_3 spectra of the catalysts. Figure 9 displays the difference DRIFT spectra in the zeolite framework region of the copper-based catalysts with different zeolite support structures. In the N-H bonds stretching region, there are three strong absorption peaks related to NH_3 adsorption at 3000–3400 cm^{-1}, which are the characteristic peak of NH_4^+ ions (3350 and 3275 cm^{-1}) and coordinated ammonia (3180 cm^{-1}) [38]. In the N-H bonds bending vibration region, 1625 and 1255 cm^{-1} can be assigned to the NH_3 coordinated to the Lewis acid sites, while 1480 cm^{-1} is due to the bending vibration of NH_4^+ ions on Brønsted acid sites [39]. Hence, the ammonia adsorbed on Brønsted acid sites were much stronger than that on Lewis acid sites. In addition, there are two negative bands in the 3500–3800 cm^{-1} and 800–1000 cm^{-1} regions. According to the literature report [40], 3780, 3675, 3625 and 3540 cm^{-1} are the characteristic peaks of the O-H bonds. Among them, 3780, 3675 and 3625 cm^{-1} are caused by NH_3 adsorbed on Si-OH, Cu-OH and Al-OH, respectively [41–43], and 3540 cm^{-1} corresponds to the bending vibration of the Si-OH-Al bond, which is a feature of Brønsted acid sites on the zeolite skeleton [44]. In the 800–1000 cm^{-1} region, it represents the characteristic peaks on the copper ion sites. The Cu-O bond formation in copper-exchanged zeolites can perturb the T-O-T vibrations of the zeolite support framework, thus being useful in the identification of different types of intrazeolitic Cu sites [45,46]. KWAK et al. [47] believe that NH_3 can be adsorbed on two types of copper ions in Cu-SSZ-13, corresponding to two characteristic peaks of 940 and 875 cm^{-1}, respectively. The first 940 cm^{-1} is attributed to the NH_3 adsorbed on the isolated copper ions in the CHA cage, while 875 cm^{-1} is due to the adsorption of NH_3 on the copper ions in the 6-MR. It is well known that the Brønsted acid sites are substances that can give protons (H^+), while the Lewis acid sites can accept electron pairs. Consequently, compared to Cu-SSZ-13, similar features of ammonia adsorption are also observed during the adsorption process on Cu-ZSM-5 and Cu-Beta. Four Bønsted acid sites exist on the zeolite catalysts' surface, namely Si-OH, Cu-OH, Al-OH and Al-OH-Si, and the Lewis acid sites are mainly two types of Cu exchange sites, which are isolated copper ions in CHA cage and 6-MR.

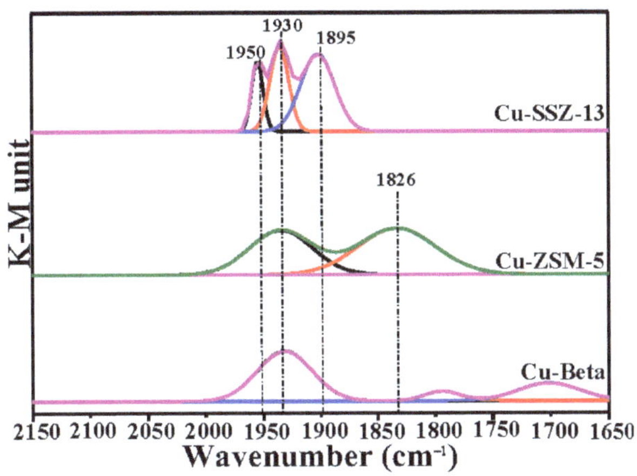

Figure 8. DRIFT Spectra of NO adsorption on the copper-based zeolites (Prior to NO adsorption, each catalyst was oxidized at 500 °C for 60 min).

Surface acidity is one of the most important parameters that evaluates the extent of NO reduction with NH_3 over copper-based zeolites, and it is usually used to determine the acid sites of the zeolite material for the adsorption and activation of NH_3, and the type and quantity of the acid sites will affect the catalytic activity [48,49]. Figure 10a shows the NH_3-TPD profiles of the Cu based zeolites. Four NH_3 desorption peaks appear for the Cu based zeolites: the low-temperature A and B peaks at 126~196 °C, the medium-temperature C peak at about 370 °C and the high-temperature D peak above 500 °C, which correspond to the four types of acid sites in the zeolite catalysts [50,51]. Among them, the desorption peaks (A and B peaks) before 200 °C are attributed to the weak adsorption of NH_3, mainly including physical adsorbed NH_3 (A peak) and NH_3 chemically adsorbed on the weak Lewis acid sites (B peak). The C peak at about 370 °C is assigned to the NH_3 adsorbed on the strong Lewis acid sites, while the D peak above 500 °C is related to NH_3 adsorbed on the Brønsted acid sites. According to the literature report [52] and NH_3-DRIFT analysis results, four Brønsted acid sites exist on the zeolite catalysts' surface, namely Si-OH, Cu-OH,

Al-OH and Al-OH-Si, and the Lewis acid sites are mainly two types of Cu ion exchange sites, which are isolated copper ions in CHA cage and 6-MR.

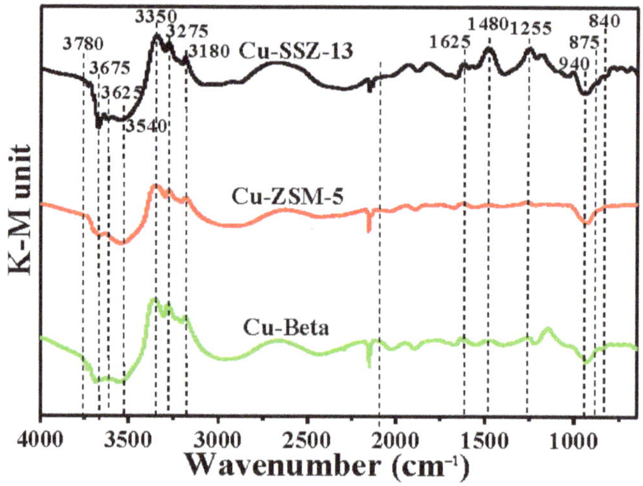

Figure 9. DRIFTS Spectra of NH_3 adsorption on the copper-based zeolites (Prior to NH_3 adsorption, each catalyst was oxidized at 500 °C for 60 min).

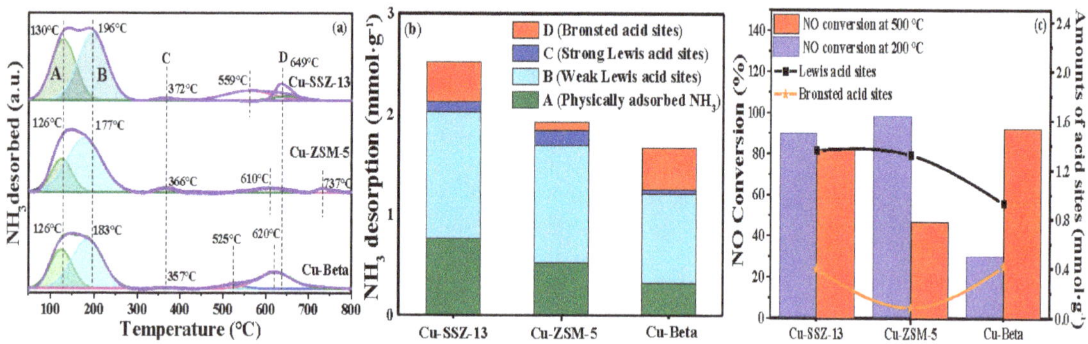

Figure 10. NH_3-TPD patterns (**a**), amount of NH_3 desorption (**b**) and relationship between acid sites and catalytic activity (**c**) for the copper-based zeolites.

After quantitative calculation of the peak separation fitting, the acid site distribution and the acid amount (desorption amount of NH_3) in the Cu based zeolites are shown in Figure 10b. The order of total acidity of the Cu based zeolites is Cu-SSZ-13 (2.61 mmol·g^{-1}) > Cu-ZSM-5 (2.13 mmol·g^{-1}) > Cu-Beta (2.05 mmol·g^{-1}), indicating that more acid sites can be generated on the Cu-SSZ-13' surface to adsorb NH_3. Moreover, the amount of Lewis acid sites in Cu-SSZ-13 (1.36 mmol·g^{-1}) and Cu-ZSM-5 (1.32 mmol·g^{-1}) is clearly higher than in Cu-Beta (0.93 mmol·g^{-1}), illustrating that the amount of isolated Cu^{2+} ions in Cu-Beta is the least, which has been confirmed in the EPR results. For the Brønsted acid site, if the reaction temperature exceeds 500 °C, the NH_3 adsorbed on the acid site can be released. Thus, the more Brønsted acid sites in the catalysts, the stronger its storage capacity for NH_3. Compared with Cu-SSZ-13 (0.40 mmol·g^{-1}) and Cu-Beta (0.42 mmol·g^{-1}), the amount of Brønsted acid sites in the Cu-ZSM-5 (0.08 mmol·g^{-1}) is the lowest, explaining that the Cu-ZSM-5 has the weakest NH_3 storage capacity at high temperature. Figure 10c reveals the relationship between acid sites and catalytic activity. It can be seen from the figure that the amount of Lewis acid sites in the zeolite catalysts is in the same order as the low-temperature SCR activity (200 °C), while the amount of Brønsted acid sites is related to the high-temperature SCR activity (500 °C). Consequently, the strong Lewis acid sites may be beneficial to the low-temperature SCR reaction, while the Brønsted acid sites perhaps promote the high-temperature SCR reaction. The Cu-SZZ-13 contains enough Lewis and Brønsted acid sites, so it shows the best catalytic activity.

2.4. Reaction Intermediates

To more completely understand the differences in the catalytic properties of Cu-SSZ-13, Cu-ZSM-5 and Cu-Beta catalysts for NH_3-SCR, the formation of reaction intermediates over these catalysts during exposure to a mixture of 500 ppm NO and 10% O_2 (N_2 as balance gas) at 150 °C was studied by in situ DRIFT spectra. As shown in Figure 11a–c, when the Cu-SSZ-13 catalyst is exposed to the mixed gas of NO and O_2 for up to 30 min, there is a continuous increase in the intensity of the absorption peak in the 1250–1750 cm^{-1} region, which is resolved into three peaks at 1625, 1575 and 1500 cm^{-1}. These peaks can be attributed to nitrate ions (NO_3^-) gradually generated on the Cu sites with increasing exposure time to 30 min, while the peak round 1575 cm^{-1} can be assigned to bidentate nitrate [53], and the amount of nitrate ion will increase with the increase of exposure time. A similar trend can be observed in the in situ DRIFT spectra of Cu-ZSM-5 and Cu-Beta catalysts compared to the Cu-SSZ-13, except for the latter two only appear two peaks around 1625 and 1575 cm^{-1}. According to DRIFTS-NO adsorption results, NO is mainly adsorbed on isolated Cu^{2+} ions and forms the Cu^{2+}-NO active species. Hence, it can be considered that NO_3^- is also mainly formed on the active Cu^{2+} ion sites. The intensity sequence of NO_3^- absorption peaks in the 1250–1750 cm^{-1} region is Cu-ZSM-5 > Cu-SSZ-13 > Cu-Beta, indicating that more NO_3^- will be formed on the Cu-ZSM-5′ surface, which may be caused by more isolated Cu^{2+} contained in the Cu-ZSM-5 catalyst. Therefore, NO_3^- is known as one of the key reaction intermediates for the NH_3-SCR reaction over metal-exchanged zeolite catalysts.

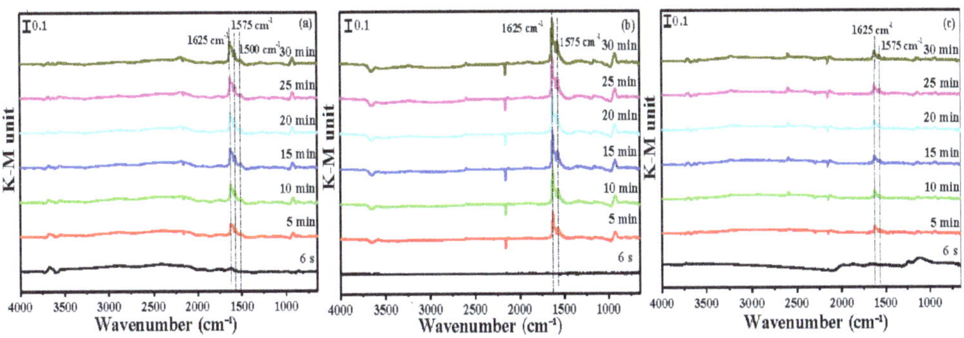

Figure 11. In situ DRIFT spectra of NO + O_2 adsorption on the copper-based zeolites at 150 °C: (**a**) Cu-SSZ-13; (**b**) Cu-ZSM-5 and (**c**) Cu-Beta.

The mixture of NO + O_2 is pre-adsorbed on the copper-based zeolites' surface at 150 °C for 30 min to generate NO_3^- followed by turning off NO + O_2 and purging with N_2 for 30 min. Figure 12 shows the in situ DRIFT spectra of NH_3 reacted with pre-adsorbed NO + O_2 on the copper-based zeolites at 150 °C for 30 min. The DRIFT spectra with a reaction time of 0 min is recorded after exposure to a mixture of 500 ppm NO and 10% O_2 (N_2 as balance gas) at 150 °C for 30 min followed by purging with N_2 for 30 min. It is easy to see in Figures 11 and 12 that no noticeable changes in the DRIFT spectra of the catalysts were caused by N_2 purging for 30 min after NO_3^- formation. The intensity of the NO_3^- absorption peak of the catalysts in the 1250–1750 cm^{-1} region will gradually weaken with the increase of exposure time in NH_3 until it disappears. Among them, the NO_3^- absorption peaks of

Cu-SSZ-13 and Cu-ZSM-5 catalysts at 1625 and 1500 cm^{-1} disappear after 20 min exposure in NH_3, while the NO_3^- absorption peaks of Cu-Beta catalyst disappears completely after 15 min. This indicates that the rate of NO_3^- disappearance is much faster in Cu-Beta than in Cu-SSZ-13 and Cu-ZSM-5. Simultaneously, two new peaks appear in Cu-SSZ-13 and Cu-ZSM-5 catalysts around 1620 and 1470 cm^{-1}, while the Cu-Beta catalyst only has a weak absorption peak at 1620 cm^{-1}. These two peaks are attributed to the adsorption of NH_3 onto the strong Lewis acid sites generated by the exchange of isolated Cu^{2+} ions and the Brønsted acid sites generated by zeolite protons, respectively [54]. This distinctly shows that the NO_3^- species produced in Cu-SSZ-13, Cu-ZSM-5 and Cu-Beta catalysts can be reduced by NH_3 [55]. It is worth noting that the disappearance rate of the reaction intermediates NO_3^- in the Cu based zeolites is Cu-Beta > Cu-SSZ-13 > Cu-ZSM-5, which may be due to the fact that the concentration of the NO_3^- ions generated in the Cu-Beta catalyst is less than in Cu-SSZ-13 and Cu-ZSM-5, rather than its higher reactivity with NH_3, which has been well confirmed in Figure 11. Consequently, the most likely reason is that the amount of isolated Cu^{2+} ions in the Cu-Beta catalyst is the least compared to Cu-SSZ-13 and Cu-ZSM-5 catalysts, which leads to the weakest adsorption capacity of Cu-Beta for NO, thus affecting the amount of the Cu^{2+}-NO active species, and ultimately limiting the formation of the NO_3^- species that play an important role in the low-temperature SCR reaction. This well explains that the standard SCR activity of the Cu-Beta catalyst at low temperature ($\leq$300 °C) is significantly lower than Cu-SSZ-13 and Cu-ZSM-5.

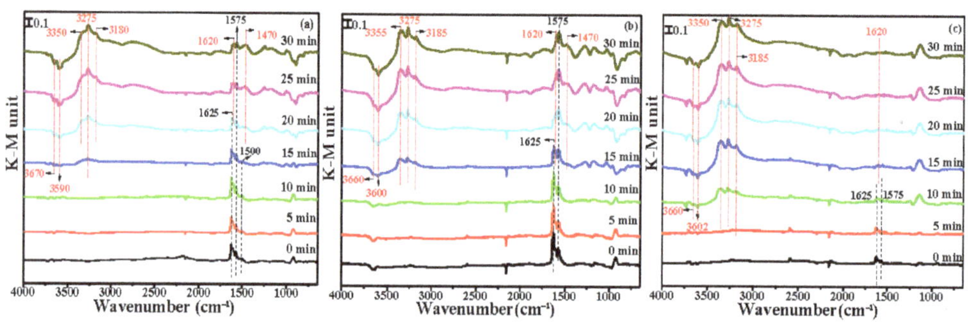

Figure 12. In situ DRIFT spectra of NH_3 reacted with pre-adsorbed NO + O_2 on the copper-based zeolites at 150 °C for 30 min: (**a**) Cu-SSZ-13; (**b**) Cu-ZSM-5 and (**c**) Cu-Beta.

In addition, it can be observed from Figure 12 that the Cu-SSZ-13, Cu-ZSM-5 and Cu-Beta catalysts appear as two negative absorption peaks in the 3500–3700 cm^{-1} region and as three positive absorption peaks with high peak intensity in the 3000–3400 cm^{-1} region after exposure to NH_3 for 10 min. According to DRIFTS-NH_3 analysis results, the two negative absorption peaks at 3500–3700 cm^{-1} can be attributed to the weak NH_3 adsorption of OH bands on the Cu-based zeolites' surface, while the three positive absorption peaks at 3000–3400 cm^{-1} are ascribed to the vibration of the N-H bonds. Therefore, it can be seen from the above analysis that the reaction intermediates NO_3^- produced in the Cu-SSZ-13, Cu-ZSM-5 and Cu-Beta catalysts are reduced by NH_3 until it disappears, the excess NH_3 mainly exists on the catalysts' surface in the form of the weak NH_3 adsorption at low temperature.

2.5. *Effect of Copper and Acid Sites on NH_3-SCR Performance*

From the above analysis results, it can be seen that different zeolite skeleton structures can affect the amount of isolated Cu^{2+} ions, the distribution of Cu species, the surface acidity and the amount of reaction intermediates in the catalysts, thus determining the catalytic activity and hydrothermal stability of the copper-based zeolites. First of all, Cu-SSZ-13 and Cu-ZSM-5 have similar catalytic activities in the low temperature region (150~300 °C), and their NO conversion is obviously better than Cu-Beta in this region. However, the

Cu-ZSM-5 has the lowest catalytic activity in the high temperature region (400~600 °C), especially as the reaction temperature reaches 600 °C, its NO conversion can be reduced to below 15%. Secondly, the chemical state, distribution and amount of the Cu species are one of the important reasons for the difference of NH_3-SCR performance of the zeolite catalysts. From the XPS and H_2-TPR results, it can be found that there are two types of copper cations in these catalysts: isolated Cu^{2+} and Cu^+, and the latter one mainly locates on the catalyst ($Cu^+/Cu > 70\%$). Although the content of the isolated Cu^{2+} ions in the copper-based zeolites is small, the reduction temperature of Cu^{2+} ions is significantly lower than Cu^+ ions, indicating that Cu^{2+} ions are easier to reduce than Cu^+ ions. Therefore, the isolated Cu^{2+} ions are generally considered as the active sites of the NH_3-SCR reaction, and their content often has a significant impact on the low-temperature SCR activity of the catalyst [18]. Moreover, the Cu content in the Cu-SSZ-13, Cu-ZSM-5 and Cu-Beta catalysts is basically the same (2.3~2.5%), but the EPR and XPS results show that the amount of isolated Cu^{2+} ions on the skeleton or surface of Cu-Beta is the least, suggesting that the microporous structure (i.e., pore size) of the copper-based zeolites has an impact on the amount of isolated Cu^{2+} ions in the catalyst. Consequently, the lack of enough isolated Cu^{2+} ions is one of the reasons for the poor SCR activity of Cu-Beta in the low temperature region (150~300 °C). For Cu^+ ions, it can be found from the H_2-TPR results that there are highly stable Cu^+ ions that can be reduced at high temperatures (>500 °C) and unstable Cu^+ ions with relatively low reduction temperatures (<500°C) in Cu-zeolites. Gao et al. [30,31] believed that the reduction temperature of highly stable Cu^+ ions is usually the temperature at which the framework structure of zeolite begins to collapse. According to the results in Figure 7c, the proportion of stable Cu^+ to Cu in Cu-ZSM-5 (8%) is significantly lower than Cu-SSZ-13 (65%) and Cu Beta (42%). Hence, this may also be one of the reasons for the poor high-temperature SCR performance of Cu-ZSM-5. Thirdly, the type and number of acid sites are also the key factors affecting the catalytic activity. It can be seen from the NH_3-TPD results that there are mainly two types of acid sites on the surface of the copper-based zeolites, namely Lewis and Brønsted acid sites. The total acidity of the Cu-SSZ-13 catalyst is significantly higher than Cu-ZSM-5 and Cu-Beta, and the differences in the physical adsorbed ammonia related to the pore structure of these three catalysts is the main reason for the difference in the total acidity. Compared with Cu-SSZ-13 and Cu-ZSM-5, Cu-Beta has the least number of the Lewis acid sites generated by isolated Cu^{2+} ions, which further proves that the amount of isolated Cu^{2+} ions in Cu-Beta is the least, thus affecting its low-temperature SCR performance. For the Brønsted acid site, Wang et al. [56] found that if the reaction temperature exceeds 500 °C, the NH_3 adsorbed on the acid site can be released, which plays an important role in storing and providing NH_3 reaction molecules to the active Cu sites in the NH_3-SCR reaction. According to Figure 10b, the amount of Brønsted acid sites in Cu-ZSM-5 is apparently lower than Cu-SSZ-13 and Cu-Beta. Therefore, the lack of sufficient Brønsted acid sites may also be the cause of poor high-temperature SCR performance of Cu-ZSM-5 [57,58]. Finally, NO_3^- is known as one of the key reaction intermediates for the NH_3-SCR reaction over metal-exchanged zeolite catalysts. According to the in situ DRIFT spectra analysis results, the amount of isolated Cu^{2+} ions in the copper-based zeolites is the basic reason to limit the effective formation of NO_3^- ions, which play an important role in the low-temperature SCR reaction. This well explains that the standard SCR activity of the Cu-Beta catalyst at low temperature (≤300 °C) is clearly lower than Cu-SSZ-13 and Cu-ZSM-5. In conclusion, the amount of isolated Cu^{2+} ions in the catalysts directly determines the formation of Lewis acid sites and reaction intermediate NO_3^- ions, thus affecting the low-temperature SCR performance, while the amount of highly stable Cu^+ ions and Brønsted acid sites is related to the high-temperature SCR performance of the catalysts.

3. Materials and Methods

3.1. Catalyst Preparation

Cupric acetate ($Cu(CH_3COO)_2 \cdot H_2O$) and H-SSZ-13 (Weihai Baidexin New Materials Co., Ltd., Weihai, China), H-ZSM-5 (Tianjin Nanfang Chemical Catalyst Co., Ltd., Tianjin, China) and H-Beta (Tianjin Baotou Steel Rare Earth Research, Tianjin, China) commercial supports with the Si/Al ratio of 14 were used for the experiment. The purity of cupric acetate produced by Tianjin Fengchuan Chemical Reagent Technology Co., Ltd. (Fengchuan, Tianjin, China) was 99%, and the specific surface areas of H-SSZ-13, H-ZSM-5 and H-Beta zeolites were 805.83, 474.04 and 759.91 $m^2 \cdot g^{-1}$, respectively. A certain amount of the hydrogen-type zeolite was slowly added into a 0.01 mol L^{-1} copper acetate solution according to the ratio of solid to liquid of 1:50 (g:mL) during stirring, and then ion exchanged in a water bath at 90 °C for 6 h, filtered, washed until the solution reached neutral, dried overnight at 90 °C and calcined at 550 °C for 8 h to obtain the copper-based catalysts with different zeolite supports. The fresh catalyst was thermally aged in 10% H_2O in air at 650 °C for 100 h, 700 °C for 50 h, 800 °C for 16 h and 950 °C for 2 h, respectively, and labeled as aged-650 °C-100 h, aged-700 °C-50 h, aged-800 °C-16 h and aged-950 °C-2 h. The durability of the catalyst was expressed using the deterioration rate, which is the average value for the decrease in NO conversion for the aged samples at four temperature test points (200, 250, 450 and 500 °C).

3.2. Evaluation of the Catalytic Activity

The activity of 0.06 g of the catalysts was tested in a fixed-bed quartz reactor (inner diameter = 6 mm). The gas mixture simulates a real diesel exhaust, which contains 500 ppm NO, 500 ppm NH_3, 8 vol% CO_2, 10 vol% O_2, 5 vol% H_2O and N_2 as the balance gas. The total flow rate was 300 $mL \cdot min^{-1}$ corresponding to a gas hourly space velocity (GHSV) of 300,000 h^{-1}. The effluent gas, including NO, NO_2 and O_2 was continuously analyzed using an online Fourier infrared analyzer. The results for the steady-state activity were collected after 20 min at each temperature. The NO conversion was calculated as follows:

$$\text{NO conversion} = \frac{[NO]_{inlet} - [NO]_{outlet}}{[NO]_{inlet}} \times 100\%$$

3.3. Characterization of the Catalysts

X-ray powder diffractometry (XRD) analysis was performed on a panalytical X-ray powder diffraction analyzer (PANalytical B.V., Almelo, The Netherlands). A Cu Ka (λ = 1.5406 Å) was used as the radiation source. The test conditions used were a tube current of 40 mA and a tube voltage of 40 kV. During the test, wide-angle XRD scanning was performed at a speed of 2°/min in the range of 2θ = 5–55° with a step size of 0.02°.

The Brunauer—Emmett—Teller analysis (BET) and pore structure distribution were measured at −196 °C on a 3H-2000PM2 physical adsorption instrument manufactured by the Bester company (Beijing, China) by using the nitrogen adsorption–desorption method. The surface area and pore size distribution curve were determined using the density functional theory (DFT) method in the 0–0.3 partial pressure range. Before the test, the sample was desorbed in a vacuum at 200 °C for 3 h.

Microstructures of the catalyst samples were observed with a sigma 500 electron microscope of ZEISS (Carl Zeiss, Oberkochen, Germany). The test conditions used were a voltage of 3 KV and a working distance of 6 mm.

X-ray photoelectron spectroscopy (XPS) analysis was obtained using a Thermo ESCALAB 250Xi spectrometer (ThermoFisher Scientic, Waltham, MA, USA) with Al Ka radiation (1486.6 eV). The binding energy (B.E.) spectrum was calibrated according to the C 1s standard spectrum (B.E. = 284.6 eV). The composition on the surface of the catalyst according to the atomic ratios was calculated, and the Shirley background and Gaussian—Lorentzian methods were used for peak analysis ^{29}Si and ^{27}Al MAS NMR spectra (Bruker, Billerica, MA, USA) were recorded using a 4 mm-diameter zirconia rotor on a Bruker AVANCE III

400 MHz WB operating at 79.5 MHz and 104.3 MHz, respectively. The rotor was spun at 12 kHz for ^{29}Si and ^{27}Al MAS NMR and using single pulse sampling. SiO_2 and $Al(OH)_3$ were used as chemical shift references for ^{27}Al MAS NMR and ^{29}Si MAS NMR, respectively.

Electron paramagnetic resonance (EPR) measurements were obtained at 110 K using a Bruker EMX-10/12-type spectrometer in the X-band (Bruker, Billerica, MA, USA). The dehydrated sample was prepared by placing 50 mg of the fresh sample in a quartz tube and pretreating under dry N_2 at 550 °C for 3 h. The sample was then cooled and sealed for measurement. The choice of a lower measuring temperature (110 K) can effectively avoid signal broadening and loss caused by coupling between copper ions.

H_2 temperature programmed reduction (H_2-TPR) was measured using a Quantachrome: Chem BET chemisorption analyzer (Micromeritics, Norcross, GA, USA) and a thermal conductivity detector (TCD) detector was used. Before the test, 80 mg of the sample was heated from room temperature to 300 °C at a rate of 10 °C· min^{-1} for the drying pretreatment and purged with He gas flow (30 mL/min) for 1 h to remove the impurities adsorbed on the sample. Then the sample was cooled to 30 °C, and the gas flow was switched to 10% H_2/Ar mixture (30 mL/min). After the baseline was stabilized for 0.5 h, the sample was heated from room temperature to 800 °C at a rate of 10 °C·min^{-1}. The signal is recorded through TCD, and the H_2 consumption is calculated from the peak area.

NH_3 temperature programmed desorption (NH_3-TPD) was conducted on a Quantachrome: Chem BET chemisorption analyzer (Micromeritics, Norcross, GA, USA). First, 80 mg of the sample was pretreated at 300 °C in an He flow at 30 mL·min^{-1} for 2 h, then cooled down to 50 °C. It was adsorbed in 5 vol% NH_3/He for 30 min (20 mL·min^{-1}). After that, the NH_3 was turned off, the gas flow was switched back to pure He (20 mL·min^{-1}) again and it was purged for 30 min to remove the physically adsorbed species. Then, the desorption was completed by increasing the temperature from 50 to 800 °C at a rate of 10 °C·min^{-1}. After calibration with standard NH_3, the desorption amount of NH_3 through the peak area was calculated.

The in situ diffuse reflectance infrared transform spectroscopy (in situ DRIFTS) was applied on a Nicolet 6700 FTIR spectrometer (ThermoFisher Scientic, Waltham, MA, USA) with an in situ diffuse reflectance pool and a highly sensitive MCT detector cooled by liquid N_2. Prior to IR measurements, a sample of 10 mg mounted in a ceramic holder was heated at 500°C under 21% O_2 in N_2 flow for 60 min and cooled to the desired temperature. Then, the difference DRIFT spectra of adsorbed NH_3 or NO were obtained after exposure of 500 ppm NH_3 or 500 ppm NO in N_2 flow at room temperature for 60 min followed by purging with N_2 for 30 min, respectively. For in situ DRIFT experiments, the difference DRIFT spectra were recorded as a function of exposure time during the stepwise introduction of reactant gases (e.g., first 500 ppm NO and 10% O_2 and then 500 ppm NH_3) balanced with N_2 at 150 °C, respectively.

4. Conclusions

In this paper, the NH_3-SCR performance differences of Cu-SSZ-13, Cu-ZSM-5 and Cu-Beta catalysts prepared by the ion-exchange method were systematically studied. It is found that there are two copper cations (isolated Cu^{2+} and Cu^+) and two acid sites (Lewis and Brønsted acid sites) in these catalysts, and they play different catalytic roles in NH_3-SCR reaction. The amount of isolated Cu^{2+} ions in the catalysts directly determines the formation of Lewis acid sites and reaction intermediate NO_3^- ions, thus affecting the low-temperature SCR performance; while the amount of highly stable Cu^+ ions and Brønsted acid sites is related to the high-temperature SCR performance of the catalysts. The Cu-SSZ-13 catalyst contains enough isolated Cu^{2+} ions, highly stable Cu^+ ions and Brønsted acid sites, so it has excellent NH_3-SCR performance. However, the Cu-ZSM-5 has the worst catalytic activity in the high temperature region (400–600 °C) due to the lack of sufficient stable Cu^+ ions and Brønsted acid sites. For Cu-Beta, the lack of enough isolated Cu^{2+} ions will lead to a significant reduction in the formation of Lewis acid sites and reaction intermediates NO_3^- ions, making it the worst catalytic activity in the low temperature region (150–300 °C).

Supplementary Materials: The following supporting information can be downloaded at: https://www.mdpi.com/article/10.3390/catal13020301/s1, Figure S1: N_2 adsorption-desorption isotherms of the copper-based zeolites; Figure S2: ^{27}Al (a) and ^{29}Si (b) NMR Spectra of the copper-based zeolites.

Author Contributions: Conceptualization, W.Z.; formal analysis, W.Z.; investigation, W.Z., Y.Z.; resources, M.S.; data curation, W.Z.; writing—original draft preparation, W.Z.; writing—review and editing, X.L.; supervision, X.R.; project administration, X.R.; funding acquisition, M.S. All authors have read and agreed to the published version of the manuscript.

Funding: This research received no external funding.

Data Availability Statement: Data is contained within the article.

Acknowledgments: First of all, I would like to show my deepest gratitude to Baotou Research Institute of Rare Earths for providing the necessary laboratory, experimental materials, funds and other support during my doctoral period. Secondly, I shall extend my thanks to Zhao (Dongyue Zhao) for his professional guidance on this article. Finally, I would also like to my family for their encouragement and support.

Conflicts of Interest: The authors declare no conflict of interest.

References

1. Beale, A.M.; Gao, F.; Lezcano-Gonzalez, I.; Peden, C.H.F.; Szanyi, J. Recent advances in automotive catalysis for NO_x emission control by small-pore microporous materials. *Chem. Soc. Rev.* **2015**, *44*, 7371–7405. [CrossRef]
2. Wang, J.H.; Zhao, H.W.; Haller, G.; Li, Y.D. Recent advances in the selective catalytic reduction of NO_x with NH_3 on Cu-Chabazite catalysts. *Appl. Catal. B Environ.* **2017**, *202*, 346–354. [CrossRef]
3. Liu, F.D.; Yu, Y.B.; He, H. Environmentally-benign catalysts for the selective catalytic reduction of NO_x from diesel engines: Structure-activity relationship and reaction mechanism aspects. *Chem. Commun.* **2014**, *50*, 8445–8463. [CrossRef]
4. Iwamoto, M.; Furukawa, H.; Mine, Y.; Uemura, F.; Mikuriya, S.I.; Kagawa, S. Copper (II) ion-exchanged ZSM-5 zeolites as highly active catalysts for direct and continuous decomposition of nitrogen monoxide. *J. Chem. Soc.-Chem. Commun.* **1986**, *16*, 1272–1273. [CrossRef]
5. Iwamoto, M.; Yahiro, H.; Torikai, Y.; Yoshioka, T.; Mizuno, N. Novel preparation method of highly copper ion-exchanged ZSM-5 zeolites and their catalytic activities for NO decomposition. *Chem. Lett.* **1990**, *11*, 1967–1970. [CrossRef]
6. Burch, R. Knowledge and know-how in emission control for mobile applications. *Catal. Rev.* **2004**, *46*, 271–334. [CrossRef]
7. Brandenberger, S.; Kröcher, O.; Tissler, A.; Althoff, R. The state of the art in selective catalytic reduction of NO*x* by ammonia using metal-exchanged zeolite catalysts. *Catal. Rev.* **2008**, *50*, 492–531. [CrossRef]
8. Kwak, J.H.; Tran, D.; Burton, S.D.; Szanyi, J.; Lee, J.H.; Peden, C.H.F. Effects of hydrothermal aging on NH_3-SCR reaction over Cu-Zeolites. *J. Catal.* **2012**, *287*, 203–209. [CrossRef]
9. Ryu, T.; Kim, H.; Hong, S.B. Nature of active sites in Cu-LTA NH_3-SCR catalysts: A comparative study with Cu-SSZ-13. *Appl. Catal. B Environ.* **2019**, *245*, 513–521. [CrossRef]
10. Liu, Q.L.; Fu, Z.C.; Ma, L.; Niu, H.J.Y.; Liu, C.X.; Li, J.H.; Zhang, Z.Y. MnOx-CeO_2 supported on Cu-SSZ-13: A novel SCR catalyst in a wide temperature range. *Appl. Catal. A* **2017**, *547*, 146–154. [CrossRef]
11. Wang, J.C.; Peng, Z.L.; Qiao, H.; Yu, H.F.; Hu, Y.F.; Chang, L.P.; Bao, W.R. Cerium-Stabilized Cu-SSZ-13 Catalyst for the Catalytic Removal of NOx by NH_3. *Ind. Eng. Chem. Res.* **2016**, *55*, 1174–1182. [CrossRef]
12. Zhao, Z.C.; Yu, R.; Shi, C.; Gies, H.; Xiao, F.S.; De Vos, D.; Yokoi, T.; Bao, X.H.; Kolb, U.; McGuire, R.; et al. Rare-earth ion exchanged Cu-SSZ-13 zeolite from organotemplate-free synthesis with enhanced hydrothermal stability in NH3-SCR of NOx. *Catal. Sci. Technol.* **2019**, *9*, 241–251. [CrossRef]
13. Chen, Z.; Guo, L.; Qu, H.; Liu, L.; Xie, H.; Zhong, Q. Controllable positions of Cu^{2+} to enhance low-temperature SCR activity on novel Cu-Ce-La-SSZ-13 by a simple one-pot method. *Chem. Commun.* **2020**, *56*, 2360–2363. [CrossRef]
14. Luo, J.Y.; Gao, F.; Kamasamudram, K.; Currier, N.; Peden, C.H.F.; Yezerets, A. New insights into Cu/SSZ-13 SCR catalyst acidity. Part I: Nature of acidic sites probed by NH_3 titration. *J. Catal.* **2017**, *348*, 291–299. [CrossRef]
15. De-La-Torre, U.; Pereda-Ayo, B.; Gutiérrez-Ortiz, M.A.; González-Marcos, J.A.; González-Velasco, J.R. Steady-state NH_3-SCR global model and kinetic parameter estimation for NO*x* removal in diesel engine exhaust aftertreatment with Cu/chabazite. *Catal. Today* **2017**, *296*, 95–104. [CrossRef]
16. Gao, C.; Shi, J.W.; Fana, Z.Y.; Wang, B.R.; Wang, Y.; He, C.; Wang, X.B.; Li, J.; Niu, C.M. "Fast SCR" reaction over Sm-modified MnO*x*-TiO_2 for promoting reduction of NO_X with NH_3. *Appl. Catal. A* **2018**, *564*, 102–112. [CrossRef]
17. Mercier, F.; Alliot, C.; Bion, L.; Thromat, N.; Toulhoat, P. XPS study of Eu(III) coordination compounds:Core levels binding energies in solid mixed-oxo-compounds $Eu_mX_xO_y$. *J. Electron. Spectrosc.* **2006**, *150*, 21–26. [CrossRef]

18. Han, S.; Ye, Q.; Cheng, S.Y.; Kang, T.F.; Dai, H.X. Effect of the hydrothermal aging temperature and Cu/Al ratio on the hydrothermal stability of CuSSZ-13 catalysts for NH_3-SCR. *Catal. Sci. Technol.* **2017**, *7*, 703–717. [CrossRef]
19. Xue, J.; Wang, X.; Qi, G.; Wang, J.; Shen, M.; Li, W. Characterization of copper species over Cu/SAPO-34 in selective catalytic reduction of NO*x* with ammonia: Relationships between active Cu sites and de-NO_x performance at low temperature. *J. Catal.* **2013**, *297*, 56–64. [CrossRef]
20. Zhang, T.; Qiu, F.; Li, J. Design and synthesis of core-shell structured meso-Cu-SSZ-13@mesoporous aluminosilicate catalyst for SCR of NO with NH_3: Enhancement of activity, hydrothermal stability and propene poisoning resistance. *Appl. Catal. B Environ.* **2016**, *195*, 48–58. [CrossRef]
21. Luo, M.M.; Cheng, Y.; Peng, X.Z.; Pan, W. Copper modified manganese oxide with tunnel structure as efficient catalyst for low-temperature catalytic combustion of toluene. *Chem. Eng. J.* **2019**, *369*, 758–765. [CrossRef]
22. Liu, D.; Zhou, W.; Wu, J. Effect of Ce and La on the activity of CuO/ZSM-5 and MnO*x*/ZSM-5 composites for elemental mercury removal at low temperature. *Fuel* **2017**, *194*, 115–122. [CrossRef]
23. Albarracin-Caballero, J.D.; Khurana, I.; Iorio, J.R.D.; Shih, A.J.; Schmidt, J.E.; Dusselier, M.; Davis, M.E.; Yezerets, A.; Miller, J.T.; Ribeiro, F.H.; et al. Structural and kinetic changes to small-pore Cu-zeolites after hydrothermal aging treatments and selective catalytic reduction of NO_x with ammonia. *React. Chem. Eng.* **2016**, *2*, 168–179. [CrossRef]
24. Li, J.R.; Zhang, W.P.; Li, C.; Xiao, H.; He, C. Insight into the catalytic performance and reaction routes for toluene total oxidation over facilely prepared Mn-Cu bimetallic oxide catalysts. *Appl. Surf. Sci.* **2021**, *550*, 149179. [CrossRef]
25. Zhang, Y.D.; Li, C.T.; Zhu, Y.C.; Du, X.Y.; Lyu, Y.; Li, S.H.; Zhai, Y.B. Insight into the enhanced performance of toluene removal from simulated flue gas over Mn-Cu oxides modified activated coke. *Fuel* **2020**, *276*, 118099. [CrossRef]
26. Wang, D.; Jangjou, Y.; Liu, Y.; Sharma, M.K.; Luo, J.Y.; Li, J.H.; Kamasamudram, K.; Epling, W.S. A comparison of hydrothermal aging effects on NH_3-SCR of NOx over Cu-SSZ-13 and Cu-SAPO-34 catalysts. *Appl. Catal. B Environ.* **2015**, *165*, 438–445. [CrossRef]
27. Chen, B.H.; Xu, R.N.; Zhang, R.D.; Liu, N. Economical way to synthesize SSZ-13 with abundant ion-exchanged Cu^+ for an extraordinary performance in selective catalytic reduction (SCR) of NO_x by ammonia. *Environ. Sci. Technol.* **2014**, *48*, 13909–13916. [CrossRef]
28. Zhang, Q.; Chen, G.R.; Dong, G.P.; Zhang, G.; Liu, X.F.; Qiu, J.R.; Zhou, Q.L.; Chen, Q.X.; Chen, D.P. The reduction of Cu^{2+} to Cu^+ and optical properties of Cu^+ ions in Cu-doped and Cu/Al-codoped high silica glasses sintered in an air atmosphere. *Chem. Phys. Lett.* **2009**, *482*, 228–233. [CrossRef]
29. Bulánek, R.; Wichterlová, B.; Sobalík, Z.; Tichý, J. Reducibility and oxidation activity of Cu ions in zeolites Effect of Cu ion coordination and zeolite framework composition. *Appl. Catal. B Environ.* **2001**, *31*, 13–25. [CrossRef]
30. Putluru, S.S.R.; Riisager, A.; Fehrmann, R. Alkali resistant Cu/zeolite deNO*x* catalysts for flue gas cleaning in biomass fired applications. *Appl. Catal. B Environ.* **2011**, *101*, 183–188. [CrossRef]
31. Gao, F.; Walter, E.D.; Karp, E.M.; Luo, J.Y.; Tonkyn, R.G.; Kwak, J.H.; Szanyi, J.; Peden, C.H.F. Structure-activity relationships in NH_3-SCR over Cu-SSZ-13 as probed by reaction kinetics and EPR studies. *J. Catal.* **2013**, *300*, 20–29. [CrossRef]
32. Ma, L.; Cheng, Y.; Cavataio, G.; McCabe, R.W.; Fu, L.; Li, J. In situ DRIFTS and temperature-programmed technology study on NH_3-SCR of NO_x over Cu-SSZ-13 and Cu-SAPO-34 catalysts. *Appl. Catal. B Environ.* **2014**, *156–157*, 428–437. [CrossRef]
33. Wang, D.; Zhang, L.; Li, J.; Kamasamudram, K.; Epling, W.S. NH_3-SCR over Cu/SAPO-34 -Zeolite acidity and Cu structure changes as a function of Cu loading. *Catal. Today* **2014**, *231*, 64–74. [CrossRef]
34. Cheung, T.; Bhargava, S.K.; Hobday, M.; Foger, K. Adsorption of NO on Cu Exchanged Zeolites, an FTIR Study: Effects of Cu Levels, NO Pressure, and Catalyst Pretreatment. *J. Catal.* **1996**, *158*, 301–310. [CrossRef]
35. Bordiga, S.; Lamberti, C.; Bonino, F.; Travert, A.; Thibault-Starzyk, F. Probing zeolites by vibrational spectroscopies. *Chem. Soc. Rev.* **2015**, *44*, 7262–7341. [CrossRef]
36. De la Torre, U.; Urrutxua, M.; Pereda-Ayo, B.; Gonzalez-Velasco, J.R. On the Cu species in Cu/beta catalysts related to $DeNO_x$ performance of coupled NSR-SCR technology using sequential monoliths and dual-layer monolithic catalysts. *Catal. Today* **2016**, *273*, 72–82. [CrossRef]
37. Yi, X.F.; Wang, J.X.; Liu, Y.Q.; Chen, Y.T.; Chen, J.S. Promotional effect of Fe and Ce co-doping on a V_2O_5-WO_3/TiO_2 catalyst for SCR of NO_x with high K and Pb resistance. *Catal. Sci. Technol.* **2022**, *12*, 4169–4180. [CrossRef]
38. Luo, J.; Oh, H.; Henry, C.; Epling, W. Effect of C_3H_6 on selective catalytic reduction of NOx by NH_3 over a Cu/zeolite catalyst: A mechanistic study. *Appl. Catal. B Environ.* **2012**, *123–124*, 296–305. [CrossRef]
39. Wang, L.; Li, W.; Qi, G.S.; Weng, D. Location and nature of Cu species in Cu/SAPO-34 for selective catalytic reduction of NO with NH_3. *J. Catal.* **2012**, *289*, 21–29. [CrossRef]
40. Shan, Y.L.; Du, J.P.; Yu, Y.B.; Shan, W.P.; Shi, X.Y.; He, H. Precise control of post-treatment significantly increases hydrothermal stability of in-situ synthesized Cu-zeolites for NH_3-SCR reaction. *Appl. Catal. B Environ.* **2020**, *266*, 118655. [CrossRef]
41. Zhu, H.; Kwak, J.H.; Peden, C.H.F.; Szanyi, J. In situ DRIFTS-MS studies on the oxidation of adsorbed NH_3 by NOx over a Cu-SSZ-13 zeolite. *Catal. Today* **2013**, *205*, 16–23. [CrossRef]
42. Sommer, L.; Mores, D.; Svelle, S.; Stöcker, M.; Weckhuysen, B.M.; Olsbye, U. Mesopore formation in zeolite H-SSZ-13 by desilication with NaOH. *Micropor. Mesopor. Mat.* **2010**, *132*, 384–394. [CrossRef]
43. Gao, F.; Washton, N.M.; Wang, Y.; Kollár, M.; Szanyi, J.; Peden, C.H.F. Effects of Si/Al ratio on Cu/SSZ-13 NH_3-SCR catalysts: Implications for the active Cu species and the roles of Brønsted acidity. *J. Catal.* **2015**, *331*, 25–38. [CrossRef]

44. Zhang, T.; Qiu, F.; Chang, H.Z.; Li, X.; Li, J.H. Identification of active sites and reaction mechanism on low-temperature SCR activity over Cu-SSZ-13 catalysts prepared by different methods. *Catal. Sci. Technol.* **2016**, *6*, 6294–6304. [CrossRef]
45. Luo, J.; Wang, D.; Kumar, A.; Li, J.; Kamasamudram, K.; Currier, N.; Yezerets, A. Identification of two types of Cu sites in Cu/SSZ-13 and their unique responses to hydrothermal aging and sulfur poisoning. *Catal. Today* **2016**, *267*, 3–9. [CrossRef]
46. Di Iorio, J.R.; Bates, S.A.; Verma, A.A.; Delgass, W.N.; Ribeiro, F.H.; Miller, J.T.; Gounder, R. The dynamic nature of bronsted acid sites in Cu-Zeolites during NOx selective catalytic reduction: Quantification by gas-phase ammonia titration. *Top. Catal.* **2015**, *58*, 424–434. [CrossRef]
47. Kwak, J.H.; Zhu, H.Y.; Lee, J.H.; Peden, C.H.F.; Szanyi, J. Two different cationic positions in Cu-SSZ-13? *Chem. Commun.* **2012**, *48*, 4758–4760. [CrossRef]
48. Ren, L.M.; Zhang, Y.B.; Zeng, S.J.; Zhu, L.F.; Sun, Q.; Zhang, H.Y.; Yang, C.G.; Meng, X.J.; Yang, X.G.; Xiao, F.S. Design and Synthesis of a Catalytically Active Cu-SSZ-13 Zeolite from a Copper-Amine Complex Template. *Chin. J. Catal.* **2012**, *33*, 92–105. [CrossRef]
49. Rodríguez-González, L.; Hermes, F.; Bertmer, M.; Rodríguez-Castellón, E.; Jiménez-López, A.; Simon, U. The acid properties of H-ZSM-5 as studied by NH_3-TPD and ^{27}Al-MAS-NMR spectroscopy. *Appl. Catal. A* **2007**, *328*, 174–182. [CrossRef]
50. Borfecchia, E.; Lomachenko, K.A.; Giordanino, F.; Falsig, H.; Beato, P.; Soldatov, A.V.; Bordiga, S.; Lamberti, C. Revisiting the nature of Cu sites in the activated Cu-SSZ-13 catalyst for SCR reaction. *Chem. Sci.* **2015**, *6*, 548–563. [CrossRef]
51. Gao, Y.; Zheng, B.H.; Wu, G.; Ma, F.W.; Liu, C.T. Effect of the Si/Al ratio on the performance of hierarchical ZSM-5 zeolites for methanol aromatization. *RSC Adv.* **2016**, *6*, 83581–83588. [CrossRef]
52. Wang, D.; Zhang, L.; Kamasamudram, K.; Epling, W.S. In situ-DRIFTS study of selective catalytic reduction of NOx by NH_3 over Cu-exchanged SAPO-34. *ACS Catal.* **2013**, *3*, 871–881. [CrossRef]
53. Yang, S.J.; Wang, C.Z.; Li, J.H.; Yan, N.Q.; Ma, L.; Chang, H.Z. Low temperature selective catalytic reduction of NO with NH_3 over Mn-Fe spinel: Performance, mechanism and kinetic study. *Appl. Catal. B Environ.* **2011**, *110*, 71–80. [CrossRef]
54. Elzey, S.; Mubayi, A.; Larsen, S.C.; Grassian, V.H. FTIR study of the selective catalytic reduction of NO_2 with ammonia on nanocrystalline NaY and CuY. *J. Mol. Catal. A Chem.* **2008**, *285*, 48–57. [CrossRef]
55. Lee, H.; Nuguid, R.J.G.; Jeon, S.W.; Kim, H.S.; Hwang, K.H.; Krocher, O.; Ferri, D.; Kim, D.H. In situ spectroscopic studies of the effect of water on the redox cycle of Cu ions in Cu-SSZ-13 during selective catalytic reduction of NOx. *Chem. Commun.* **2022**, *58*, 6610–6613. [CrossRef] [PubMed]
56. Wang, L.; Li, W.; Schmieg, S.J.; Weng, D. Role of Brønsted acidity in NH_3 selective catalytic reduction reaction on Cu/SAPO-34 catalysts. *J. Catal.* **2015**, *324*, 98–106. [CrossRef]
57. Ma, Y.; Wu, X.D.; Cheng, S.Q.; Cao, L.; Liu, L.P.; Xu, Y.F.; Liu, J.B.; Ran, R.; Si, Z.C.; Weng, D. Relationships between copper speciation and Bronsted acidity evolution over Cu-SSZ-13 during hydrothermal aging. *Appl. Catal. A-Gen.* **2020**, *602*, 117650. [CrossRef]
58. Lezcano-Gonzalez, I.; Deka, U.; Arstad, B.; Van Yperen-De Deyne, A.; Hemelsoet, K.; Waroquier, M.; Van Speybroeck, V.; Weckhuysen, B.M.; Beale, A.M. Determining the storage, availability and reactivity of NH_3 within Cu-Chabazite-based ammonia selective catalytic reduction systems. *Phys. Chem. Chem. Phys.* **2014**, *16*, 1639–1650. [CrossRef]

catalysts

Article

Three-Dimensional Graphene with Preserved Channeling as a Binder Additive for Zeolite 13X for Enhanced Thermal Conductivity, Vapor Transport, and Vapor Adsorption Loading Kinetics

Evan Gildernew, Syed Tareq and Sungwoo Yang *

Department of Civil and Chemical Engineering, University of Tennessee at Chattanooga, Chattanooga, TN 37403, USA; nmx363@mocs.utc.edu (E.G.); zhd777@mocs.utc.edu (S.T.)
* Correspondence: sungwoo-yang@utc.edu; Tel.: +1-(423)-425-4366

Abstract: Atmospheric water vapor extraction through adsorption to highly porous materials holds promise for its incorporation into broader technologies, including potable water generation. These technologies require breakthroughs in synthesis and design. Here, we demonstrate a composite of zeolite 13X sorbent for high adsorption capacity infiltrated with a light-weight three-dimensional graphene binder, which effectively networks a substrate structure into the sorbent. The composites described maintained fidelity when passing through the pore structure. This was accomplished by the utilization of a sacrificial polymer for safeguarding channel networking during sorbent infiltration of the binder for the extension of substrate networking. The performance measures for adsorbate loadings and thermal flux are evaluated with additional measurements taken for considering compactions of sorbent/substrates. Graphene/Zeolite 13X with preserved channeling demonstrated specific heat flux at 7664 W/kg, while samples without preserved channeling measured 4206 W/kg. A 0.6 g/cm^3 compaction resulted in a 412% and a 368% improvement in mass transport while compaction at 1.2 g/cm^3 resulted in a 333% and a 290% improvement in mass transport.

Keywords: zeolite; vapor adsorption; water harvesting; heat and mass transfer

Citation: Gildernew, E.; Tareq, S.; Yang, S. Three-Dimensional Graphene with Preserved Channeling as a Binder Additive for Zeolite 13X for Enhanced Thermal Conductivity, Vapor Transport, and Vapor Adsorption Loading Kinetics. *Catalysts* **2022**, *12*, 292. https://doi.org/10.3390/catal12030292

Academic Editors: Maja Milojević-Rakić and Danica Bajuk-Bogdanović

Received: 29 January 2022
Accepted: 2 March 2022
Published: 4 March 2022

1. Introduction

Much of our world is built with the expectation that water will return to us through natural systems and in abundance. Population growth, untreated industrial wastewater, and climate change have resulted in daunting cleanup challenges, rivers with reduced flow, and reservoirs with depleting supplies [1]. In a growing number of regions, the expectation of unstrained access to potable water requiring minimal energy input is untenable for agricultural, residential, and industrial planning [2]. Atmospheric Water Harvesting (AWH) is a wide field of research with a long past. Beautiful ancient rainwater collecting structures known as Khadins, Ahars, and Kattas effectively supported large populations in India. In the 14th century, Chittorgarh Water Fort, utilizing 84 reservoirs and water bodies, had enough capacity to endure a year without rain [3]. In many parts of the world, rain- and dew-collecting structures can be found atop residences with markedly little improvement in the centuries-old engineering underlying their design. AWH dew-collector research, such as the WarkaWater tower, inverted pyramid designs [4], multiscaled curvatures [5], pore-controlled synthesis [6], metal–organic framework adsorbent synthesis [7,8], and macrotexturing [9], seek game-changing performance improvements by incorporating modern advances. Decades of water restrictions in the American West are seeing the installation of Atmospheric Water Extraction (AWE) by way of optimized electric chillers with accompanying cisterns, a modern option for those who can afford it [10]. Such water chiller technologies require a highly resourced infrastructure and could not be used alongside energy conservation efforts. Other techniques of AWE, such as cloud seeding and solar stills, are active areas of research [11]. The holy grail of AWE is a low-energy input device with sorbents designed for rapid vapor uptake/release cycling across a wide range of humiditys and temperature environs. A sorbent-based AWE device meeting such requirements designed in consideration of advances in passive radiative cooling has the potential to be a breakthrough technology [12]. Sorbent-based AWE relies upon the surface attractive forces of highly porous materials wherein an optimized sorbent/substrate structure presents the limiting design hurdle. The structure of a sorbent/substrate composite will directly affect the vapor loadings, the mass diffusivities, and the thermal conductivities. All of which require precise engineering to achieve the rapid uptake/release cycling required for AWE. Compositing of Graphene for the formation of sorbent/substrate structures with Zeolite was demonstrated for the removal of various pollutants from water; for removal of Methyl Orange and Cu(II) [13], for removal of Arsenic [14], for removal of Lead [15], and for removal of Fe(III) [16]. Compositing of Graphene–Zeolite has received attention from a wide array of adsorption-driven applications, for the adsorption of Ethane [17], for Hydrogen [18], and for Carbon capture [19]. The use of functionalized Zeolites for compositing was demonstrated to improve adsorption kinetics with water vapor [20,21]. Our prior research effort in compositing includes Zeolite-driven adsorption with binder additives for climate control and thermal energy storage systems [22], with this publication directly building on a functionalized Graphene–Zeolite compositing synthesis protocol for heat pump applications, which investigates the Zeolite–Water dynamic. Efforts to advance adsorption-based heat pump performance by compositing Graphene with Zeolite has continued to receive attention by researchers [23]. In this publication, the performance of an integrated material "composite" with preserved channeling by sorbent/substrate synthesis with steps for utilization of a sacrificial polymer is considered for use in AWE. The sorbate–sorbent pair were selected in the hope of demonstrating this accelerated uptake/release. The substrate/binder's higher thermal conductivity and micro-channel pore dimensions are intended to desirably manage the enthalpies of adsorption while facilitating vapor transport. In this study, we demonstrate enhanced heat and mass transfers which function to keep high heat and mass transfer with increasing volumetric adsorption capacity via compaction. We synthesized zeolite composites with three-dimensional graphene-embedded hollow micro-channels (3dGRCh) that enable adsorption stacks to overcome inherent thermal and mass transport limitations simultaneously.

2. Materials and Methods

Three-dimensional graphene (3dGR) was synthesized by way of Chen's template-directed chemical vapor deposition (CVD). Nickel foam (American Elements, PPI 110, Los Angeles, CA, USA) acted as catalyst [24]. Methane/hydrogen gasses were used as carbon sources for growth. For the pre-annealing process Ni foam was annealed for about 24 h at 1100 °C under flowing H_2 (40 standard cubic centimeters per minute, sccm) and cooled down to room temperature at 0.5 °C min^{-1} before conducting CVD growth to enlarge Ni grain size resulting in higher thermal conductivity of 3dGR than without the pre-annealing process. The Ni foam was cut into strips of the desired size (25.4 × 25.4 × 5 mm^3), and then placed in a furnace (Thermo Scientific™, BlueM®/Lindburg, Waltham, MA, USA). The temperature of the furnace was ramped to 1000 °C in 1 h under a flowing Ar (40 sccm) and H_2 (10 sccm) mixture and held at 1000 °C for an additional 30 min. To coat graphene on the Ni foam, the mixture of CH_4 (50 sccm), H_2 (50 sccm) and Ar (400 sccm) was introduced for 1 h at 1000 °C. After the growth, the furnace was quickly cooled to room temperature (20 °C min^{-1}). After the growth of 3dGR on nickel foam in a furnace at 1000 °C, the nickel foam template was removed by placing the sample in a diluted HCl acid solution for 3 days at 50 °C, followed by washing with water. As the synthesized 3dGR is hydrophobic, we functionalized the surface of 3dGR to maximize the interaction with zeolite 13X (NaX, Sigma Aldrich, St. Louis, MO, USA 283,592, molecular sieves, 13X, 2 μm), which is hydrophilic. The synthesized 3dGR was oxidized by a mixture of HNO_3 and H_2SO_4 solutions which was described by Menna et al. [25] As shown in Figure 1a, we used poly(methyl methacrylate)

(PMMA) to protect channels inside functionalized 3dGR (f3dGR) while we integrated with zeolites. f3dGR was placed in PMMA solution (PMMA 950, MicroChem, Round Rock, TX, USA) for 30 min, followed by drying with Kimwipes to remove unnecessary PMMA between macro-pores of f3dGR. After curing of PMMA at 170 °C for one-hour, aqueous zeolite solution (50 wt%) was slowly poured. Zeolite particle macro-pores were infiltrated with f3dGR and solidified after 12 h. After drying at 100 °C for overnight, the composite was placed in acetone to remove the PMMA in the micro-channels of the f3dGR. To fabricate the 3dGR-zeolite composite without micro-channels (3dGR/NaX), the zeolite suspension (50 wt% zeolite in H_2O) was directly dropped on the top of the wet f3dGR in a beaker, followed by similar solidifying and drying procedures described above.

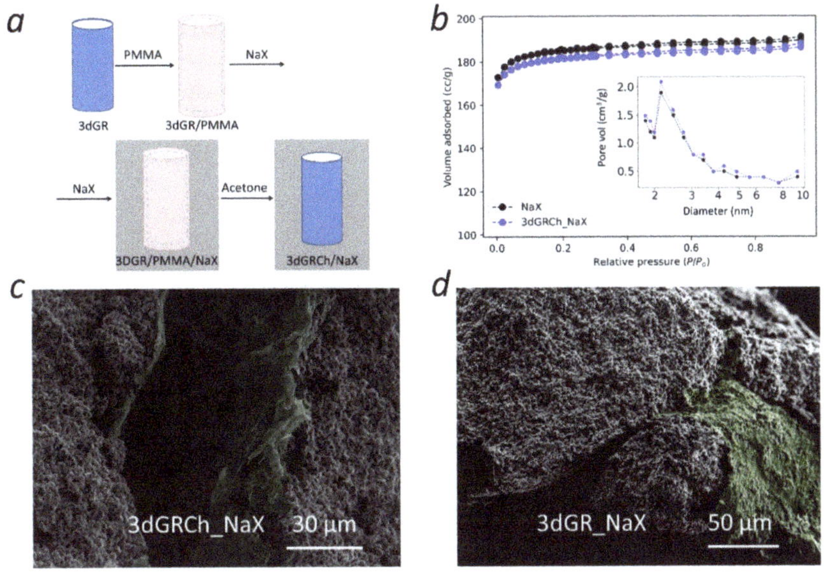

Figure 1. (**a**) Schematic of the synthesis of 3dGRCh-NaX using PMMA as sacrificial polymer to protect the micro-channels of 3dGR. SEM images of 3dGRCh-NaX and 3dGR-NaX composites; (**b**) N_2 adsorption–desorption isotherms at 77 K for NaX (black-round) and 3dGRCh-NaX (blue-round) composite. Inset: pore size distribution based on N_2 adsorption–desorption isotherm. y-axis is incremental pore volume (cm^3/g) and x-axis is diameter (nm). (**c**) NaX with infiltrated 3dGRCh thermal additive binder coating, and (**d**) 3dGR-NaX without preserved channeling. The 3dGR was falsely colorized in green for better contrast.

A composite of zeolite with 3dGR and 3dGRCh of desired dimension ($25 \times 25 \times 5$ mm^3) was mounted on a heat-flux sensing adsorption platform (HFS) in a pressure-controlled vacuum chamber, which was built to measure the heat released from an adsorption stack via conduction through a thin film heat flux sensor to a temperature-controlled sample holder. To measure heat generated by the adsorption stack, we fabricated the HFS rig shown in the inset of Figure 3c. The HFS was located in the custom environmental chamber where vapor pressure and wall temperature were precisely controlled. A thin film heat flux sensor (HFS-4, OMEGA, Norwalk, CT, USA) was located between the adsorption stack and a copper sample holder. A thin layer of thermal paste (silicone heat sink compound, 340, DOW CORNING, Midland, TX, USA) was applied to bond the adsorption stack and the heat flux sensor. To minimize thermal resistance, it was important to apply a thin but uniform layer of thermal paste. A thermal pad (A15896-02, Laird Technologies, Chesterfield, CT, USA) was placed between the heat flux sensor and the copper sample holder. The use of a thermal pad, rather than additional thermal paste, ensured equivalent thermal contact resistance from the sample to the sample holder between subsequent trials. Two resistive heaters (OMEGALUX, Kapton Insulated Flexible Heater, OMEGA, Norwalk, CT, USA) were located in the middle of the copper sample holder to heat the adsorption during the desorption process. A U-shaped tube was inserted in the copper sample holder to cool or maintain its temperature at a constant ~22 °C. The U-shaped tube was connected with a water pump, and a flow meter was installed to monitor the flow of cooling water. During the adsorption process, heat released by the adsorption stack was passed through the heat flux sensor, which generated electrical signals collected by a data acquisition system (34972A LXI Data Acquisition, Agilent, Santa Clara, CA, USA). To convert the electrical signals (mV) to heat flux (W), calibration was conducted using the HFS and a resistive heater with the same dimensions (25.4 × 25.4 mm^2). The power (W) provided by the resistive heater was simply the product of voltage (V) and current (I), and corresponding voltage signal from the HFS was recorded. The calibration factor, converting the electrical signal to heat flux, was 349.1 W $m^{-1}V^{-1}$. A replacement in the HFS showed a small variation in the calibration factor. Therefore, the calibration experiments were conducted again upon replacing the HFS. The reported uncertainty is based on the standard deviation of multiple measurements. The specific surface area and pore size distribution of the solid were based on the multipoint Brunner–Emmett–Teller theory (BET) and nonlocal density functional theory (NLDFT). SEM images were obtained on a JEOL 6010LA SEM. The thermal conductivity of NaX and NaX/3dGRCh composite were measured by laser flash method (NETZSCH, LFA 457 MicroFlash). Pellets of NaX or NaX/3dGRCh were prepared in 12.5 mm diameter for use of laser flash method. Adsorption isotherms were obtained on a DVS Vacuum Surface Measurement System (0.1 μg sensitivity), in which the vapor pressure and temperature was precisely controlled. Typical loading mass of absorbent was ~30 mg. Before adsorption on the DVS Vacuum, all adsorbents were fully dried. The desorption conditions used for zeolite (NaX) were 400 °C and 0.1 Pa for 6 h. To achieve an accurate isotherm, the adsorbent was exposed to vapor at the desired relative pressure until it reached its absorption plateau.

3. Results and Discussion

We created micro-channels inside of the 3dGR as vapor transport channels to enhance vapor transport through adsorption composite stack, while a percolated graphene network of 3dGR ensures effective thermal transport without compromising adsorption uptake due to light weight of 3dGR. Figure 1a shows a schematic of the synthesis of the 3dGR with micro-channels (3dGRCh). Details of the procedure can be found in the experimental section above. Figure 1c,d show SEM images of (3dGRCh-NaX) composite with micro-channels and (3dGR-NaX) composite without micro-channels. Figure 1a shows the purpose of the sacrificial polymer (PMMA) in preserving the 3dGR microchannels, which is critical to maximize the mass transport of water vapor while enhancing thermal transport through a maximized binder 3dGR network to the substrate. Once the NaX particles were successfully incorporated with 3dGRCh, the sacrificial polymer was removed via acetone washing. Figure 1c,d show the 3dGRCh and 3dGR networking formed on NaX. The 3dGR was falsely colorized in green for better contrast. The BET specific surface area of the NaX and 3dGRCh-NaX composite were estimated using the multipoint BET method based on adsorption–desorption isotherms at 77 K, as shown in Figure 1b. The surface area of the NaX and 3dGRCh-NaX composite was 657.5 m^2/g and 644.8 m^2/g, respectively. In addition, the inset of Figure 1b shows the pore-size distribution. Micropore volumes of NaX and 3dGRCh-NaX composite were estimated at 0.282 cm^3/g and 0.285 cm^3/g, respectively. This isotherm analysis demonstrated that the surface area and micropore volume of the samples remained relatively constant, even after the addition of 3dGRCh. The relatively unaffected surface area and pore-volume suggests that the effect of the addition of thermal additive (3dGRCh) was not significant and the pores of the NaX particles stayed open during the synthetic process of 3dGRCh-NaX. The vapor adsorption dynamics (Figure 2) and the heat-flux sensing adsorption results below (Figure 3c,d) also support preserved micropores in the 3dGRCh-NaX composite.

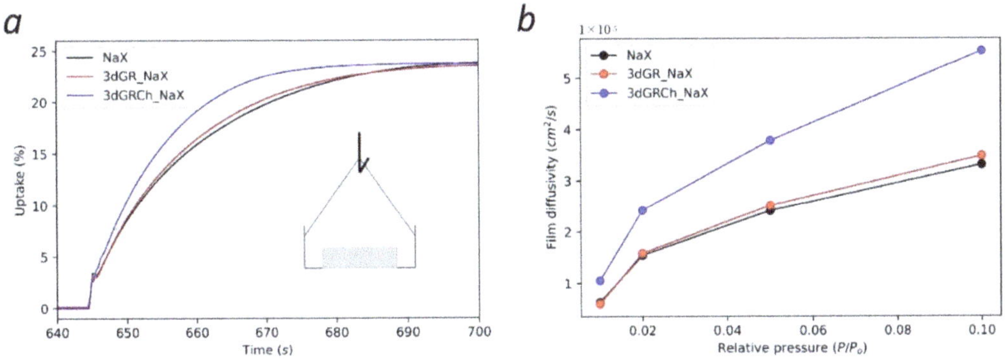

Figure 2. (**a**) Adsorption dynamics comparison between NaX powder (black), and compacted 3dGR-NaX (red) and 3dGRCh-NaX (blue) composites in time. The inset shows a schematic of sample loading in DVS. (**b**) Thermal conductivity comparison between NaX (black squares), 3dGR-NaX (red rounds), and 3dGRCh-NaX (blue triangles) in different density.

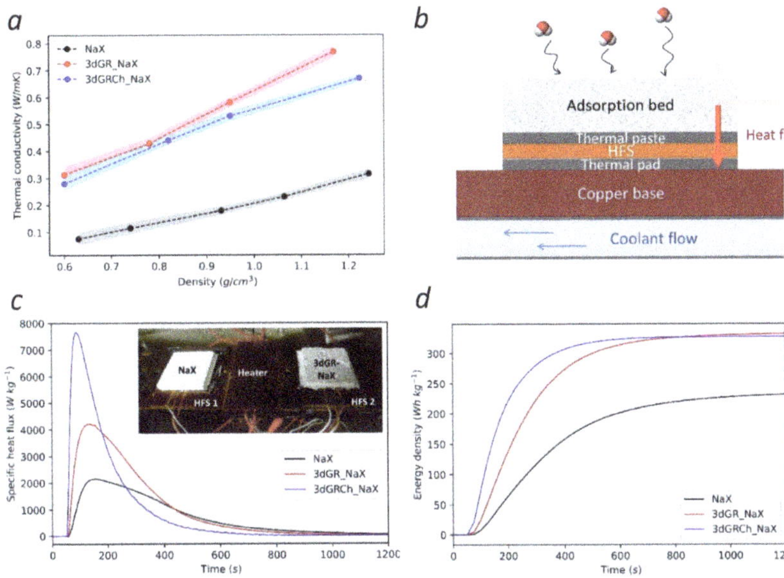

Figure 3. (**a**) Thermal conductivity comparisons of NaX (black), 3dGR-NaX (red) and 3dGRCh-NaX (blue). (**b**) A schematic side view of the heat-flux sensing adsorption platform (HFS) to measure specific heat flux responses and gravimetric energy density of the adsorption heat. (**c**) Specific heat flux responses and (**d**) gravimetric energy density adsorption stacks with 3dGRCh (blue triangle) and 3dGR (red rounds), and without binder (black square). The inset of (**c**). shows an image of experimental set up. The 3dGRCh-NaX composite exhibits increased thermal power density as its higher thermal conductivity increases the proportion of heat that is delivered to the substrate, rather than lost to the vapor ambient. The inset is an image of the experimental setup.

To compare the adsorption dynamics of the 3dGRCh-NaX to the composite 3dGR-NaX, 3dGR-NaX and 3dGRCh-NaX composites were pressurized at 50 Mpa and samples were taken ($3 \times 3 \times 1$ mm^3) which were then loaded into a vapor adsorption analyzer (DVS Vacuum, Surface Measurement Systems). The desorption condition was characterized at ~10^{-4} Pa and 400 °C and the adsorption condition was characterized at 25 °C with 1% relative pressure. Figure 2a shows that adsorption dynamics of the 3dGR-NaX composite after densification was similar to that of the NaX disk. However, vapor transport of the 3dGRCh-NaX was faster than that of 3dGR-NaX. This is attributed to the small weight fraction (<2%) of 3dGR in the composite of 3dGRCh-NaX, minor reduction of adsorption capacity was observed after adding 3dGR. Figure 2b presents the moisture diffusion coefficients in the composites by measuring adsorption isotherms on the thin films in a DVS instrument. The diffusion constants for the composites were estimated by utilizing the diffusion equation employed by Crank and Park [26]. The initial kinetics of sorption into a bulk adsorbent can be described as:

$$\frac{M_t}{M_\infty} = \frac{4}{d}\sqrt{\frac{Dt}{\pi}}$$

where M_t is amount adsorbed at time t, M_∞ is amount adsorbed at thermodynamic equilibrium, d is the thickness of a sample, and D is diffusion constant. The equation is generally accepted for values of $\frac{M_t}{M_\infty} < 0.4$, where a plot of $\frac{M_t}{M_\infty}$ against $\frac{\sqrt{t}}{d}$ is linear. The diffusion constant, D, was calculated from the slop of the line. Sorption data points were collected every 2 s for the experiment, and a minimum R^2 value of 99.9% was applied for the diffusion constant characterization.

Figure 2b presents a general trend for diffusion coefficients calculated for each discrete step in humidity. The diffusion constants from the composites increased with increasing partial vapor pressure of water. In addition, the diffusion constants of the composite with micro-channels (3dGRCh-NaX) were higher than those without micro-channels (3dGR-NaX). The diffusion constants of the NaX disk were the lowest. The diffusion data suggest enhanced diffusion coefficient by adding micro-channels.

While 3dGRCh showed enhanced mass transport, it exhibits comparable enhancement in thermal conductivity to that of 3dGR, as shown in Figure 3a. The improvements of 3dGRCh and 3dGRCh were approximately 412% and 368%, respectively, at a low composite density (~0.6 g/cm^3), and approximately 333% and 290%, respectively, at a high composite density (~1.2 g/cm^3). The amount of enhancement decreased with densification as the thermal conductivities of the zeolite without the binders increased. In addition, the composite of 3dGR and NaX without channels shows higher thermal conductivity compared to those with channels. Figure 3b shows a schematic side view of the heat-flux sensing (HFS) adsorption platform to measure specific heat flux responses and gravimetric energy density of the adsorption heat. A thermal paste and a thermal pad were applied to ensure the heat flow of the adsorption heat generated by the adsorption bed. Figure 3c presents thermal responses from a sample without any thermal additive (NaX) and composites with micro-channels (3dGRCh-NaX) and without (3dGR-NaX). A water vapor step input was from ~1 Pa to ~2500 Pa, comparable to the operating conditions in practical adsorption chillers. The released heat was normalized by the total weight of the adsorption stack. The sample without any thermal additive (NaX) not only showed slowest thermal energy release among the three samples but also experienced significant heat loss to its surroundings due to low thermal conductivity. This is because the thermal resistance from the adsorbent stack without thermal additive to the sample base was comparable to the thermal resistance to the surroundings, resulting in thermal loss to the ambient. The 3dGRCh-NaX showed higher specific heat flux (7664.0 W kg^{-1}) than that of 3dGR-NaX (4206.7 W kg^{-1}). Figure 3d shows that the 3dGRCh-NaX and 3dGR-NaX composites released 99% of their thermal energy in 632 s and 1046 s, respectively, while energy release from the NaX sample without additives was clearly slower. These results demonstrate the enhancement in thermal transport using the micro-channels of the 3dGRCh.

4. Conclusions

In summary, 3dGR with micro-channels (3dGRCh) appears to be a promising thermal and mass transport additive to simultaneously enhance thermal and mass transports. The percolated structure of the 3dGR provides effective heat transfer through the adsorbent stack while the reduction of adsorption capacity was minimal due to it being light weight. The micro-channels inside of the 3dGR, created by PMMA as a sacrificial material, enhanced mass transport through the adsorption stacks compared to the samples without the micro-channels. This work demonstrates that heat and mass transports were simultaneously improved by the 3dGR, light weight thermal additive, and the micro-channels of 3dGRCh. Significant improvement in thermal additives and mass transport channels is expected to result in enhanced water harvesting. While the micro-channels demonstrated a way for simultaneous improvement of thermal and mass transports, the investigation of a new class of thermal and mass additives is clearly needed in consideration of improving its flexibility, weight, and cost.

Author Contributions: S.Y. designed the experiments with inputs from all authors. E.G., S.T. and S.Y. conducted the designed experiments and analyzed collected data. E.G. and S.Y. wrote the manuscript with inputs from all authors. S.Y. guided the overall project. All authors have read and agreed to the published version of the manuscript.

Funding: This research was funded by the Center of Excellence for Applied Computational Science and Engineering (CEACSE) program at the University of Tennessee. The funding number is CEACSE-FY2021.

Acknowledgments: S.Y. gratefully acknowledges the internal funding from the College of Engineering and Computer Science (CECS) at The University of Tennessee at Chattanooga (UTC). This work made use of essential experimental equipment that were purchased by the internal funding.

Conflicts of Interest: The authors declare that they have no known competing financial interests or personal relationships that could have appeared to influence the work reported in this paper.

References

1. Kummu, M.; Guillaume, J.H.A.; De Moel, H.; Eisner, S.; Flörke, M.; Porkka, M.; Siebert, S.; Veldkamp, T.I.E.; Ward, P.J. The world's road to water scarcity: Shortage and stress in the 20th century and pathways towards sustainability. *Sci. Rep.* **2016**, *6*, 1–16. [CrossRef] [PubMed]
2. van Vliet, M.T.H.; Jones, E.R.; Flörke, M.; Franssen, W.H.P.; Hanasaki, N.; Wada, Y.; Yearsley, J.R. Global water scarcity including surface water quality and expansions of clean water technologies. *Environ. Res. Lett.* **2021**, *16*, 024020. [CrossRef]
3. Möller, D. On the history of the scientific exploration of fog, dew, rain and other atmospheric water. *Erde* **2008**, *139*, 11–44.
4. Beysens, D.; Broggini, F.; Milimouk-Melnytchouk, I.; Ouazzani, J.; Tixier, N. New architectural forms to enhance dew collection. *Chem. Eng. Trans.* **2013**, *34*, 79–84. [CrossRef]
5. Li, C.; Yu, C.; Zhou, S.; Dong, Z.; Jiang, L. Liquid harvesting and transport on multiscaled curvatures. *Proc. Natl. Acad. Sci. USA* **2020**, *117*, 23436–23442. [CrossRef]
6. Wright, A.M.; Rieth, A.J.; Yang, S.; Wang, E.N.; Dincă, M. Precise control of pore hydrophilicity enabled by post-synthetic cation exchange in metal–organic frameworks. *Chem. Sci.* **2018**, *9*, 3856–3859. [CrossRef]
7. Kim, H.; Rao, S.R.; Kapustin, E.A.; Zhao, L.; Yang, S.; Yaghi, O.M.; Wang, E.N. Adsorption-based atmospheric water harvesting device for arid climates. *Nat. Commun.* **2018**, *9*, 1191. [CrossRef]
8. Kim, H.; Yang, S.; Rao, S.R.; Narayanan, S.; Kapustin, E.A.; Furukawa, H.; Umans, A.S.; Yaghi, O.M.; Wang, E.N. Water harvesting from air with metal-organic frameworks powered by natural sunlight. *Science* **2017**, *356*, 430–434. [CrossRef]
9. Cheng, Y.; Liu, Y.; Ye, X.; Liu, M.; Du, B.; Jin, Y.; Wen, R.; Lan, Z.; Wang, Z.; Ma, X. Macrotextures-enabled self-propelling of large condensate droplets. *Chem. Eng. J.* **2021**, *405*, 126901. [CrossRef]
10. Magrini, A.; Cattani, L.; Cartesegna, M.; Magnani, L. Water production from air conditioning systems: Some evaluations about a sustainable use of resources. *Sustainability* **2017**, *9*, 1309. [CrossRef]
11. Xu, Z.; Zhang, L.; Zhao, L.; Li, B.; Bhatia, B.; Wang, C.; Wilke, K.L.; Song, Y.; Labban, O.; Lienhard, J.H.; et al. Ultrahigh-efficiency desalination: Via a thermally-localized multistage solar still. *Energy Environ. Sci.* **2020**, *13*, 830–839. [CrossRef]
12. Chen, H.; Ran, T.; Zhang, K.; Chen, D.; Gan, Y.; Wang, Z.; Jiang, L. Highly Efficient Multiscale Fog Collector Inspired by Sarracenia Trichome Hierarchical Structure. *Glob. Chall.* **2021**, *5*, 2100087. [CrossRef]
13. Liu, H.; Yu, A.; Liu, H.; Chu, S.; Tan, S. Preparation of graphene/zeolite composites and the adsorption of pollutants in water. *Russ. J. Appl. Chem.* **2017**, *90*, 1171–1180. [CrossRef]
14. Khatamian, M.; Khodakarampoor, N.; Saket-Oskoui, M. Efficient removal of arsenic using graphene-zeolite based composites. *J. Colloid Interface Sci.* **2017**, *498*, 433–441. [CrossRef]

15. Manjili, M.; Silva, M.R.; Garman, D.; Zhang, H.F. Graphene oxide and thiol functionalized natural zeolite for the removal of lead from water. *Water Sci. Technol. Water Supply* **2020**, *20*, 2577–2588. [CrossRef]
16. Ali, I.O.; El-Sheikh, S.M.; Salama, T.M.; Abdel-Khalek, E.K.; Thabet, M.S.; Bakr, M.F.; Fodial, M.H. Novel Composites of Multifunctional NaP Zeolite/Graphene Oxide for Highly Efficient Removal of Fe(III) from Aqueous Solution. *J. Inorg. Organomet. Polym. Mater.* **2021**, *31*, 577–590. [CrossRef]
17. Lee, S.K.; Park, H.; Yoon, J.W.; Kim, K.; Cho, S.J.; Maurin, G.; Ryoo, R.; Chang, J.S. Microporous 3D Graphene-like Zeolite-Templated Carbons for Preferentia l Adsorption of Ethane. *ACS Appl. Mater. Interfaces* **2020**, *12*, 28484–28495. [CrossRef]
18. Soldatov, A.P. Mechanism of Hydrogen Adsorption in Graphene Nanostructures Synthesized in Membrane Pores and on Zeolites. *Russ. J. Phys. Chem.* **2019**, *93*, 494–500. [CrossRef]
19. Firdaus, R.M.; Desforges, A.; Rahman Mohamed, A.; Vigolo, B. Progress in adsorption capacity of nanomaterials for carbon dioxide capture: A comparative study. *J. Clean. Prod.* **2021**, *328*, 129553. [CrossRef]
20. Silva, M.R.; Lecus, A.; Gajdardziska-Josifovska, M.; Schofield, M.; Virnoche, M.; Chang, J.; Chen, J.; Garman, D. Graphene-oxide loading on natural zeolite particles for enhancement of adsorption properties. *RSC Adv.* **2020**, *10*, 4589–4597. [CrossRef]
21. Yang, S.; Huang, X.; Chen, G.; Wang, E.N. Three-dimensional graphene enhanced heat conduction of porous crystals. *J. Porous Mater.* **2016**, *23*, 1647–1652. [CrossRef]
22. Narayanan, S.; Yang, S.; Kim, H.; Wang, E.N. Optimization of adsorption processes for climate control and thermal energy storage. *Int. J. Heat Mass Transf.* **2014**, *77*, 288–300. [CrossRef]
23. Rocky, K.A.; Pal, A.; Rupam, T.H.; Nasruddin; Saha, B.B. Zeolite-graphene composite adsorbents for next generation adsorption heat pumps. *Microporous Mesoporous Mater.* **2021**, *313*, 110839. [CrossRef]
24. Chen, Z.; Ren, W.; Gao, L.; Liu, B.; Pei, S.; Cheng, H.-M. Three-dimensional flexible and conductive interconnected graphene networks grown by chemical vapour deposition. *Nat. Mater.* **2011**, *10*, 424–428. [CrossRef]
25. Menna, E.; Della Negra, F.; Dalla Fontana, M.; Meneghetti, M. Selectivity of chemical oxidation attack of single-wall carbon nanotubes in solution. *Phys. Rev. B* **2003**, *68*, 193412. [CrossRef]
26. Frisch, H.L. "Diffusion in polymers" edited by J. Crank and G. S. Park, Academic Press, London and New York, 1968; 452 pg. *J. Appl. Polym. Sci.* **1970**, *14*, 1657. [CrossRef]

catalysts

MDPI

Review

ORR Catalysts Derived from Biopolymers

Jelena Rupar [1], Danijela Tekić [2], Aleksandra Janošević Ležaić [1] and Kush K. Upadhyay [3,*]

1 Department of Physical Chemistry and Instrumental Methods, University of Belgrade—Faculty of Pharmacy, 11221 Belgrade, Serbia
2 University of Belgrade—Faculty of Physical Chemistry, 11158 Belgrade, Serbia
3 Centro de Química Estrutural-CQE, Departamento de Engenharia Química, Instituto Superior Técnico, Universidade de Lisboa, 1049-001 Lisboa, Portugal
* Correspondence: kush.upadhyay@c2cnewcap.com or kushkumar434@gmail.com

Abstract: Due to the limited reaction rate of the oxygen reduction reaction (ORR), it is considered as a limiting factor in the performance of fuel cells and metal-air batteries. Platinum is considered the benchmark catalyst for ORR; however, the scarcity of platinum, its high price, the drift phenomenon, its insufficient durability, and its susceptibility to gas poisoning are the reasons for the constant search for new ORR catalysts. Carbon-based catalysts show exceptional promise in this respect considering economic profitability and activity, and, in addition, they have favorable conductivity and often a large specific surface area. The use of chitin, cellulose, lignin, coconut shell particles, shrimp shells, and even hair for this purpose was reported, as they had similar electrochemical activity regarding Pt. Alginate, a natural polymer and a constituent of brown algae, can be successfully used to obtain carbon materials that catalyze ORR. In addition, metal atomic-level catalysts and metal N-doped porous carbon materials, obtained from sodium alginate as a precursor, have been proposed as efficient electrocatalysts for ORR. Except for alginate, other biopolymers have been reported to play an important role in the preparation of ORR catalysts. In this review, recent advances regarding biopolymer-derived ORR catalysts are summarized, with a focus on alginate as a source.

Keywords: oxygen reduction reaction; biopolymers; alginate; carbon porous catalysts

Citation: Rupar, J.; Tekić, D.; Janošević Ležaić, A.; Upadhyay, K.K. ORR Catalysts Derived from Biopolymers. *Catalysts* **2023**, *13*, 80. https://doi.org/10.3390/catal13010080

Academic Editor: Nicolas Alonso-Vante

Received: 30 November 2022
Revised: 17 December 2022
Accepted: 25 December 2022
Published: 30 December 2022

1. Introduction

In recent years, the awareness of the human influence on climate change has increased. More and more attention is being paid to factors leading to global warming, especially CO_2 emissions. As fossil fuel usage leads to CO_2 release into the atmosphere [1], in addition to their limited availability [2], research groups worldwide are dedicated to finding alternatives to the carbon cycle. Fuel cells are of extraordinary importance in this respect, as they can participate in the hydrogen cycle. Fuel cells, as devices that convert oxygen and chemical energy into electricity with a high efficiency, are considered environmentally friendly and can be used for small, portable electronic devices and military and space devices [3–5]. Metal-air batteries represent cost-effective devices that require an air atmosphere [6], and the most promising are Zn-air batteries, with a high theoretical energy density, solid rechargeability, and flat discharge voltage [7,8]. During discharge, the reduction of oxygen at the cathode occurs while the metal oxidizes and releases electrons which pass through the external circuit in metal-air batteries. In the fuel cells, the H_2 dissociation occurs at the anode, while at the cathode, O_2 reduction occurs, transforming chemical energy into electricity [9]; thus, oxygen reduction reaction (ORR) is the most important cathode reaction for both types of devices [10–12].

The ORR catalyst is the main descriptor for the performance of these devices, with conventional platinum still regarded as the best ORR catalyst in both acid and alkaline electrolytes. However, Pt has a high price, a susceptibility to time-dependent drift, and serious anode crossover, and it is easily poisoned by CO (an intermediate product of electro-oxidizing alcohol fuels), which limits its usage. The limited availability of Pt is another important fact and constrains the possibility of using devices whose operation is based on ORR [13,14]. Therefore, platinum has to be replaced with appropriate alternatives that have a similar catalytic activity and can be produced massively and cost-effectively from Earth's abundant resources [15]. Electrocatalysts should have high electrical conductivities, large specific surface areas, and electroactive properties [11,16]. Conductive polymers (CPs) and carbon materials (grapheme, carbon nanotubes (CNTs), amorphous carbon, and carbon nanofibers) can fulfil these requirements, and they have been widely explored for oxygen reduction in fuel cells, among other applications [13,17].

CPs that have been found to be useful as ORR catalysts can be classified into three groups: inherent CPs, CP-derived heteroatom-doped carbons, and CP composites. The activity of inherent CPs is a consequence of their structure, so the transfer of electrons with oxygen in molecular form occurs due to the characteristic that the neutral CP is an electron donor for O_2, while O_2 is an electron acceptor. Adsorbed oxygen on the CP surface accepts an electron, and its molecular symmetry decreases while the length of the O = O bond increases. In addition, due to the reaction with adsorbed oxygen molecules, the CPs that are mildly oxidized transform to a higher oxidation state, while the oxygen molecules are reduced into O^{2-} anions [17,18]. As many research groups directed their research interest towards the development of ORR catalysts from CPs, it was established that catalytic performances can be improved if a metal is incorporated into the CP [19] or if the CP is doped with a heteroatom. When Dai et al., in 2009, reported the strong ORR catalytic activity of carbon nanotubes doped with nitrogen, a new research field for N-doped carbon catalysts investigation was opened [17–20]. In addition to monoatomic doping, co-doping with other heteroatoms (B, O, S, P) proved to be an important direction for the synthesis of ORR catalysts, taking into account the increase in the asymmetrical spin density obtained in that case [21–23]. While B- and P-doped carbon without N-doping mainly enabled ORR via a two-electron process [24,25], B,N-doped carbon was proposed as an ORR catalyst, indicating a four-electron ORR in alkaline media [26]. In acidic media, the co-doping of N-doped material with only B influenced the increase in H_2O_2 production, and additionally doping it with P decreased the production of H_2O_2, consequently leading to it favoring the four-electron process [21]. Zhang et al. [22] designed N and P co-doped material by using polyaniline pyrolysis and developed the ORR OER dual catalyst, which enables the reduction to occur via the four-electron pathway.

For tridoped nanoporous carbon material with N-, O-, and S-, it was suggested that the O_2 molecule was reduced via a four-electron process in 0.1 M KOH solution. Compared to metal-based catalysts, this material had improved the catalytic activity; thus, it influenced the growth of research in this area [23].

Additionally, it was reported that the incorporation of transition-metal active centers had an influence on the electrocatalytic activity [27–30]. Transition metals such as Mn, Fe, Co, Ni, etc. have empty 3d orbitals; thus, they can accept an electron and have the ability to reduce the bonding energy between OOH*,O*/OH* intermediates [31,32]. As facilitating the O-O bond breaking in OOH* leads to the inhibition of H_2O_2 formation, this material quality is important for improving the four-electron ORR pathway selectivity [33].

Published materials [34] containing Fe [35], which had an electrocatalytic activity comparable to Pt/C in alkaline as well as in neutral and acidic medium, with an electron transfer number of about 4. Additionally, there is evidence that the introduction of two transition metals instead of one enhances the four-electron pathway selectivity [31].

Yang et al. [31] synthesized a binuclear catalyst containing Fe and Mn with excellent four-electron pathway selectivity in acidic media [34].

In accordance with the above, the heteroatom-doped carbon and transition metal coordinated by N in the carbon network are considered as the most promising catalysts for ORR [13].

The porous structure is an important parameter for the platinum group metal (PGM)-free ORR catalysts. Micropores host the majority of the active sites, while the mesopores primarily have the role of the mass transfer channel. Thus, the electrocatalytic activity is determined by micro- and mesoporosity [13,36].

The materials derived from biomass can inherit the macro-, meso-, and microporosity of the original biological precursors well after appropriate conversion processes [13,37–39].

The carbonization temperature presents another important factor that affects the activity of the bio-based carbon catalyst. As the temperature increases, the graphitization degree increases from about 75% to about 85%, and the electrical conductivity also increases [40,41]. In addition, the temperature increase leads to porosity development [36]. Additionally, the activity of the carbon-based catalyst depends on the relationship between the electron conductivity and the specific surface area. However, many works do not contain information about the specific surface area. The future is certainly represented by carbon foams with a system of open pores that enable the highest availability of active sites for oxygen and thus improve catalytic activity.

The defects of different sorts had a great influence on the ORR mechanism in the alkaline medium, while in the acidic medium, the influence was less pronounced [42]. Additionally, there is evidence that the rate of heteroatom influenced the catalytic activity; a higher content of heteroatom leads to an increase in structural defects and improves catalytic activity [43]. The researchers agree on the fact that the level of nitrogen influences the electrocatalytic activity. While increasing the level of nitrogen content, the ORR activity increases. The performance improvement is primarily due to the presence of free-flowing sp2-hybridized π electrons. In addition, the presence of free electron pairs of nitrogen atoms contributes to the bonding of the delocalized carbon matrix system, thereby further improving the electrocatalytic activity of ORR to a large extent. It is also known that heteroatoms doping can enhance the catalytic activity given that the doped heteroatom generally has the greater electronegativity compared to carbon atoms; this induces a partial positive charge near the carbon atoms and enhances the ORR [44,45]. The problem arises when the influence of the nitrogen type is discussed, since it is not entirely clear which nitrogen functional groups had a direct effect on the catalytic activity.

Various loadings of carbon-based materials in the range between 0.10 and 1.00 mg/cm^2 [46,47] were tried, and its influence on ORR was assessed. The results showed that low-carbon loadings were not sufficient in covering the electrode surface and consequently led to lower current densities and, predominantly, the 2e$^-$ mechanism. With an increase in loading up to around 0.25 mg/cm^2, higher currents were measured, with a shift in the mechanism towards 4e$^-$ reduction. A further increase in loading was detrimental to the ORR performance of carbon-based materials. An explanation for this occurrence is the excessively long electron part caused by the catalyst layer thickness, which results in substantial electrical resistance and mass transport losses. Both phenomena limited the access to some of the electroactive sites, independent of the sites' actual activity towards ORR. Film thickness also affects the total available surface area as well as the materials' bulk density, so there exists a trade-off between all these different effects when considering the overall ORR performance. Lastly, layer thickness seems to have an effect on the ORR mechanism itself by increasing the number of apparently exchanged electrons with the increase in loading. Namely, thicker layer intermediates, such as OOH-, remain near the carbon surface for longer periods of time and participate in the second 2e$^-$ reduction to OH- [46,48], pushing the number of electrons towards four. This phenomenon is more pronounced for materials with sites of higher activity.

It is of particular importance to mention that ORR catalysts can also be synthesized using biogenic precursors. The authors Ye et al. [49] presented a synthesis process of ORR catalysts containing Fe-P active centers, After the carbonization of the biogenic precursor, i.e., the bacteria@vivianite composite, which was obtained due to the reduction of polyferric flocs by Shewanella oneidensis MR-1, the obtained material showed an excellent catalytic activity. Chitin, as the most abundant natural nitrogen-containing compound, is an ideal precursor for the synthesis of ORR catalysts using only a high temperature, without additional dopants [50,51]. Human hair, consisting of keratin, with contents of about

15–16% nitrogen and 4.5–5.5% sulfur [52], may also be used as a biogenic precursor for obtaining heteroatom-doped carbon material through the process that uses no hazardous reagents [53].

Biopolymers, substances consisting of numerous repeating monomer units, are present in natural sources. In Table 1, the principal advantages and disadvantages of natural biopolymers are presented [54,55]. These are promising candidates for different spheres of medicine and industry due to their biocompatibility and biodegradability. The biopolymers are used as edible films, emulsions, packaging materials in the food industry, drug transport materials, medical implants, tissue scaffolds, dressing materials in pharmaceutical industries, etc. [55].

Table 1. The positive and negative aspects of natural biopolymers.

Positive Aspects	Negative Aspects
biologically renewable biodegradable biocompatible non-toxic biofunctional	low melting point less stable structurally more complex high surface tension

In this review, we will discuss the use of some biopolymers-based materials as promising candidates for ORR catalysis, whose performance was most often tested in alkaline conditions. Those that are applied most often are alginate, cellulose, chitin, and lignin due to their broad availability and low price [56–61].

Figure 1 represents the biopolymer categorization [55] with respect to the biopolymers considered in this review paper.

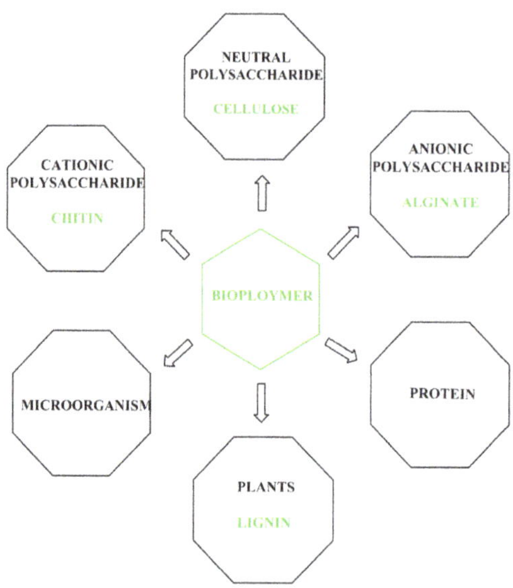

Figure 1. A pictorial depiction of several natural, renewable biopolymers categorized according to their source [55].

In Table 2, the structures of the biopolymers reviewed in this paper are presented.

Table 2. Main biopolymers and their chemical structures [62].

alginate	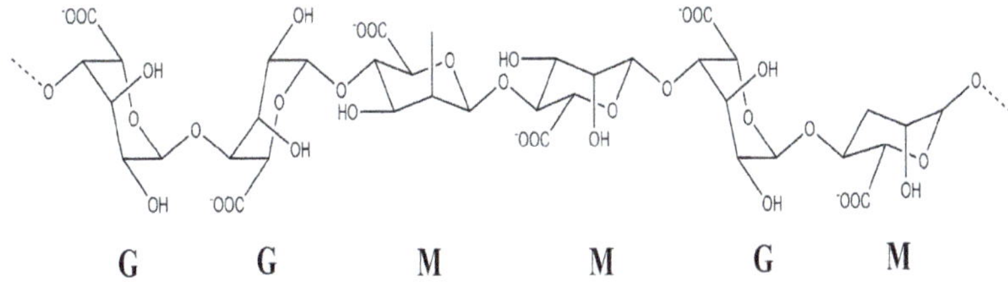
alginate monomer	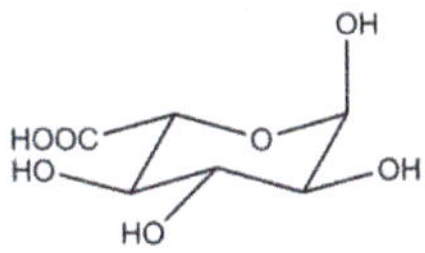 α-D guluronic acid 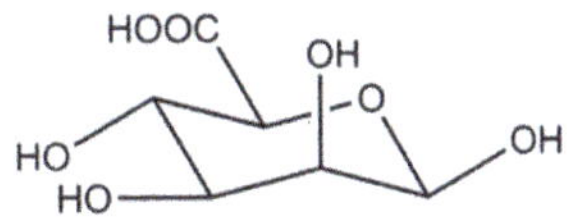 β-D mannuronic acid
cellulose	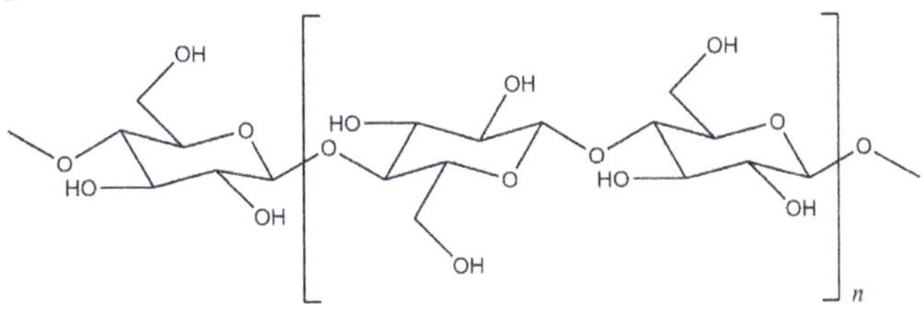
cellulose monomer	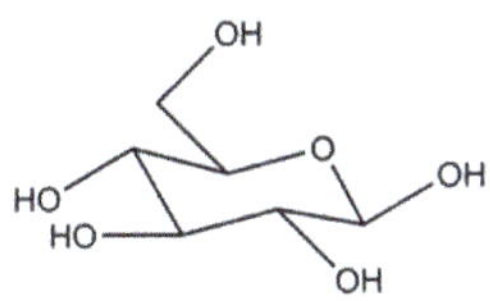 glucose
chitin	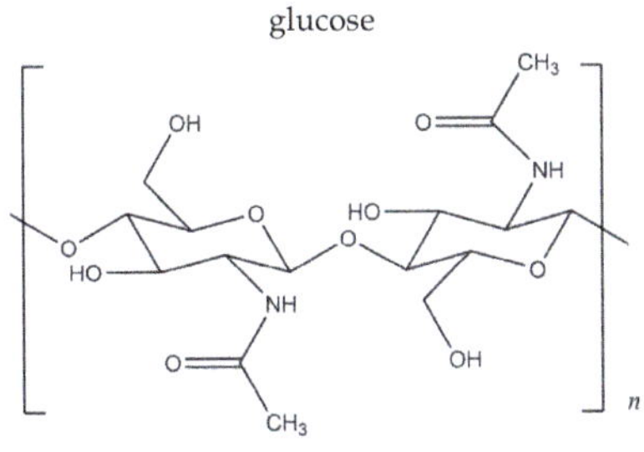 N-acetylglucosamine

Table 2. *Cont.*

lignin	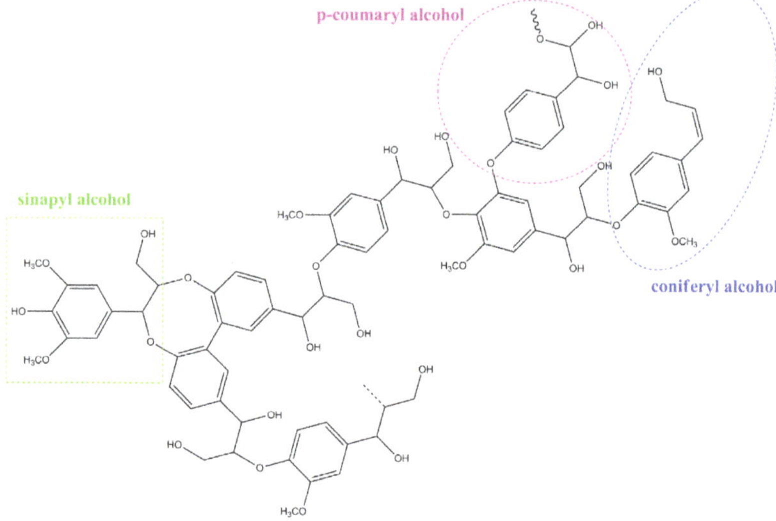
lignin monomer	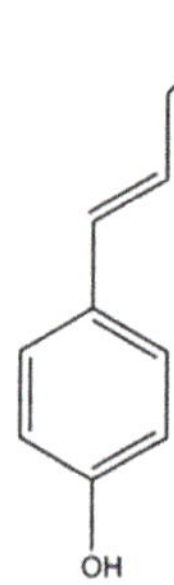 p-coumaryl alcohol 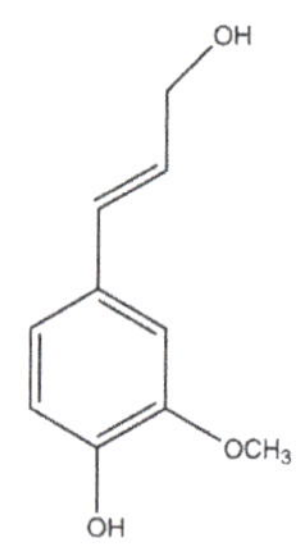 coniferyl alcohol 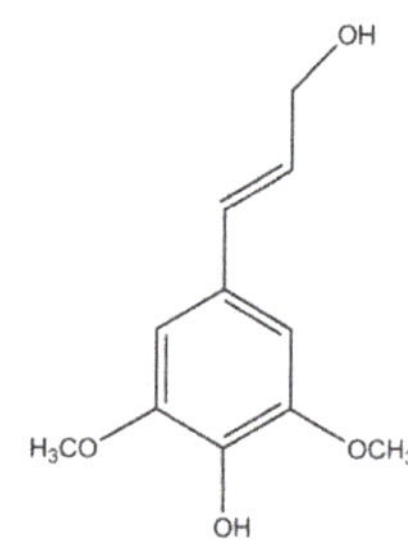 synapil alcohol

Major challenges during the synthesis of ORR catalysts include the efforts to extract biopolymers from the biomass, along with methods for cleaning and decomposition, simplifying the conversion processes with the aim of ensuring the low cost of synthesis and production on a bigger scale. Knowledge of active sites and the relationship between the structure and mass transfer can facilitate the selection of materials and the methodology for ORR synthesis. Additionally, it is necessary to intensively transfer the half-cells tests of synthesized ORR catalysts to corresponding devices, such as fuel cells and metal-air batteries [13].

2. Alginate as a Source of Carbon Material

Biopolymers, as renewable and non-toxic precursors, are important for the synthesis of carbon materials. Abundant sea resources represent one of the most important sources, considering that the ocean occupies three-quarters of the Earth's surface [63].

The wide alginates availability, with the simplicity of their extraction and synthesis via "green" processes, makes them exceptionally favorable materials, both commercially and environmentally. The field of their applications and significance is expanding, and one of the main challenges is their use as precursors of carbon materials production on a large scale.

Brown algae are a valuable source of alginate, a natural polymer composed of β-D-mannuronate and α-L guluronate. Due to the structure, as alginate consists of a large number of carboxyl and hydroxyl groups in the polymer chain, it is possible to obtain porous carbon material after alginate carbonization [64,65]. However, the different sequence and structure that vary depending on the source may be a disadvantage when this biopolymer is used [66]. Alginate interacts with metal ions and can chelate divalent and trivalent ions such as Ca^{2+}, Co^{2+}, Ni^{2+} Zn^{2+}, and Fe^{3+}, consequently forming an "egg-box" structure that can be used for the synthesis of metal-doped/free three-dimensional carbon nanomaterials with multimodal pores. This represents one of the main advantages of using the alginate as a precursor considering that the 1D structure has a very pronounced limited mass transfer of electrolyte ions. Multimodal pores formation was reported as a result of the material acid rinsing after thermal treatment in an inert gas atmosphere [67,68]. Large pores corresponded to the elimination of metal chelated into alginate, while smaller pores were the result of the release of H_2O and CO_2 during the thermal treatment.

Li et al. developed N-doped porous graphitic carbon nanofibers (N-PCNFs) with multimodal pores by the pyrolysis of electrospun alginate nanofibers at different temperatures (Figure 2). The result of the annealing treatment at 600 °C in ammonia was the formation of a metal-free catalyst for ORR with a higher stability and methanol tolerance for ORR in comparison to Pt/C. The mechanism revealed the four-electron pathway in an alkaline solution, with a capacity of 625 mAh g^{-1} [63]. Porous graphitic carbon nanofibers without N were also prepared via the pyrolysis of alginate. However, N-doping influenced the formation of defective structures in the carbon framework with a specific area of 283 m^2 g^{-1}, implying increased active sites and an enhanced electrochemical performance. Based on the Koutecky–Levich (K-L) plots, the electron transfer number was calculated. At 0.4 V, the n-value was 3.95, indicating that the catalyst ensures that the reduction mechanism takes place as a four-electron mechanism. Voltametric studies were employed, and cyclic voltammograms of N-PCNFs showed a well-defined reduction peak (at about 0.76 V vs. RHE), which indicated oxygen reduction. Linear sweep voltammetry (LSV) was also applied in an O_2-saturated 0.1 M KOH solution with a rotating disk electrode (RDE), and the ORR onset potential and half-wave potential of N-PCNFs-600 were 0.953 and 0.810 V, respectively. In 0.5 M H_2SO_4, the onset potential was 0.55 V, while the half-wave potential was 0.35 V. Chronoamperometric measurements were conducted at 0.70 V with a rotating speed of 1600 rpm with the aim of evaluating the methanol tolerance, whereby N-PCNF-600 had a better fuel selectivity toward ORR than Pt/C considering that the current density was stable after the addition of methanol. The durability was examined at a constant voltage of 0.3 V for 15 h in an 0.1 M KOH solution at a rotation rate of 1600 rpm. The synthesized catalyst had a slower decrease of 16% than Pt/C (25%) and, thus, a better durability.

The high charge and discharge performance, based on the specific structure and bimodal shape of the pore size distribution, is one of the important advantages of this catalyst.

Zhao et al. used calcium ions to form a network with alginate, after which wet spinning was applied with the aim of obtaining calcium alginate fibers (CAFs). CAFs were treated with an acid solution to remove metal ions [69,70], and the remaining carbon fibers were used for further experiments. N and S atoms from thioacetamide were doped into carbon fibers during pyrolysis, and after annealing at 1000 °C, the recombinant carbon atoms formed a metal-free structure without S and N atoms with defects (D-CFs), which aimed to improve the catalytic activity for ORR [71–73]. D-CFs had micro- and mesopores which enabled the availability of active sites and facilitated the transport of the reaction participants with a specific surface area of 485.2 m^2 g^{-1} and a 795 mAh g^{-1} capacity [74]. The initial potential was 0.92 V vs. RHE in 0.1 KOH, and the limiting current density was 5.38 mA cm^{-2}, which were comparable to those of Pt/C. The half-wave potential was 0.84 V vs. RHE, equivalent to that of Pt/C in 0.1 M KOH. The reaction process catalyzed by D-CFs was determined to take place as a four-electron pathway at 0.25, 0.35, and 0.45 V vs. RHE. By comparing D-CFs and Pt/C, the chronoamperometric results confirmed the better methanol tolerance that was expected. After 10 h, the current density for D-CFs was 94%, compared to 87% for Pt/C [70].

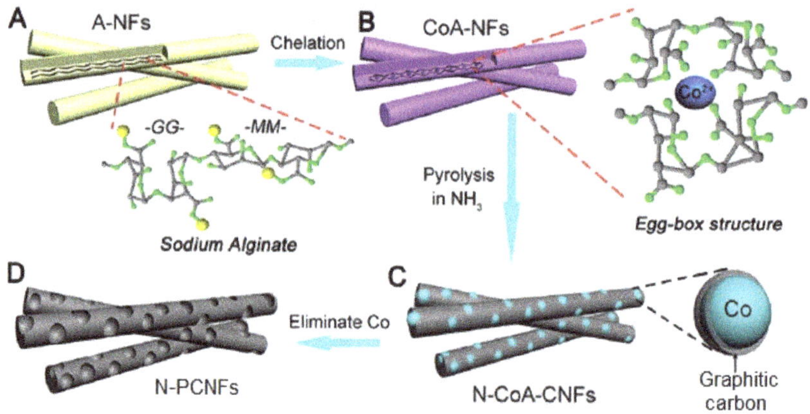

Figure 2. Schematic illustration of the synthesis process of N-doped porous graphitic carbon nanofibers (N-PCNFs). Reprinted from [63] with permission. Copyright 2015 American Chemical Society.

Considering the characteristics mentioned in the paper and the low cost, the obtained material could find its usage in large-scale applications of zinc-air batteries.

Alginate-Based Materials Doped with Metal Ions

Shu et al. [75] reported a novel strategy, thus broadening the methodology for the synthesis of Co/N-doped hierarchical porous carbon microspheres (Co/NHPCMS) with ZIF-67 (zeolite imidazolate framework), which were grown on the carbon frameworks. During the synthesis, ZIF-67 nanocrystals were grown on a carbon framework, and they improved the catalytic activity considering that they represent an in situ source of nitrogen for the material doping [76]. Sodium alginate/$NaHCO_3$ (SA/$NaHCO_3$) solution as a precursor and the electrospinning technique were used to form Co-SA/$NaHCO_3$ hydrogel microspheres. The specific structure of the alginate enabled the formation of a 3D network which improved the mass/charge transfer and increased the density of active sites. In such a structure, Co^{2+} formed an "egg-box" model and acted as the metal source of ZIF-67. ZIF-67@SA/$NaHCO_3$ microspheres were pyrolyzed in the nitrogen atmosphere, and Co/NHPCMS was obtained, with a specific surface area of 252 m^2 g^{-1}. For comparison, the same procedure was performed without $NaHCO_3$, and the ORR activity test showed that $NaHCO_3$ contributes to better ORR catalytic activity considering that the decomposition of $NaHCO_3$ results in the formation of more pores and defects, which increase the density of active sites. According to the XPS spectrum, the pyridinic and graphitic-N were considered as active sites. Additionally, it was suggested that pyridinic nitrogen could coordinate with cobalt ion. The Co nanoparticles influenced the aggregation process and corrosion and improved the stability of the catalyst [75,77].

The LSV-obtained data included the half-wave potential, which was 0.827 V, close to that of commercial Pt/C (0.851 V). Such positive values indicated a great advantage of the hierarchical porous structure. The K-L plots gave a linear dependence, and the electron transfer number was greater than 4, probably as a result of the adopted values for the K-L equation being different from the ideal. According to the chronoamperometric measurement, in O_2-saturated 0.1 M KOH at a potential of 0.6 V (vs. RHE), at a rotation rate of 1600 rpm, the relative current decreased to 79.10% after 18,000 s. Thus, compared to Pt/C, the newly synthesized catalyst showed greater ORR stability. Co/NHPCMS was also more tolerant to methanol compared to commercial Pt/C, as no change in the current density was observed after 200 s, when 1 mL of methanol was added [75].

Ma et al. [78] developed a novel hybrid material, a three-dimensional nanostructured electrocatalyst, for ORR and OER by the pyrolysis of (Ni,Co)/carbon nanotubes (CNTs) alginate hydrogels. SA/CNTs aqueous solution (Figure 3a,d) was exposed to a solution containing Co^{2+} and Ni^{2+} ions to form an "egg-box structure" (Figure 3b,e). Upon obtaining hydrogels, the freeze-drying process was used to form (Ni, Co)-alginate/CNT aerogels (Figure 3c,f). Carbonization in the NH_3 atmosphere was conducted in order to obtain the final form of the catalyst, whereby the catalyst Ni/NiO/NiCO_2O_4-N-CNT-As (As-aerogels) showed a significant advantage in terms of catalytic properties in comparison with commercial Pt/C, with a surface area of 222 m^2 $g-^1$. The high activity of CNTs was explained as a result of the possibility of pyridinic and graphitic nitrogen exerting an electron on the encapsulated Co nanoparticles [14,65,79]. Accordingly, the catalytic activity was attributed to the ternary Ni/NiO/NiCO_2O_4 active site for electron transfer and the 3D hierarchical mesoporous hybrid network for mass transport. The onset potential and half-wave potential of Ni/NiO/NiCO_2O_4-N-CNT-As (obtained by LSV) were 0.89 V and 0.74 V vs. RHE, respectively, compared to 0.92 V and 0.78 V for Pt/C. Based on the reviewed results, newly synthesized catalysts were recommended as ORR catalysts, especially as bifunctional for ORR and OER. According to the N_2 adsorption-desorption isotherms and BJH pore size distribution, the synthesized material had a bimodal shape in the pore size distribution, large mesopores as a consequence of interconnected voids from tangled CNTs in the aerogels, and small mesopores as a consequence of H_2O and CO_2 release [79].

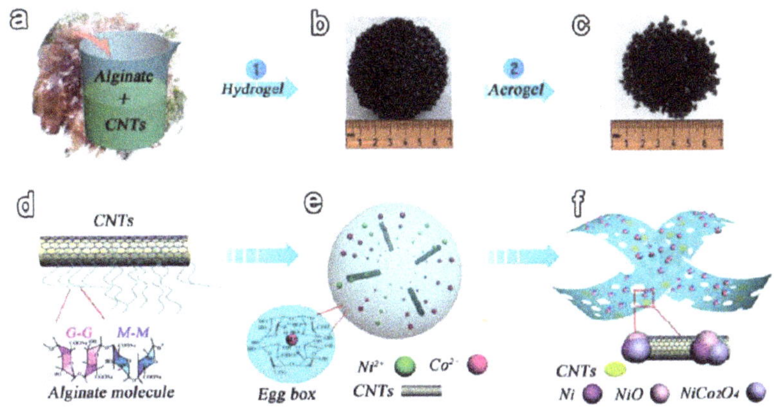

Figure 3. Schematic synthesis of Ni/NiO/NiCo2O4/N-CNT-As electrocatalysts; (**a**,**d**) shematic representation of SA with CNs, (**b**,**e**) hydrogels formed upon addition of aqueous solution containing Co^{2+} and Ni^{2+} ions; (**c**,**f**) freeze-dryied hydrogels formed aerogels. Reprinted with permission from [78]. Copyright 2015 The Royal Society of Chemistry.

According to the K–L equation, as in other reported materials in this review, the number of electrons was 3.8, a value desirable for ORR. The durability was tested at a constant voltage of 0.69 V vs. RHE for 22,000 s at a rotation rate of 1600 rpm, where the current loss was 9%, proving the improved durability compared to commercial Pt/C. Additionally, the tolerance to methanol was better for the Ni/NiO/NiCO_2O_4-N-CNT-As, as the negative current appeared after the introduction of O_2 in the KOH saturated with N_2 at 1000 s, while subsequently, after the addition of 3M methanol, the newly synthesized catalyst's current had no changes [78].

The proposed method may be easily large-scaled, so it presents a valuable reference for the formation of dual ORR and OER catalysts.

Zhang et al. [80] developed a catalyst consisting of a 3D N-doped porous carbon matrix with Co_3O_4/Co active species–Co_3O_4/Co-NPC and a uniform porous nanostructure with a surface area of 496.4 m^2 g^{-1}. The great advantage of the developed method was certainly the simple and low-cost synthesis, as they used hydrothermal reaction, unlike methods that use an electrospun technique or complex procedures. This method is one of the most promising methods for the synthesis of ORR catalysts [81–85]. In addition to catalyzing the ORR, the catalyst can be used for anodic oxygen evolution reaction (OER) as well; thus, it possibly has practical applications for energy conversion.

Carbon materials contributed to a better stability and conductivity and an increased surface area compared to the nanostructures of cobalt oxides and/or hydroxides [86–89]. Nevertheless, hybrid material consisting of metal at the nanoscale and heteroatom co-doped carbon showed better performances as ORR catalysts and bifunctional ORR/OER catalysts [90,91]. As biopolymers provide a cost-effective and renewable source, they attracted the attention of these authors as well [80]. After the chemical reaction of SA with Co ions, SA formed a network such that, after carbonization, a catalyst containing both Co and Co_3O_2 species-Co_3O_4/Co-NPC was obtained. Melamine was used as a source of nitrogen, and after the carbonization at different temperatures, the product obtained at 800 °C had the optimal characteristics [92–94]. The synergistic effect of the Co_3O_4/Co active species and NPC was responsible for the desired mass transport and charge transfer and the specific dual activity towards ORR and OER.

CV measurements were conducted to evaluate the ORR activity. In 0.1 M KOH solution, Co_3O_4/Co-NPC had a peak at 0.738 V, which was at more positive potentials compared to Pt/C. The onset potential obtained by LSV was 0.91 V due to its value being smaller than that for Pt/C (0.987 V) and its half-wave potential being 0.806 V vs. RHE (less than the half-wave potential for Pt/C 0.8875 V). This result encourages the application of the synthesized catalyst for ORR catalysis. According to K–L plots, the average electron transfer number at the potential from 0.25 to 0.45 V was 3.9, indicating a four-electron mechanism for ORR. The chronoamperometric curves of Co_3O_4/Co-NPC and Pt/C were compared to define the tolerance to methanol, where the synthesized catalyst had a negligible response after the addition of methanol and thus showed better tolerance. O_2-saturated 0.1 M KOH at 0.5 V (vs. RHE), with a rotating speed of 1600 rpm, was used to determine the long-term stability, where, after 7 h, Co_3O_4/Co-NPC had a current value at 94.5% compared to Pt/C (86.5%). Thus, the porous N-doped carbon hybrid with Co and Co_3O_4 was suggested as the promising ORR and OER agent [80].

Alginate fibers (AF) which formed an "egg-box" structure with Co^{2+} ions were used as growing sites for the coating of Co-based ZIF (Zeolitic imidazolate frameworks)-MOF (metal organic framework), and by applying the solvothermal process, the AF@ZIF-67 was obtained [95]. Synthesized material was carbonized at different temperatures, and optimal characteristics were shown by the Co-CF obtained at temperatures of 900 °C in an N_2 atmosphere. Due to the specific structure formed, Co and N, which were uniformly distributed on the one-dimensional CFs, were active sites and had the effect of accelerating the process of mass transfer and the transmission of electrons [96]. As ZIFs were composed of molecules with a large number of N atoms, after carbonization, the CoNC-CF material was obtained, which was shown to be an efficient catalyst for the ORR reaction. In 0.1 KOH, the half-wave potential was 0.833 V vs. RHE, and the onset potential was 0.926 V (vs. RHE) compared to 0.827 V and 0.942 V vs. RHE, respectively, for 20 wt.% Pt/C. The limiting current density was -5.56 mA^{-2}, and, compared with catalysts without N and Co active sites, the aforementioned catalyst had obvious advantages, as stated. The authors confirmed that the reduction reaction occurred through a four-electron pathway according to K–L plots. After 20,000 s, at 1600 rpm, CoNC-CF pyrolyzed at 900 °C had 94.8% of relative current in contrast to Pt/C, with a retention of 80.9% [95].

Compared to previously reported materials, the greatest disadvantage of this material would be its 1D structure. Despite that, the obtained material has been shown to be an excellent ORR catalyst.

The synthesis of Fe_2N/C ORR catalysts via a simple, eco-friendly route, where Fe^{3+} cations formed a novel structure of an "egg box" with alginate in the excellent yields, was described in work of Liu et al. [14]. Fe_2N nanoaerogels had outstanding ORR electrocatalytic activity, stability, and methanol tolerance in alkaline as well as in an acidic medium.

The (SA) was mixed with graphene; thus, the dispersion was obtained. The mixture was added to the $FeCl_3$ solution, and hydrogel was formed [63,65,97]. Upon a freeze-drying process, 3D Fe-alginate/graphene aerogels (GAs) were made, and as the size of the aerogel was only limited by the size of the container where the ion-exchange process took place, the proposed method may be a facile strategy for scaling up the production of Fe_2N/N-GAs from brown algae. After calcination at different temperatures, the material obtained at 700 °C in the NH_3 atmosphere, i.e., the optimal 3D Fe_2N/N-GAs, was formed. The 3D Fe_2N/N-GAs had a high-density N-doped amorphous carbon shell (N-AC) which encapsulated Fe_2/N NPs. This characteristic presents the great advantage of this material, considering that some Fe/N/C electrocatalysts for ORR did not have the controlled porous structure; thus, active sites were not exposed adequately, and transport properties were poor. The resulting aerogel was bimodal, as 3D mesoporous networks of about 30 nm and small mesopores of 3 nm were achieved, where the specific surface area was 465 $m^2\ g^{-1}$ [14]. The molecular transport throughout the entire 3D architecture was supposed/achieved, and small mesopores were beneficial for the O_2 adsorption in the ORR [98]. The rate-determining step in the ORR, i.e., the splitting of O-O bonds as a consequence of the transfer of two electrons from active sites to adsorbed O_2, was influenced by quaternary-type and pyridine-like nitrogen, while the Fe was probably the site for O_2 adsorption.

The Fe_2N-based nanoaerogels had electrocatalytic activity for ORR in 0.1 M KOH and 1 M $HClO_4$ solution. In an alkaline medium, the onset potential and half-wave potentials were 1.02 V and 0.93 V, respectively, while in an acidic medium, those values were 0.82 V for the onset and 0.71 V for the half-wave potential vs. RHE. The electron transfer number was four, and the current density was -4.5 mA cm^{-2}. By comparing the durability of Fe_2N/N-GAs with that of Pt/C, the authors showed that the durability in acidic and alkaline media was better for Fe_2N/N-GAs. After 9 h, there was only a 10% loss of the initial current density in KOH (for Pt/C, 20%), while in $HClO_4$, there was a 19% loss after 10.5 h (for Pt/C, 45%). Fe_2N/N-GAs also had a good ability to resist the crossover effect; thus, there was no noticeable change in the current density after the addition of methanol [14]. This excellent catalytic activity was probably a consequence of an optimal balance of the surface area, the density of the active sites, and the electrical conductivity [99].

Yu. et al. [100] developed a simple and universal method for obtaining single-atom catalysts (SACs) with a surface area of 1551 $m^2\ g^{-1}$ using Cu ion as a representative of transition metals that can form a network with alginate (i.e., "egg-box" structure). As one of the main problems for the investigation of SACs applications is the lack of a universal method for synthesis, these authors have significantly contributed to the development of SACs. In addition, they avoided physical and traditional chemical approaches which usually result in low production, complex equipment, a cumbersome process, and a high cost [101,102].

After the calcium from Ca-alginate (derived from sodium alginate) was replaced with H^+ ions, the chemical reaction with $CuCl_2$ was enabled. Hydrogel consisting of Cu and alginate was dried and carbonized at 900 °C under an NH_3 atmosphere. With the aim of eliminating the metal ions, the obtained material was washed with an acidic solution. The obtained product, Cu-SAC/N, was examined for ORR activity, as well as SACs synthesized using the same procedure but with different transition metals, such as Fe-, Co-, and Mn. For catalysts obtained with Cu, in 0.1 M KOH, the onset potential was 0.90 V, and the half-wave potential was 0.80 V vs. RHE. Compared to the onset and half-wave potentials for the Pt/C catalyst (0.93 and 0.83 V, respectively), the newly synthesized material had an improved electrocatalytic activity, primarily owing to the catalytic site, which was Cu-N_4. As in the majority of experimentally obtained materials, only surface atoms acted as active sites, and the SACs have a great atom utilization efficiency. The experimentally obtained transfer electron number was about four. After the addition of methanol, the current remained stable. The same trend was achieved by analyzing the current density after 3000 cycles of the CV, where the decrease was not more than 10%.

To summarize, a comparative table with the indicated method for obtaining the type of carbon material and the characteristics of the ORR catalysts derived from alginate is given in Table 3.

Table 3. ORR catalysts derived from alginate; characteristics, methods for obtention, and type of carbon materials.

Reference	[63]	[70]	[75]	[78]	[80]	[95]	[14]	[100]
Specific area	283 $m^2\ g^{-1}$	485.2 $m^2\ g^{-1}$	252 $m^2\ g^{-1}$	222 $m^2\ g^{-1}$	496.4 $m^2\ g^{-1}$	/	465 $m^2\ g^{-1}$	1551 $m^2\ g^{-1}$
Heteroatom	N	N and S eliminated upon annealing	N	N	N	N	N	N
Dopant	/	/	Co	Co, Ni	Co	Co	Fe	Cu
Half-wave potential of catalyst/Pt/C	0.810V vs. RHEl	0.84 V vs. RHE	0.827 V vs. RHE	0.74 V vs. RHE	0.806 V vs. RHE	0.833 V vs. RHE	0.93 V vs. RHE	0.80 V vs. RHE
Optimal preparation temperature	600 °C	1000 °C	1000 °C	400 °C	800 °C	900 °C	700 °C	900 °C
Derived catalyst	N-doped porous graphitic carbon nanofibers	defective carbon fibers	3D Co/N-doped hierarchical porous carbon mi-crospheres	carbon nanotubes nanoaero-gels	Co_3O_4/Co species incorporated into the N-doped carbon matrix	carbon fiber-coated Co@N-doped porous car-bondoped	3D Fe-alginate/ graphene aerogels	single-atom catalyst
Synthesis technique	electrospun	wet spinning	electrospinning	freeze-drying	hydrothermal reaction	chemical reaction	freeze-drying	chemical reaction

3. Other Widely Used Biopolymers

Cellulose, lignin, and chitin are widely distributed polymer materials that are increasingly used as sustainable precursors for the synthesis of carbonaceous materials since they are non-toxic, biodegradable, and biocompatible [103,104].

3.1. Cellulose

Cellulose is a homopolymer consisting of glucopyranose [105]; therefore, to obtain materials with good ORR electrocatalytic activity, doping or co-doping with heteroatoms is desirable [51]. The chains of cellulose are connected in parallel, and the crystalline structure, via hydrogen and van der Waals interactions, forming microfibrils [51]. The degree of polymerization varies according to its source [106] Besides many advantages, such as the large amounts, biodegradability, low cost, etc., they also have some disadvantages; the low moisture resistance may be the most important [107].

Kim et al. used cellulose as the biomaterial for the high-value-added N-doped hierarchical porous carbon (NHPC) [108].

They developed a simple, inexpensive, and efficient process in which they mixed cotton cellulose with magnesium nitrate hexahydrate or magnesium acetate tetrahydrate and urea in different proportions. As previously reported methods included high-cost materials and low outputs, this method presents a great improvement. Upon drying, the mixture was pre-pyrolyzed at 500 °C in an Ar atmosphere. Afterwards, the obtained material was carbonized at different temperatures. The authors explained in detail the influence of the preparation parameters and contributed to the improvement of the methodology of making carbon-based ORR catalysts. Pyrolysis and carbonization were separated in order to characterize the material after pyrolysis and to avoid the pollution of high-temperature furnaces due to a large amount of gas emission. The treatment with an acidic solution was applied after the carbonization with the aim of obtaining a final material without metal ions.

The novelty of the proposed method was that the authors used accelerated pyrolysis, where the exothermic reactions were caused by the application of nitrates, which encouraged pyrolysis at low temperatures and the rapid exfoliation of cellulose fibers. In addition, urea was used as an additional source of nitrogen but also as a reactant that supports the total exfoliation of cellulose when applied in the optimal content (Figure 4). Consequently, the sample synthesized with magnesium nitrate and urea had a highly 3D porous structure with macropores, mesopores, and abundant micropores and a specific surface area of 1173 m^2 g^{-1}. Graphitic and pyridinic N, i.e., the N-C parts of the catalysts, were considered as active sites, and the high specific surface area combined with the hierarchical open pore structure contributed to the increase in the exposed active sites, thus enabling efficient mass transport and fast ion transport. According to the electrochemical measurements, including CV and LSV, the onset potential was 0.94 V, the half-wave potential was 0.83 V, and the estimated number of transferred electrons was in the range of 3.5–4 for the sample carbonized at a temperature of 1000 °C. As the temperature of carbonization influences the doping amount and doping species of N, as the graphitization degree of the carbon, according to the stated results, the optimal temperature was 1000 °C. The durability of the optimal material was better compared to that of the Pt/C catalyst, as the current obtained by the chronoamperometric measurements was 88% after 10 h in KOH, while for the Pt/C, it was 83% at 0.6 V vs. RHE in the O_2-saturated KOH solution at a rotating speed of 1600 rpm. The material was methanol-tolerant, as the current had no significant changes after the addition of 2% methanol, while for the Pt/C, the current decreased [108].

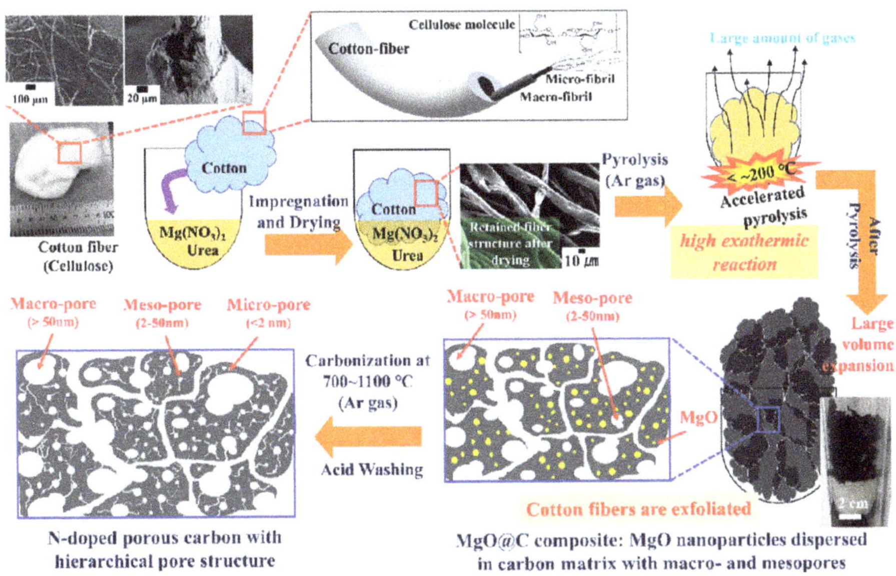

Figure 4. Schematic diagram of the preparation of N-doped porous carbon particles with a hierarchical pore structure from cellulose. Reprinted with permission from [108]. Copyright 2019 American Chemical Society.

3.2. Chitin

Chitin is, besides cellulose, the most widespread biopolymer [109,110]. It has a high content of nitrogen. Before using from the biomass, the chitin has to be extracted. This process includes mechanical grinding, chemical demineralization, and deprotonation, techniques that may be time-demanding. In addition, the direct carbonization of chitin could not provide a defined morphology and porosity of the newly obtained material [111]. Chitosan, the deacetylated form of chitin with a different degree of deacetylation, may also be used for different purposes [110,112]. As chitosan contains about 7% of N, it presents a valuable resource of material containing carbon and nitrogen from amine and acetamide functional groups [57,103,113,114], which, upon thermal treatment, may exclude an additional step of N-doping in the synthesis process of functionalized carbons.

The specific surface area and the type of functional groups [115], both dependent on the carbonization temperature, are the main factors favoring an increase in the catalytic activity for N-doped porous carbons obtained from chitin and chitosan. The urea treatment of these materials had a great benefit for their ORR performance, since it further increased the specific surface area and enhanced the N-doping [116].

Wang et al. designed cobalt- and nitrogen-co-doped carbon material (CoNC), using a mixture of chitin and cobalt as a precursor, through a one-step pyrolysis process at 800 °C [117]. The obtained CoNC showed a specific area of 165 $m^2\ g^{-1}$, and it was reported as a good material for ORR electrocatalysts in alkaline media for Al-air batteries. The onset potential was 0.86 V vs. RHE, and the high-limiting current density of 4.91 mA cm^{-2} was comparable to those obtained for Pt/C 0.9 V and 5.67 mA cm^{-2}. The electron-transfer number was 3.73. The long-time durability was examined at 900 rpm in 0.1 M KOH saturated with O_2, and after 15,000 s, the current density was 94.82%.

3.3. Lignin

Lignin is the by-product obtained in the process of pulping. It is the constituent of grass, trees, and plants [118].

Lignin presents an amorphous aromatic polymer [58] which consists of p-hydrophenyl (H), syringyl (S), and guaiacyl (G) components, which form a three-dimensional structure. As lignin has a high content of benzene rings and phenolic functional groups, it is widely used as a precursor for the synthesis of carbon material (carbon spheres, nanofibers, nanosheets, 3D-porous carbon, carbon composites, etc.). The availability, low cost, the unique designability and controllability of the lignin structure are certainly its great advantages for its usage in different spheres [58]. The disadvantages of lignin include the fact that its usage may be a complicated procedure that disables the large-scale production. Accordingly, the development of an environmentally friendly process for the production of carbon materials from lignin is mandatory, with special attention given to the fact that the relationship between the structure of carbonized lignin and its catalytic activity is not absolutely understood yet.

The reported carbon materials derived from lignin include nitrogen-doped and nitrogen/sulfur- and nitrogen-phosphorus-co-doped catalysts [119–121]. Li et al. [122] used lignosulfonate and synthesized a robust multifunctional carbon catalyst for ORR, OER, and hydrogen evolution reaction (HER), which contained iron, nitrogen, phosphorus, and sulfur. Although different methods for carbon-doping were reported, the catalytic activity for the ORR, OER, and HER of such materials was not satisfactory. The improvement was achieved by the synthesis of the catalyst with a larger specific surface area, where active sites were exposed to the surface. The highly porous structure improved the rate of mass/electron transfer, the doping with heteroatoms upgraded the electroconductibility and charge transfer, and the FeN_x and FeP_x species had an impact on the prevention of the agglomeration of active sites and thus provided great durability. Finally, the obtained catalyst had the intrinsic activity of each active site and, consequently, great activity toward these three different reactions. The material was synthesized during the process that consisted of the preparation of lignin-Fe, and then the obtained substance was mixed with hypophosphite (inorganic molten salt template), annealed at 500 °C, and carbonized at 800 °C. After treatment with HCl solution, the material was again annealed at 800 °C and tested for catalytic activity [122]. The obtained catalyst had a specific surface area of 782 m^2 g^{-1}, which is larger than that of some reported catalysts [121]. The half-wave potential was 0.9 V, and the electron transfer number was 3.9. After injecting 3 M methanol, the current density had a negligible change, indicating a robust tolerance to methanol [122]. After 86,400 s, the current changed by 9.1%, while for the Pt/C, the change was 17.1%. Shen et al. [121] used lignin with melamine to form lignin carbon nanosheets of a surface area of 1208 m^2 g^{-1} co-doped with nitrogen and sulfur. The material may be used in an acid medium as well as in alkaline due to its better performances compared to Pt/C (a more positive half-wave potential and (nearly) current density and a high graphitic N ratio including four electron mechanisms for ORR).

4. Biomass Used for the ORR Catalysts Preparation

The carbonization of waste biomass, including adequate pre- and post-treatment processes, can be used for the preparation of ORR catalysts. Biobased N-doped carbons have been synthetized from coconut shell particles, shrimp shell, and even hair.

Due to the specificity of the structure formed by cross-linked cellulose, hemicellulose, and lignin molecules, coconut shells have a high density and low porosity; they therefore represent a good precursor material whose carbonization produces catalysts of high carbon contents and low ash residues [123]. Borghei et al. [124] developed an environmentally friendly method, without using hazardous chemicals to synthesize N/P-doped porous carbon, where, after the single-step activation of coconut shells with phosphoric acid, urea was the source of nitrogen doping. New material obtained after carbonization at 1000 °C with a specific surface area of 1260 m^2 g^{-1} had a comparable electrocatalytic activity with Pt/C, a better tolerance to methanol crossover, and an improved long-term durability in alkaline media.

Although it is shown that the tailoring of the pore size distribution is better achieved through physical activation [125], the authors performed the activation chemically using phosphoric acid, which led to the high-yield synthesis of carbons with a three-dimensional structure of a high porosity. The long-term stability was evaluated by chronoamperometry at 0.7 V and at 900 rpm after 50,000 s, and the results showed that the functionalization with urea improved the stability by reaching a plateau of about 75% after 13 h. The tolerance to methanol was examined, where the current intensity was the same before and after the addition of methanol, which indicated a good selectivity of the material toward the ORR. The excellent electrocatalytic activity of this material may be the result of different factors, including the contribution of a large fraction of mesopores, which enables enhanced electrolyte mass transport and the easy access of oxygen molecules to the active sites due to the 3D open pore structure, the synergic effect of N and P doping, and the large proportion of N-graphitic and N-pyridinic species. The authors suggested that the O_2 adsorption probably occurred on N-graphitic sites, while N-pyridinic sites enabled fast decomposition or an electroreduction of hydroperoxide, resulting in a fast conversion of oxygen molecules [126]. Nevertheless, the influence of the N-doping level and ORR activity still remains absolutely defined [127,128]

Jahran et al. [123] also reported the synthesis of N-doped activated carbon from coconut shells with bifunctional electrocatalytic activities towards ORR and OER, showing a good stability in alkaline electrolytes. First, they pulverized the coconut shells, soaked them overnight with 50 wt.% HNO_3, and exposed them to a temperature of 550 °C under the N_2 atmosphere. After the functionalization of the material, the N-doping was accomplished with urea, and the obtained product was subsequently pyrolyzed at 900 °C.

N-doped material had a three-dimensional structure with pores, and upon the presence of nitrogen, it had an enhanced electrocatalytic performance, a higher current density, and a more positive onset potential compared to raw material. The authors used a K–L plot to determine the number of electrons transferred in ORR, where a direct four-electron reduction pathway was obtained for N-doped activated carbon [123]. The LSV analysis of the material confirmed that the presence of nitrogen enhances the electrocatalytic activity, i.e., enables a more positive onset potential and a higher current density. The durability of the prepared material was better compared to that of Pt/C since there was a slow decrease in the current after 1000 s in 0.1 M KOH. The authors claimed that the method was optimal for ORR catalyst synthesis, but no data for the onset and half-wave potential were presented.

Liu et al. presented a simple template-assisted method for ORR catalyst development using shrimp-shell-based carbons [129]. Shrimp waste is seafood production waste [110,130], and chitin is one of the most important parts of a shrimp shell, as it makes up as much as 40% of the shrimp mass [110,131]. Shrimp shells were ground to powder and exposed to a hydrothermal reaction such that N-doped carbon nanodots (N-CN) were formed. N-CN reacted with SiO_2 spheres under experimental conditions that were optimal for obtaining N-CNs@SiO_2, which can be pyrolyzed in an N_2 atmosphere at 800 °C. At the end of the process, SiO_2 was removed, and the material with an intact three-dimensional N-doped porous carbon network was obtained. A three-modal-pore structure was noticed, with micro-, meso-, and macropores; thus, the material contained many defects in the carbon structure and, consequently, active sites for electrocatalysis predominantly consisting of pyridinic and graphitic-N [132–134]. The onset potential in 0.1 M KOH saturated with O_2 was 0.905 V, which was close to that of Pt/C. The limiting current was 5.3 mA cm^{-2} at 0.6 V vs. RHE. The electron transfer number was between 3.75 and 3.95, over the potential range from 0.56 V to 0.32 V. The NPC material had no crossover effect, as the current change after the addition of 3.0 M methanol was negligible.

The authors also examined the durability, and after 20,000 s, the NPC exhibited a 12% decrease in current at a rotation speed of 1600 rpm and under an applied potential of 0.62 V, which certainly recommends the obtained material for use as an ORR catalyst [129].

Some studies suggested that carbon nanodots have high water adsorption properties, especially in an alkaline solution, due to the presence of O- and N-containing hydrophilic functional groups; thus, they could not be used as catalysts owing to the low uniformity and stability of the film made of the catalyst on the electrode [135–137]. The solution to this problem may involve the immobilization of carbon nanodots onto conductive carbon with the aim of overcoming the falling-off process [135–137]. Additionally, the low graphitization degree as a result of the low-temperature hydrothermal process, as one of the problems, may be solved by increasing the pyrolysis temperature, but agglomeration with a low surface area and an unacceptably low electrical conductivity may still be a problem [138]. Taking all of this into account, Li et al. contributed significantly to the field of efficient catalysts synthesis using biomass as a source material.

As human and animal hair can be classified as organic waste and represents the source of carbon, it can be used for the synthesis of carbon-based materials with different characteristics, including porous structures with a great surface area [53,139,140]. Chaudhari et al. [53] synthesized porous carbon material by using hair as a source of carbons and heteroatoms.

Hair consists of the α-keratin protein and thus contains nitrogen and sulfur from amino acids which are interconnected by peptide bonds [52]. Keratin is known as a protein with high content of cysteine, an amino acid rich in sulfur, making hair a good precursor material for heteroatom-doped carbon material synthesis. The specific structure of hair makes it an excellent single precursor for N- and S-doped material, consequently avoiding the usage of hazardous chemicals. Chaudhari et al. used carbonization, mild activation, and graphitization with the aim of obtaining the high-porosity material that consists of N- and S- doped carbons, which may be used as ORR catalysts. After carbonization at 250 °C in the N_2 atmosphere, the obtained char was activated at 600 °C using NaOH under a constant N_2 flow. As the graphitization was conducted at different temperatures, the temperature of 900 °C was proved to be the best for optimal ORR characteristics. The combination of heteroatom doping, a high surface area, and electrical conductivity makes the hair a promising precursor for the synthesis of ORR catalysts. The micropores and mesopores ranging from 0.5 to 4.0 nm were dominant. The specific surface area was 1813.95 m^2 g^{-1}. The electrochemical parameters were as follows: the onset potential was

0.949 V vs. RHE, and the reaction current was −4.92 mA cm^{-2} at 0.4 V. The electron transfer number was between 3.80 and 3.90. According to the chronoamperometry results, there was no noticeable change in the current response after 300 s of the addition of 3.0 M methanol. The durability and stability were tested at a constant potential of 0.7 V for 14 h, and after 50,000 s, the current density in 0.1 M KOH saturated with O_2 decreased by 14% at 1600 rpm.

5. Conclusions

Biopolymers contribute inestimable value towards the preparation of ORR catalysts. After appropriate treatment (mainly carbonization at an optimized temperature), these compounds give highly porous carbon material with/without dopants, making them excellent candidates for ORR catalysts. The porous carbon materials obtained from alginate are of great interest due to their low cost, wide availability, and relatively simple fabrication procedure. The inherent ability of alginate to form the "egg-box" structure after cross-linking with metal ions makes it possible to obtain materials whose electrocatalytic activity is comparable to that of Pt/C. Apart from alginate, cellulose, chitin, and lignin are excellent precursors for the preparation and production of large quantities of catalysts. As the application of Pt/C catalysts is limited by its high production costs, poor tolerance to methanol, and low durability, the limits of biopolymer applications are expanding.

Undoubtedly, natural polymers represent a wide range of compounds, which, apart from being available and affordable, have a specific structure which makes them an excellent starting compound for obtaining ORR catalysts with better durability, an improved tolerance to methanol, and better electrocatalytic properties compared to commercial ORR catalysts. Various biopolymer-derived, metal-free ORR catalysts have been studied and shown to possess excellent electrocatalytic activity, a remarkable tolerance to CO poisoning, fuel crossover effects, and a concomitant, good operational stability that is comparable to the best among non-precious metal catalysts. Porosity in biopolymer-derived carbons can ensure efficient oxygen transport and even alter the reaction mechanism, making them a viable component in PEMFCs. Despite the tremendous advancements made in the development, with heteroatom doping, a pore structure design, and surface modification, some inherent drawbacks still remain. Changes in the conductivity with higher doping levels, the lack of catalytic activity in acid media, and the difficulty of the preparation of catalysts with an exact structure and surface groups are still some of the issues that remain unresolved. The exact active site for ORR is still under debate among many researchers, many of whom turned to theoretical calculation, mainly DFT, to explore the main descriptor guiding the catalysis. Predictions of the electronic structure, active sites, reaction pathways, and intermediates provided by DFT present a part of the solution, with further research being needed to guide and connect the predictions from theoretical studies to the results from the laboratory. Biopolymer-derived, carbon-centered electrocatalysts have proved very versatile, with a great potential for improvement through doping, the pore structure, and active site design, making them an unavoidable part of the solution for many energy conversion and storage devices.

Author Contributions: J.R.: Conceptualization, Writing—Original Draft, Visualization, Writing—Review and Editing. D.T.: Visualization, Writing—Review and Editing, A.J.L.: Writing-Original Draft, Writing—Review and Editing. K.K.U.: Conceptualization, Writing—Original Draft, Supervision, Writing—Review and Editing. All authors have read and agreed to the published version of the manuscript.

Funding: This research was funded by the Ministry of Education, Science, and Technological Development, the Republic of Serbia, through a Grant Agreement with the University of Belgrade—Faculty of Pharmacy No. 451-03-68/2022-14/200161 and the University of Belgrade—Faculty of Physical Chemistry No. 451-03-68/2022-14/200146.

Data Availability Statement: Not applicable.

Conflicts of Interest: The authors declare no conflict of interest.

References

1. Dühnen, S.; Betz, J.; Kolek, M.; Schmuch, R.; Winter, M.; Placke, T. Toward Green Battery Cells: Perspective on Materials and Technologies. *Small Methods* **2020**, *4*, 2000039. [CrossRef]
2. Kwak, W.-J.; Rosy; Sharon, D.; Xia, C.; Kim, H.; Johnson, L.R.; Bruce, P.G.; Nazar, L.F.; Sun, Y.-K.; Frimer, A.A.; et al. Lithium-Oxygen Batteries and Related Systems: Potential, Status, and Future. *ACS Appl. Mater. Interfaces* **2020**, *120*, 6626–6683. [CrossRef] [PubMed]
3. Wang, Y.; Chu, F.; Zeng, J.; Wang, Q.; Naren, T.; Li, Y.; Cheng, Y.; Lei, Y.; Wu, F. Single Atom Catalysts for Fuel Cells and Rechargeable Batteries: Principles, Advances, and Opportunities. *ACS Nano* **2021**, *15*, 210–239. [CrossRef]
4. Carrette, L.; Friedrich, K.A.; Stimming, U. Fuel Cells: Principles, Types, Fuels, and Applications. *ChemPhysChem* **2000**, *1*, 162–193. [CrossRef]
5. Dyer, C.K. Fuel Cells for Portable Applications. *Low-Power Electron. Des.* **2002**, *106*, 31–34. [CrossRef]
6. Hilal, M.E.; Aboulouard, A.; Akbar, A.R.; Younus, H.A.; Horzum, N.; Verpoort, F. Progress of MOF-Derived Functional Materials toward Industrialization in Solar Cells and Metal-Air Batteries. *Catalysts* **2020**, *10*, 897. [CrossRef]
7. He, Y.; Yang, X.; Li, Y.; Liu, L.; Guo, S.; Shu, C.; Liu, F.; Liu, Y.; Tan, Q.; Wu, G. Atomically Dispersed Fe-Co Dual Metal Sites as Bifunctional Oxygen Electrocatalysts for Rechargeable and Flexible Zn-Air Batteries. *ACS Catal.* **2022**, *12*, 1216–1227. [CrossRef]
8. Sun, J.; Wang, N.; Qiu, Z.; Xing, L.; Du, L. Recent Progress of Non-Noble Metal Catalysts for Oxygen Electrode in Zn-Air Batteries: A Mini Review. *Catalysts* **2022**, *12*, 843. [CrossRef]
9. Slavova, M.; Mihaylova-Dimitrova, E.; Mladenova, E.; Abrashev, B.; Burdin, B.; Vladikova, D. Zeolite Based Carbon-Free Gas Diffusion Electrodes for Secondary Metal-Air Batteries. *J. Electroanal. Sci. Eng.* **2020**, *10*, 229–234. [CrossRef]
10. Wang, J.; Huang, Z.; Liu, W.; Chang, C.; Tang, H.; Li, Z.; Chen, W.; Jia, C.; Yao, T.; Wei, S.; et al. Design of N-Coordinated Dual-Metal Sites: A Stable and Active Pt-Free Catalyst for Acidic Oxygen Reduction Reaction. *J. Am. Chem. Soc.* **2017**, *139*, 17281–17284. [CrossRef]
11. Debe, M.K. Electrocatalyst Approaches and Challenges for Automotive Fuel Cells. *Nature* **2012**, *486*, 43–51. [CrossRef] [PubMed]
12. Wu, G.; More, K.L.; Johnston, C.M.; Zelenay, P. High-Performance Electrocatalysts for Oxygen Reduction Derived from Polyaniline, Iron, and Cobalt. *Science* **2011**, *332*, 443–447. [CrossRef] [PubMed]
13. Du, L.; Zhang, G.; Liu, X.; Hassanpour, A.; Dubois, M.; Tavares, A.C.; Sun, S. Biomass-Derived Nonprecious Metal Catalysts for Oxygen Reduction Reaction: The Demand-Oriented Engineering of Active Sites and Structures. *Carbon Energy* **2020**, *2*, 561–581. [CrossRef]
14. Liu, L.; Yang, X.; Ma, N.; Liu, H.; Xia, Y.; Chen, C.; Yang, D.; Yao, X. Scalable and Cost-Effective Synthesis of Highly Efficient Fe2N-Based Oxygen Reduction Catalyst Derived from Seaweed Biomass. *Small* **2016**, *12*, 1295–1301. [CrossRef]
15. Popadić, D.; Gavrilov, N.; Ignjatović, L.; Krajišnik, D.; Mentus, S.; Milojević-Rakić, M.; Bajuk-Bogdanović, D. How to Obtain Maximum Environmental Applicabilitz from Natural Silicates. *Catalysts* **2022**, *12*, 519. [CrossRef]
16. Yun, S.; Hagfeldt, A.; Ma, T. Pt-Free Counter Electrode for Dye-Sensitized Solar Cells with High Efficiency. *Adv. Mater.* **2014**, *26*, 6210–6237. [CrossRef]
17. Zhou, Q.; Shi, G. Conducting Polymer-Based Catalysts. *J. Am. Chem. Soc.* **2016**, *138*, 2868–2876. [CrossRef]
18. Winther-Jensen, B.; Winther-Jensen, O.; Forsyth, M.; MacFarlane, D.R. High Rates of Oxygen Reduction over a Vapor Phase-Polymerized PEDOT Electrode. *Science* **2008**, *321*, 671–674. [CrossRef]
19. Bashyam, R.; Zelenay, P. A Class of Non-Precious Metal Composite Catalysts for Fuel Cells. *Nature* **2006**, *443*, 63–66. [CrossRef]
20. Gong, K.; Du, F.; Xia, Z.; Durstock, M.; Dai, L. Nitrogen-Doped Carbon Nanotube Arrays with High Electrocatalytic Activity for Oxygen Reduction. *Science* **2009**, *323*, 760–764. [CrossRef]
21. Choi, C.H.; Park, S.H.; Woo, S.I. Binary and Ternary Doping of Nitrogen, Boron, and Phosphorus into Carbon for Enhancing Electrochemical Oxygen Reduction Activity. *ACS Nano* **2012**, *6*, 7084–7091. [CrossRef] [PubMed]
22. Zhang, J.; Zhao, Z.; Xia, Z.; Dai, L. A Metal-Free Bifunctional Electrocatalyst for Oxygen Reduction and Oxygen Evolution Reactions. *Nat. Nanotechnol.* **2015**, *10*, 444–452. [CrossRef] [PubMed]
23. Meng, Y.; Voiry, D.; Goswami, A.; Zou, X.; Huang, X.; Chhowalla, M.; Liu, Z.; Asefa, T. N-, O-, and S-Tridoped Nanoporous Carbons as Selective Catalysts for Oxygen Reduction and Alcohol Oxidation Reactions. *J. Am. Chem. Soc.* **2014**, *136*, 13554–13557. [CrossRef] [PubMed]
24. Liu, Z.-W.; Peng, F.; Wang, H.-J.; Yu, H.; Zheng, W.-X.; Yang, J. Phosphorus-Doped Graphite Layers with High Electrocatalytic Activity for the Oxygen Reduction Reaction in an Alkaline Medium. *Angew. Chem. Int. Ed.* **2011**, *50*, 3257–3261. [CrossRef] [PubMed]
25. Yang, L.; Jiang, S.; Zhan, Y.; Zhu, L.; Chen, S.; Wang, X.; Wu, Q.; Ma, J.; Ma, Y. Boron-Doped Carbon Nanotubes as Metal-Free Electrocatlysts for the Oxygen Reduction Reaction. *Angew. Chem. Int. Ed.* **2011**, *50*, 7132–7135. [CrossRef]
26. Wang, S.; Iyyamperumal, E.; Roy, A.; Xue, Y.; Yu, D.; Dai, L. Vertically Aligned BCN Nanotubes as Efficient Metal-Free Electrocatalysts for the Oxygen Reduction Reaction: A Synergistic Effect by Co-Doping with Boron and Nitrogen. *Angew. Chem. Int. Ed.* **2011**, *50*, 11756–11760. [CrossRef]
27. Liang, H.; Wei, W.; Wu, Z.; Feng, X.; Mullen, K. Mesoporous Metal—Nitrogen-Doped Carbon Electrocatalysts for Highly Efficient Oxygen Reduction Reaction. *J. Am. Chem. Soc.* **2013**, *135*, 16002–16005. [CrossRef]

28. Ding, W.; Li, L.; Xiong, K.; Wang, Y.; Li, W.; Nie, Y.; Chen, S.; Qi, X.; Wei, Z. Shape Fixing via Salt Recrystallization: A Morphology-Controlled Approach To Convert Nanostructured Polymer to Carbon Nanomaterial as a Highly Active Catalyst for Oxygen Reduction Reaction. *J. Am. Chem. Soc.* **2015**, *137*, 5414–5420. [CrossRef]
29. Li, Q.; Cao, R.; Cho, J.; Wu, G. Nanocarbon Electrocatalysts for Oxygen Reduction in Alkaline Media for Advanced Energy Conversion and Storage. *Adv. Energy Mater.* **2014**, *4*, 1301415. [CrossRef]
30. Masa, J.; Xia, W.; Muhler, M.; Schuhmann, W. On the Role of Metals in Nitrogen-Doped Carbon Electrocatalysts for Oxygen Reduction. *Angew. Chemie-Int. Ed.* **2015**, *54*, 10102–10120. [CrossRef]
31. Yang, G.; Zhu, J.; Yuan, P.; Hu, Y.; Qu, G.; Lu, B.-A.; Xue, X.; Yin, H.; Cheng, W.; Cheng, J.; et al. Regulating Fe-spin state by atomically dispersed Mn-N in Fe-N-C catalysts with high oxygen reduction activity. *Nat. Commun.* **2021**, *12*, 1734. [CrossRef] [PubMed]
32. Wang, H.; Liu, R.; Li, Y.; Lu, X.; Wang, Q.; Zhao, S.; Yuan, K.; Cui, Z.; Li, X.; Xin, S.; et al. Durable and efficient hollow porous oxide spinel microspheres for oxygen reduction. *Joule* **2018**, *2*, 337–348. [CrossRef]
33. Brezny, A.C.; Johnson, S.I.; Raugei, S.; Mayer, J.M. Selectivity-Determining Steps in O_2 Reduction Catalyzed by Iron(tetramesitylporphyrin). *J. Am. Chem. Soc.* **2020**, *142*, 4108–4113. [CrossRef] [PubMed]
34. Ma, Q.; Jin, H.; Zhu, J.; Li, Z.; Xu, H.; Liu, B.; Zhang, Z.; Ma, J.; Mu, S. Stabilizing Fe-N-C Catalysts as Model for Oxygen Reduction Reaction. *Adv. Sci.* **2021**, *8*, 2102209. [CrossRef] [PubMed]
35. Zhao, M.; Liu, H.; Zhang, H.; Chen, W.; Sun, H.; Wang, Z.; Zhang, B.; Song, L.; Yang, Y.; Ma, C.; et al. A pH-universal ORR catalyst with single-atom iron sites derived from a double-lazer MOF for superior flexible quasi-solid-state rechargeable Zn-air batteries. *Energy Environ. Sci.* **2021**, *14*, 6455–6463. [CrossRef]
36. Rupar, J.; Bajuk-Bogdanović, D.; Milojević-Rakić, M.; Krstić, J.; Upadhyay, K.; Gavrilov, N.; Janošević Ležaić, A. Tailored porosity development in carbons via Zn^{2+} monodispersion: Fitting supercapacitors. *Micropor. Mesopor. Mat.* **2022**, *335*, 111790. [CrossRef]
37. Zhou, H.; Zhang, J.; Amiinu, I.S.; Zhan, C.; Liu, X.; Tu, W.; Pan, M.; Mu, S. Transforming Waste Biomass with an Intrinsically Porous Network Structure into Porous Nitrogen-Doped Graphene for Highly Efficient Oxygen Reduction. *Phys. Chem. Chem. Phys.* **2016**, *18*, 10392–10399. [CrossRef]
38. Huang, Y.; Wu, D.; Cao, D.; Cheng, D. Facile Preparation of Biomass-Derived Bifunctional Electrocatalysts for Oxygen Reduction and Evolution Reactions. *Int. J. Hydrog. Energy* **2018**, *43*, 8611–8622. [CrossRef]
39. Calvo-Muñoz, E.M.; García-Mateos, F.J.; Rosas, J.M.; Rodríguez-Mirasol, J.; Cordero, T. Biomass Waste Carbon Materials as Adsorbents for CO_2 Capture under Post-Combustion Conditions. *Front. Mater.* **2016**, *3*, 23. [CrossRef]
40. Xu, X.; Yuan, T.; Zhoy, Y.; Li, Y.; Li, Y.; Lu, J.; Tian, X.; Wang, D.; Wang, J. Facile synthesis of boron and nitrogen-sdoped graphene asefficient electrocatalyst for the oxygen reduction reaction in alkaline media. *Int. Hydrog. Energy* **2014**, *39*, 16043–16052. [CrossRef]
41. Huang, Z.-H.; Liu, T.-Y.; Song, Y.; Li, Y.; Liu, X.-X. Balancing the electrical double layer capacitance and pseudocapacitance of heteto-atom doped carbon. *Nanoscale* **2017**, *35*, 13119–13127. [CrossRef] [PubMed]
42. Bandosz, T.J. Porous Carbons as Oxygen Reduction Electrocatalysts. In *Porous Materials*; Springer: Berlin/Heidelberg, Germany, 2021; pp. 41–77.
43. Wu, J.; Ma, L.; Yadav, R.; Yang, Y.; Zhang, X.; Vajtai, R.; Lou, J.; Ajayan, P.M. Nitrogen-Doped Graphene with Pyridinic Dominance as a Highly Active and Stable Electrocatalyst for Oxygen Reduction. *ACS Appl. Mater. Interfaces* **2015**, *7*, 14763–14769. [CrossRef]
44. Yuan, S.; Cui, L.-L.; Dou, Z.; Ge, X.; He, X.; Zhang, W.; Asefa, T. Nonprecious Bimetallic Sites Coordinated on N-Doped Carbons with Efficient and Durable Catalytic Activity for Oxygen Reduction. *Small* **2020**, *16*, 2000742. [CrossRef]
45. Zheng, Y.; Song, H.; Chen, S.; Yu, X.; Zhu, J.; Xu, J.; Zhang, K.A.I.; Zhang, C.; Liu, T. Metal-Free Multi-Heteroatom-Doped Carbon Bifunctional Electrocatalysts Derived from a Covalent Triazine Polymer. *Small* **2020**, *16*, 2004342. [CrossRef] [PubMed]
46. Zhang, G.; Wei, Q.; Yang, X.; Tavares, A.C.; Sun, S. RRDE experiments on noble-metal and noble-metal-free catalysts: Impact of loading on the activity and selectivity of oxygen reduction reaction in alkaline solution. *Appl. Catal. B Environ.* **2017**, *206*, 115–126. [CrossRef]
47. Bouleau, L.; Pérez-Rodríguez, S.; Quílez-Bermejo, J.; Izquierdo, M.; Xu, F.; Fierro, V.; Celzard, A. Best practices for ORR performance evaluation of metal-fre porous carbon electrocatalysts. *Carbon* **2022**, *189*, 349–361. [CrossRef]
48. Gabe, A.; Ruiz-Rosas, R.; Gonzales-Gaitan, C.; Morallon, E.; Cazorla-Amoros, D. Modeling of oxygen reduction reaction in porous carbon reaction in porous carbon materials in alkaline medium. Effect of microporosity. *J. Power Sources* **2019**, *412*, 451–464. [CrossRef]
49. Ye, Y.; Duan, W.; Yi, X.; Lei, Z.; Li, G.; Feng, C. Biogenic precursor to size-controlled synthesis of Fe_2P nanoparticles in heteroatom-doped graphene-like carbons and their electrocatalytic reduction of oxygen. *J. Power Sources* **2019**, *435*, 226770. [CrossRef]
50. Li, Y.; Zhang, H.; Liu, P.; Wang, Y.; Yang, H.; Li, Y.; Zhao, H. Self-supported bimodal-pore structured nitrogen-doped carbon fiber electrocatalyst for oxygen reduction reaction. *Electrochem. Commun.* **2015**, *51*, 6–10. [CrossRef]
51. Borghei, M.; Lehtonen, J.; Liu, L.; Rojas, O.J. Advanced Biomass-Derived Electrocatalysts for the Oxygen Reduction Reaction. *Adv. Mater.* **2017**, *30*, 1703691. [CrossRef]
52. Clay, R.C.; Cook, K.; Routh, J.I. Studies in the composition of human hair. *J. Am. Chem. Soc.* **1940**, *62*, 2709. [CrossRef]
53. Chaudhari, K.N.; Song, M.Y.; Yu, J.S. Transforming Hair into Heteroatom-Doped Carbon with High Surface Area. *Small* **2014**, *10*, 2625–2636. [CrossRef] [PubMed]

54. Ezeoha, S.L.; Enzenwanne, J.N. Production of Biodegradable Plastic Packaging Film from Cassava Starch. *IOSR J. Eng.* **2013**, *3*, 14–20. [CrossRef]
55. Baranwal, J.; Barse, B.; Fais, A.; Delogu, G.L.; Kumar, A. Biopolymer: A Sustainable Material for Food and Medical Applications. *Polymers* **2022**, *14*, 983. [CrossRef] [PubMed]
56. Berg, J.M.; Tymoczko, J.L.; Stryer, L. *Biochemistry*, 5th ed.; W.H. Freeman and Company: New York, NY, USA, 2002.
57. Khan, A.; Goepel, M.; Colmenares, J.C.; Gläser, R. Chitosan-Based N-Doped Carbon Materials for Electrocatalytic and Photocatalytic Applications. *ACS Sustain. Chem. Eng.* **2020**, *8*, 4708–4727. [CrossRef]
58. Wang, H.; Fu, F.; Huang, M.; Feng, Y.; Han, D.; Xi, Y.; Xiong, W.; Yang, D.; Niu, L. Lignin-Based Materials for Electrochemical Energy Storage Devices. *Nano Mater. Sci.* **2022**, *in press*. [CrossRef]
59. Wittmar, A.S.M.; Ropertz, M.; Braun, M.; Hagemann, U.; Andronescu, C.; Ulbricht, M. Preparation of N-Doped Carbon Materials from Cellulose:Chitosan Blends and Their Potential Application in Electrocatalytic Oxygen Reduction. *Polym. Bull.* **2022**, 1–19. [CrossRef]
60. Li, Y.; Liu, X.; Wang, J.; Yang, L.; Chen, X.; Wang, X.; Zhang, P. Marine Algae-Derived Porous Carbons as Robust Electrocatalysts for ORR. *Catalysts* **2019**, *9*, 730. [CrossRef]
61. Hao, Y.; Zhang, X.; Yang, Q.; Chen, K.; Guo, J.; Zhou, D.; Feng, L.; Slanina, Z. Highly Porous Defective Carbons Derived from Seaweed Biomass as Efficient Electrocatalysts for Oxygen Reduction in Both Alkaline and Acidic Media. *Carbon N. Y.* **2018**, *137*, 93–103. [CrossRef]
62. Udayakumar, G.P.; Muthusamy, S.; Selvaganesh, B.; Sivarajasekar, N.; Rambabu, K.; Sivamani, S.; Sivakumar, N.; Maran, J.P.; Hosseini-Bandegharaei, A. Ecofriendly Biopolymers and Composites: Preparation and Their Applications in Water-Treatment. *Biotechnol. Adv.* **2021**, *52*, 107815. [CrossRef]
63. Li, D.; Lv, C.; Liu, L.; Xia, Y.; She, X.; Guo, S.; Yang, D. Egg-Box Structure in Cobalt Alginate: A New Approach to Multifunctional Hierarchical Mesoporous N-Doped Carbon Nanofibers for Efficient Catalysis and Energy Storage. *ACS Cent. Sci.* **2015**, *1*, 261–269. [CrossRef]
64. Wu, X.L.; Chen, L.L.; Xin, S.; Yin, Y.X.; Guo, Y.G.; Kong, Q.S.; Xia, Y.Z. Preparation and Li Storage Properties of Hierarchical Porous Carbon Fibers Derived from Alginic Acid. *ChemSusChem* **2010**, *3*, 703–707. [CrossRef] [PubMed]
65. Li, D.; Yang, D.; Zhu, X.; Jing, D.; Xia, Y.; Ji, Q.; Cai, R.; Li, H.; Che, Y. Simple Pyrolysis of Cobalt Alginate Fibres into Co_3O_4/C Nano/Microstructures for a High-Performance Lithium Ion Battery Anode. *J. Mater. Chem. A* **2014**, *2*, 18761–18766. [CrossRef]
66. Puscaselu, R.G.; Lobiuc, A.; Dimian, M.; Covasa, M. Alginate: From Food Industry to Biomedical Applications and Management of Metabolic Disorders. *Polymers* **2020**, *12*, 2417. [CrossRef] [PubMed]
67. Fourest, E.; Volesky, B. Alginate Properties and Heavy Metal Biosorption by Marine Algae. *Appl. Biochem. Biotechnol.—Part A Enzym. Eng. Biotechnol.* **1997**, *67*, 215–226. [CrossRef]
68. Braccini, I.; Pérez, S. Molecular Basis of Ca^{2+}-Induced Gelation in Alginates and Pectins: The Egg-Box Model Revisited. *Biomacromolecules* **2001**, *2*, 1089–1096. [CrossRef] [PubMed]
69. Li, P.; Jin, Z.; Qian, Y.; Fang, Z.; Xiao, D.; Yu, G. Supramolecular Confinement of Single Cu Atoms in Hydrogel Frameworks for Oxygen Reduction Electrocatalysis with High Atom Utilization. *Mater. Today* **2020**, *35*, 78–86. [CrossRef]
70. Zhao, X.; Yu, X.; Xin, S.; Chen, S.; Bao, C.; Xu, W.; Xue, J.; Hui, B.; Zhang, J.; She, X.; et al. Enhanced Oxygen Reduction Reaction for Zn-Air Battery at Defective Carbon Fibers Derived from Seaweed Polysaccharide. *Appl. Catal. B Environ.* **2022**, *301*, 120785. [CrossRef]
71. Wei, C.; Sun, Y.; Scherer, G.G.; Fisher, A.C.; Sherburne, M.; Ager, J.W.; Xu, Z.J. Surface Composition Dependent Ligand Effect in Tuning the Activity of Nickel-Copper Bimetallic Electrocatalysts toward Hydrogen Evolution in Alkaline. *J. Am. Chem. Soc.* **2020**, *142*, 7765–7775. [CrossRef]
72. Song, Q.; Li, J.; Wang, S.; Liu, J.; Liu, X.; Pang, L.; Li, H.; Liu, H. Enhanced Electrocatalytic Performance through Body Enrichment of Co-Based Bimetallic Nanoparticles In Situ Embedded Porous N-Doped Carbon Spheres. *Small* **2019**, *15*, 1–9. [CrossRef]
73. Mun, Y.; Lee, S.; Kim, K.; Kim, S.; Lee, S.; Han, J.W.; Lee, J. Versatile Strategy for Tuning ORR Activity of a Single Fe-N4 Site by Controlling Electron-Withdrawing/Donating Properties of a Carbon Plane. *J. Am. Chem. Soc.* **2019**, *141*, 6254–6262. [CrossRef]
74. Xing, X.; Liu, R.; Anjass, M.; Cao, K.; Kaiser, U.; Zhang, G.; Streb, C. Bimetallic Manganese-Vanadium Functionalized N,S-Doped Carbon Nanotubes as Efficient Oxygen Evolution and Oxygen Reduction Electrocatalysts. *Appl. Catal. B Environ.* **2020**, *277*, 119195. [CrossRef]
75. Shu, J.; Niu, Q.; Wang, N.; Nie, J.; Ma, G. Alginate Derived Co/N Doped Hierarchical Porous Carbon Microspheres for Efficient Oxygen Reduction Reaction. *Appl. Surf. Sci.* **2019**, *485*, 520–528. [CrossRef]
76. Wang, Y.; Tao, L.; Xiao, Z.; Chen, R.; Jiang, Z.; Wang, S. 3D Carbon Electrocatalysts In Situ Constructed by Defect-Rich Nanosheets and Polyhedrons from NaCl-Sealed Zeolitic Imidazolate Frameworks. *Adv. Funct. Mater.* **2018**, *28*, 1705356. [CrossRef]
77. Zhang, Y.; Lin, Y.; Jiang, H.; Wu, C.; Liu, H.; Wang, C.; Chen, S.; Duan, T.; Song, L. Well-Defined Cobalt Catalyst with N-doped Carbon Layers Enwrapping: The Correlation between Surface Atomic Structure and Electrocatalytic Property. *Small* **2017**, *14*, 1702074. [CrossRef] [PubMed]
78. Ma, N.; Jia, Y.; Yang, X.; She, X.; Zhang, L.; Peng, Z.; Yao, X.; Yang, D. Seaweed Biomass Derived (Ni,Co)/CNT Nanoaerogels: Efficient Bifunctional Electrocatalysts for Oxygen Evolution and Reduction Reactions. *J. Mater. Chem. A* **2016**, *4*, 6376–6384. [CrossRef]

79. Zhao, W.; Yuan, P.; She, X.; Xia, Y.; Komarneni, S.; Xi, K.; Che, Y.; Yao, X.; Yang, D. Sustainable Seaweed-Based One-Dimensional (1D) Nanofibers as High-Performance Electrocatalysts for Fuel Cells. *J. Mater. Chem. A* **2015**, *3*, 14188–14194. [CrossRef]
80. Zhan, T.; Lu, S.; Liu, X.; Teng, H.; Hou, W. Alginate Derived Co_3O_4/Co Nanoparticles Decorated in N-Doped Porous Carbon as an Efficient Bifunctional Catalyst for Oxygen Evolution and Reduction Reactions. *Electrochim. Acta* **2018**, *265*, 681–689. [CrossRef]
81. Zhang, C.; Antonietti, M.; Fellinger, T.P. Blood Ties: Co_3O_4 Decorated Blood Derived Carbon as a Superior Bifunctional Electrocatalyst. *Adv. Funct. Mater.* **2014**, *24*, 7655–7665. [CrossRef]
82. Wan, L.; Lin, C.; Huang, D.; Zhang, F.; Wang, M.; Jin, J. A Comparative Study of Composition and Morphology Effect of Ni XCo1-x$(OH)_2$ on Oxygen Evolution/Reduction Reaction. *ACS Appl. Mater. Interfaces* **2014**, *6*, 10172–10180. [CrossRef]
83. Wang, D.; Chen, X.; Evans, D.G.; Yang, W. Well-Dispersed Co_3O_4/Co_2MnO_4 Nanocomposites as a Synergistic Bifunctional Catalyst for Oxygen Reduction and Oxygen Evolution Reactions. *Nanoscale* **2013**, *5*, 5312–5315. [CrossRef]
84. Song, F.; Hu, X. Ultrathin Cobalt-Manganese Layered Double Hydroxide Is an Efficient Oxygen Evolution Catalyst. *J. Am. Chem. Soc.* **2014**, *136*, 16481–16484. [CrossRef] [PubMed]
85. Ryu, W.H.; Yoon, T.H.; Song, S.H.; Jeon, S.; Park, Y.J.; Kim, I.D. Bifunctional Composite Catalysts Using Co_3O_4 Nanofibers Immobilized on Nonoxidized Graphene Nanoflakes for High-Capacity and Long-Cycle Li-O_2 Batteries. *Nano Lett.* **2013**, *13*, 4190–4197. [CrossRef] [PubMed]
86. Navalon, S.; Dhakshinamoorthy, A.; Alvaro, M.; Garcia, H. Carbocatalysis by Graphene-Based Materials. *Chem. Rev.* **2014**, *114*, 6179–6212. [CrossRef] [PubMed]
87. Li, W.; Zhang, F.; Dou, Y.; Wu, Z.; Liu, H.; Qian, X.; Gu, D.; Xia, Y.; Tu, B.; Zhao, D. A Self-Template Strategy for the Synthesis of Mesoporous Carbon Nanofibers as Advanced Supercapacitor Electrodes. *Adv. Energy Mater.* **2011**, *1*, 382–386. [CrossRef]
88. You, B.; Yin, P.; Zhang, J.; He, D.; Chen, G.; Kang, F.; Wang, H.; Deng, Z.; Li, Y. Hydrogel-Derived Non-Precious Electrocatalysts for Efficient Oxygen Reduction. *Sci. Rep.* **2015**, *5*, 11739. [CrossRef] [PubMed]
89. You, B.; Kang, F.; Yin, P.; Zhang, Q. Hydrogel-Derived Heteroatom-Doped Porous Carbon Networks for Supercapacitor and Electrocatalytic Oxygen Reduction. *Carbon N. Y.* **2016**, *103*, 9–15. [CrossRef]
90. Li, Y.; Zhou, W.; Wang, H.; Xie, L.; Liang, Y.; Wei, F.; Idrobo, J.C.; Pennycook, S.J.; Dai, H. An Oxygen Reduction Electrocatalyst Based on Carbon Nanotube-Graphene Complexes. *Nat. Nanotechnol.* **2012**, *7*, 394–400. [CrossRef]
91. Zhao, Z.; Wu, H.; He, H.; Xu, X.; Jin, Y. A High-Performance Binary Ni-Co Hydroxide-Based Water Oxidation Electrode with Three-Dimensional Coaxial Nanotube Array Structure. *Adv. Funct. Mater.* **2014**, *24*, 4698–4705. [CrossRef]
92. He, W.; Jiang, C.; Wang, J.; Lu, L. High-Rate Oxygen Electroreduction over Graphitic-N Species Exposed on 3D Hierarchically Porous Nitrogen-Doped Carbons. *Angew. Chemie—Int. Ed.* **2014**, *53*, 9503–9507. [CrossRef]
93. Shibuya, R.; Kondo, T.; Nakamura, J. Active Sites of Nitrogen-Doped Carbon Materials for Oxygen Reduction Reaction Clarified Using Model Catalysts. *Carbon-Based Met. Catal. Des. Appl.* **2016**, *351*, 361–365. [CrossRef]
94. Wu, G.; Hu, Y.; Liu, Y.; Zhao, J.; Chen, X.; Whoehling, V.; Plesse, C.; Nguyen, G.T.M.; Vidal, F.; Chen, W. Graphitic Carbon Nitride Nanosheet Electrode-Based High-Performance Ionic Actuator. *Nat. Commun.* **2015**, *6*, 7258. [CrossRef]
95. Jia, N.; Li, D.; Huang, G.; Sun, J.; Lu, P.; Wan, L.; Hui, B.; She, X.; Zhao, X. Carbon Fibers-Coated Co@N-Doped Porous Carbon Derived from ZIF-67/Alginate Fibers for Efficient Oxygen Reduction Reaction. *J. Photonics Energy* **2020**, *10*, 023507. [CrossRef]
96. Guo, J.; Shu, J.; Nie, J.; Ma, G. Fe/Ni Bimetal and Nitrogen Co-Doped Porous Carbon Fibers as Electrocatalysts for Oxygen Reduction Reaction. *J. Colloid Interface Sci.* **2020**, *560*, 330–337. [CrossRef] [PubMed]
97. Lv, C.; Yang, X.; Umar, A.; Xia, Y.; Jia, Y.; Shang, L.; Zhang, T.; Yang, D. Architecture-Controlled Synthesis of MxOy (M = Ni, Fe, Cu) Microfibres from Seaweed Biomass for High-Performance Lithium Ion Battery Anodes. *J. Mater. Chem. A* **2015**, *3*, 22708–22715. [CrossRef]
98. Xiao, M.; Zhu, J.; Feng, L.; Liu, C.; Xing, W. Meso/Macroporous Nitrogen-Doped Carbon Architectures with Iron Carbide Encapsulated in Graphitic Layers as an Efficient and Robust Catalyst for the Oxygen Reduction Reaction in Both Acidic and Alkaline Solutions. *Adv. Mater.* **2015**, *27*, 2521–2527. [CrossRef]
99. Liang, J.; Zhou, R.F.; Chen, X.M.; Tang, Y.H.; Qiao, S.Z. Fe-N decorated hybrids of CNTs grown on hierarchically porous carbon for high-performance oxygen reduction. *Adv. Mater.* **2014**, *26*, 6074–6079. [CrossRef]
100. Yu, L.Q.; Xia, W.J.; Ma, W.J.; Wen, T.E.; Chen, S.L.; Jin, F.; Huang, B.C.; Jin, R.C. Universal Method to Fabricate Transition Metal Single-Atom-Anchored Carbon with Excellent Oxygen Reduction Reaction Activity. *ACS Appl. Mater. Interfaces* **2021**, *13*, 13534–13540. [CrossRef]
101. Kaden, W.E.; Wu, T.; Kunkel, W.A.; Anderson, S.L. Electronic structure controls reactivity of size-selected Pd clusters adsorbed on TiO_2 surfaces. *Science* **2009**, *326*, 826–829. [CrossRef]
102. Liu, P.; Zhao, Y.; Qin, R.; Mo, S.; Chen, G.; Gu, L.; Chevrier, D.M.; Zhang, P.; Qing, G.; Zang, D.; et al. Photochemical rooute for synthesizing atomically dispersed palladium catalysts. *Science* **2016**, *352*, 797–801. [CrossRef]
103. Varma, R.S. Biomass-Derived Renewable Carbonaceous Materials for Sustainable Chemical and Environmental Applications. *ACS Sustain. Chem. Eng.* **2019**, *7*, 6458–6470. [CrossRef]
104. Zahedifar, M.; Es-Haghi, A.; Zhiani, R.; Sadeghzadeh, S.M. Synthesis of Benzimidazolones by Immobilized Gold Nanoparticles on Chitosan Extracted from Shrimp Shells Supported on Fibrous Phosphosilicate. *RSC Adv.* **2019**, *9*, 6494–6501. [CrossRef] [PubMed]
105. Klemm, D.; Philipp, B.; Heinze, T.; Heinze, U.; Wagenknecht, W. General Considerations on Structure and Reactivity of Cellulose: Section 2.3–2.3.7. In *Comprehensive Cellulose Chemistry*; Wiley-VCH Verlag GmbH: New York, NY, USA, 1998; Volume 1, pp. 83–129. ISBN 3527294139.

106. Abe, M.M.; Ribeiro Martins, J.; Bertolino Sanvezzo, P.; Macedo, J.V.; Branciforti, M.C.; Halley, P.; Botaro, V.R.; Brienzo, M. Advantages and Disadvantages of Bioplastics Production from Sttarch and Lignocellulosic Components. *Polymers* **2021**, *13*, 2484. [CrossRef] [PubMed]
107. Sahu, P.; Gupta, M.K. Water absorption behavior of cellulosic fibres polymer composites: A review on its effectsand remedies. *J. Ind. Text.* **2020**, *51*, 7480S–7512S. [CrossRef]
108. Kim, C.; Zhu, C.; Aoki, Y.; Habazaki, H. Exothermically Efficient Exfoliation of Biomass Cellulose to Value-Added N-Doped Hierarchical Porous Carbon for Oxygen Reduction Electrocatalyst. *Ind. Eng. Chem. Res.* **2019**, *58*, 3047–3059. [CrossRef]
109. Kurita, K. Chitin and Chitosan: Functional Biopolymers from Marine Crustaceans. *Mar. Biotechnol.* **2006**, *8*, 203–226. [CrossRef]
110. Bradic, B. Isolation of Chitin from Marine Biomass Using Eutectic Solvents and Studies of N-Deacetylation Kinetics to Chitosan. Doctoral Dissertation, Univerza v Mariboru, Bojana Bradić, Slovenia, 2020.
111. De, S.; Acharya, S.; Sahoo, S.; Nayak, G.C. Chapter 12- Present status of biomass-derived carbon-based composites for supercapacitor application. In *Nanostructured, Functional, and Flexible Materials for Energy Conversion and Storage Systems*; Pandikumar, A., Rameshkumar, P., Eds.; Elsevier: Amsterdam, The Netherlands, 2020; pp. 373–415. [CrossRef]
112. Anitha, A.; Sowmya, S.; Kumar, P.T.S.; Deepthi, S.; Chennazhi, K.P.; Ehrlich, H.; Tsurkan, M.; Jayakumar, R. Chitin and Chitosan in Selected Biomedical Applications. *Prog. Polym. Sci.* **2014**, *39*, 1644–1667. [CrossRef]
113. Verma, S.; Nadagouda, M.N.; Varma, R.S. Porous Nitrogen-Enriched Carbonaceous Material from Marine Waste: Chitosan-Derived Carbon Nitride Catalyst for Aerial Oxidation of 5-Hydroxymethylfurfural (HMF) to 2,5-Furandicarboxylic Acid. *Sci. Rep.* **2017**, *7*, 13596. [CrossRef]
114. Primo, A.; Atienzar, P.; Sanchez, E.; Delgado, J.M.; García, H. From Biomass Wastes to Large-Area, High-Quality, N-Doped Graphene: Catalyst-Free Carbonization of Chitosan Coatings on Arbitrary Substrates. *Chem. Commun.* **2012**, *48*, 9254–9256. [CrossRef]
115. Skorupska, M.; Ilnicka, A.; Lukaszewicz, J.P. Successful Manufacturing Protocols of N-Rich Carbon Electrodes Ensuring High ORR Activity: A Review. *Processess* **2022**, *10*, 643. [CrossRef]
116. Ilnicka, A.; Lukaszewicz, J.P.; Shimanoe, K.; Yuasa, M. Urea treatment of nitrogen-doped carbon leads to enhanced performance for the oxygen reduction reaction. *J. Mater. Res.* **2018**, *33*, 1612–1624. [CrossRef]
117. Wang, M.; Ma, J.; Yang, H.; Lu, G.; Yang, S.; Chang, Z. Nitrogen and Cobalt Co-Coped Carbon Materials Derived from Biomass Chitin as High-Performance Electrocatalyst for Aluminium-Air Batteries. *Catalysts* **2019**, *9*, 954. [CrossRef]
118. Hatakeyama, H.; Hatakeyama, T. Lignin Structure, Properties, and Applications. *Adv. Polym. Sci.* **2009**, *232*, 1–63. [CrossRef]
119. Zheng, Y.; Chen, S.; Zhang, K.A.I.; Guan, J.; Yu, X.; Peng, W.; Song, H.; Zhu, J.; Xu, J.; Fan, X.; et al. Template-Free Construction of Hollow Mesoporous Carbon Spheres from a Covalent Triazine Framework for Enhanced Oxygen Electroreduction. *J. Colloid Interface Sci.* **2022**, *608*, 3168–3177. [CrossRef]
120. Li, W.; Zhang, H.; Zhang, K.; Hu, W.; Cheng, Z.; Chen, H.; Feng, X.; Peng, T.; Kou, Z. Monodispersed Ruthenium Nanoparticles Interfacially Bonded with Defective Nitrogen-and-Phosphorus-Doped Carbon Nanosheets Enable PH-Universal Hydrogen Evolution Reaction. *Appl. Catal. B Environ.* **2022**, *306*, 121095. [CrossRef]
121. Shen, Y.; Peng, F.; Cao, Y.; Zuo, J.; Wang, H.; Yu, H. Preparation of Nitrogen and Sulfur Co-Doped Ultrathin Graphitic Carbon via Annealing Bagasse Lignin as Potential Electrocatalyst towards Oxygen Reduction Reaction in Alkaline and Acid Media. *J. Energy Chem.* **2019**, *34*, 33–42. [CrossRef]
122. Li, P.; Wang, H.; Fan, W.; Huang, M.; Shi, J.; Shi, Z.; Liu, S. Salt Assisted Fabrication of Lignin-Derived Fe, N, P, S Codoped Porous Carbon as Trifunctional Catalyst for Zn-Air Batteries and Water-Splitting Devices. *Chem. Eng. J.* **2021**, *421*, 129704. [CrossRef]
123. Jahan, M.; Feni, F. Environmentally Friendly Bifunctional Catalyst for ORR and OER from Coconut Shell Particles. *Adv. Mater. Phys. Chem.* **2022**, *12*, 106–123. [CrossRef]
124. Borghei, M.; Laocharoen, N.; Kibena-Põldsepp, E.; Johansson, L.S.; Campbell, J.; Kauppinen, E.; Tammeveski, K.; Rojas, O.J. Porous N,P-Doped Carbon from Coconut Shells with High Electrocatalytic Activity for Oxygen Reduction: Alternative to Pt-C for Alkaline Fuel Cells. *Appl. Catal. B Environ.* **2017**, *204*, 394–402. [CrossRef]
125. Prauchner, M.J.; Rodríguez-Reinoso, F. Chemical versus Physical Activation of Coconut Shell: A Comparative Study. *Microporous Mesoporous Mater.* **2012**, *152*, 163–171. [CrossRef]
126. Borghei, M. Novel carbon nanomaterials for the direct methanol fuel cell electrodes. Doctoral Dissertation, Aalto University, School of Science, Department of Applied Physics, NanoMaterials Group, Maryam Borghei, Espoo, Finland, 2015.
127. Nagaiah, T.C.; Kundu, S.; Bron, M.; Muhler, M.; Schuhmann, W. Nutrgen-doped carbon nanotubes as a cathode catalyst for the oxygen reduction reaction in alkaline medium. *Electrochem. Commun.* **2010**, *12*, 338–341. [CrossRef]
128. Oh, H.-S.; Oh, J.-G.; Hee Lee, W.; Kim, H.-J.; Kim, H. The influence of the structural properties of carbon on the oxygen reduction reaction of nitrogen modified carbon based catalysts. *Int. J. Hydrog. Energy* **2011**, *36*, 8181–8186. [CrossRef]
129. Liu, R.; Zhang, H.; Liu, S.; Zhang, X.; Wu, T.; Ge, X.; Zang, Y.; Zhao, H.; Wang, G. Shrimp-Shell Derived Carbon Nanodots as Carbon and Nitrogen Sources to Fabricate Three-Dimensional N-Doped Porous Carbon Electrocatalysts for the Oxygen Reduction Reaction. *Phys. Chem. Chem. Phys.* **2016**, *18*, 4095–4101. [CrossRef] [PubMed]
130. Yan, N.; Chen, X. Don't Waste Seafood Waste: Turning Cast-off Shells into Nitrogen-Rich Chemicals Would Benefit Economies and the Environment. *Nature* **2015**, *524*, 155–157. [CrossRef] [PubMed]
131. Tokatlı, K.; Demirdöven, A. Optimization of Chitin and Chitosan Production from Shrimp Wastes and Characterization. *J. Food Process. Preserv.* **2017**, *42*, e13494. [CrossRef]

132. Tuci, G.; Zafferoni, C.; Rossin, A.; Milella, A.; Luconi, L.; Innocenti, M.; Truong Phuoc, L.; Duong-Viet, C.; Pham-Huu, C.; Giambastiani, G. Chemically Functionalized Carbon Nanotubes with Pyridine Groups as Easily Tunable N-Decorated Nanomaterials for the Oxygen Reduction Reaction in Alkaline Medium. *Chem. Mater.* **2014**, *26*, 3460–3470. [CrossRef]
133. Xie, J.; Zhang, H.; Li, S.; Wang, R.; Sun, X.; Zhou, M.; Zhou, J.; Wen, X.; Lou, D.; Xie, Y. Defect-Rich MoS_2 Ultrathin Nanosheets with Additional Active Edge Sites for Enhanced Electrocatalytic Hydrogen Evolution. *Adv. Mater.* **2013**, *25*, 5807–5813. [CrossRef]
134. Yu, D.; Nagelli, E.; Du, F.; Dai, L. Metal-Free Carbon Nanomaterials Become More Active than Metal Catalysts and Last Longer. *J. Phys. Chem. Lett.* **2010**, *1*, 2165–2173. [CrossRef]
135. Zhang, H.; Wang, Y.; Wang, D.; Li, Y.; Liu, X.; Liu, P.; Yang, H.; An, T.; Tang, Z.; Zhao, H. Hydrothermal Transformation of Dried Grass into Graphitic Carbon-Based High Performance Electrocatalyst for Oxygen Reduction Reaction. *Small* **2014**, *10*, 3371–3378. [CrossRef]
136. Zhang, H.; Chen, J.; Li, Y.; Liu, P.; Wang, Y.; An, T.; Zhao, H. Nitrogen-Doped Carbon Nanodots@Nanospheres as An Efficient Electrocatalyst for Oxygen Reduction Reaction. *Electrochim. Acta* **2015**, *165*, 7–13. [CrossRef]
137. Li, Y.; Zhao, Y.; Cheng, H.; Hu, Y.; Shi, G.; Dai, L.; Qu, L. Nitrogen-doped grapheme quantum dots with oxygen-rich functional groups. *J. Am. Chem. Soc.* **2012**, *134*, 15–18. [CrossRef]
138. Liang, J.; Jiao, Y.; Jaroniec, M.; Zhang Qiao, S. Sulfur and nitrogen dual-doped mesoporous grapheme electrocatalyst for oxygen reduction with synergistically enhanced performance. *Angew. Chem. Int. Ed. Engl.* **2012**, *51*, 11496–11500. [CrossRef] [PubMed]
139. Balakrishnan, M.; Batra, V.S.; Hargreaves, J.S.J.; Pulford, I.D. Waste Materials—Catalytic Opportunities: An Overview of the Application of Large Scale Waste Materials as Resources for Catalytic Applications. *Green Chem.* **2011**, *13*, 16–24. [CrossRef]
140. Mane, G.P.; Talapaneni, S.N.; Anand, C.; Varghese, S.; Iwai, H.; Ji, Q.; Ariga, K.; Mori, T.; Vinu, A. Preparation of Highly Ordered Nitrogen-Containing Mesoporous Carbon from a Gelatin Biomolecule and Its Excellent Sensing of Acetic Acid. *Adv. Funct. Mater.* **2012**, *22*, 3596–3604. [CrossRef]

catalysts

MDPI

Article

How to Obtain Maximum Environmental Applicability from Natural Silicates

Daliborka Popadić [1], Nemanja Gavrilov [2], Ljubiša Ignjatović [2], Danina Krajišnik [3], Slavko Mentus [2], Maja Milojević-Rakić [2,*] and Danica Bajuk-Bogdanović [2]

1 National Laboratory Sector, Department of Organic Residual Analysis, Serbian Environmental Protection Agency, Žabljačka 10A, 11160 Belgrade, Serbia; daliborka.bankovic@sepa.gov.rs

2 University of Belgrade-Faculty of Physical Chemistry, Studentski Trg 12-16, 11158 Belgrade, Serbia; gavrilov@ffh.bg.ac.rs (N.G.); ljignjatovic@ffh.bg.ac.rs (L.I.); slavko@ffh.bg.ac.rs (S.M.); danabb@ffh.bg.ac.rs (D.B.-B.)

3 Department of Pharmaceutical Technology and Cosmetology, University of Belgrade-Faculty of Pharmacy, Vojvode Stepe 450, 11221 Belgrade, Serbia; danina.krajisnik@pharmacy.bg.ac.rs

* Correspondence: maja@ffh.bg.ac.rs; Tel.: +381-1121-87133

Abstract: Unmodified natural silicates (bentonite, kaolin, clinoptilolite and diatomites) were tested as adsorbents for the organic pollutants in water tables using Methylene Blue (MB) as the model adsorbate. Among the selected materials, bentonite adsorbed as much as 237 mg/g, confirming its excellent suitability for pollutant removal. Spectral evidence confirmed successful MB immobilization at the bentonite surface. Furthermore, the thermal treatment of MB-saturated adsorbent in an inert atmosphere at 700 °C produced a carbon/silicate composite. EDX confirmed the formation of the nitrogen-doped carbon overlay on the silica scaffold and the obtained composite material was probed as an electrode material for oxygen reduction in an alkaline solution. Reduction proceeded via a two-electron mechanism with the main product being HO_2^-, a known nucleophile, which was subsequently used to degrade/demethylate MB. The composite showed a considerable 70% MB removal rate after an hour of electrochemical treatment. The synergy between the processes of adsorption of MB and the surface-generated HO_2^- dictates the efficiency of the method and points to a possible route for spent adsorbent reuse in the form of a durable product for environmental protection.

Keywords: bentonite; methylene blue; oxygen reduction reaction; spent adsorbent

Citation: Popadić, D.; Gavrilov, N.; Ignjatović, L.; Krajišnik, D.; Mentus, S.; Milojević-Rakić, M.; Bajuk-Bogdanović, D. How to Obtain Maximum Environmental Applicability from Natural Silicates. *Catalysts* **2022**, *12*, 519. https://doi.org/10.3390/catal12050519

Academic Editor: Roman Bulánek

Received: 12 April 2022
Accepted: 4 May 2022
Published: 5 May 2022

1. Introduction

The search for advanced materials that have favorable properties for the adsorption/degradation of various pollutants from water is still ongoing. Costly, complicated and time-consuming synthesis procedures may not always produce materials with high adsorption capacity/degradation efficiency and therefore the researcher's attention is drawn to the use/modification of abundant and low-cost natural materials [1,2]. In the environmental application field, aluminosilicates represent the most important class of minerals and Earth's crust constituents. Clinoptilolite, hydrated alkali aluminosilicate, is one of the most abundant minerals and is readily available and implemented as incipients/carriers [3–6]. Other crust ingredients are clay minerals with several candidates for targeted applications in adsorption studies. Bentonite is a highly applicable clay [7] mainly composed of montmorillonite, a clay mineral from the smectite group [8]. The characteristics of bentonite depend on the amount of smectite and exchangeable cations in the interlayer space. Due to their structural characteristics of 2:1 tetrahedral SiO_2/octahedral Al_2O_3 layers with substantial voids, bentonites are perceived as good adsorbents. One of the main characteristics of these materials is the swelling ability and high cation exchange capacity. Kaolin's (white clay) main component is kaolinite, a clay mineral of a 1:1 kaolinite group, with tightly distributed tetrahedral and octahedral layers. Diatomites are sedimentary rock silicates that mainly consist of opal and cristobalite. These materials are light, finely porous rocks consisting mainly of small opal skeletons (or parts thereof) of diatomaceous algae with interesting physico-chemical and biomedical applications [9,10].

Natural material does not necessarily imply efficiency in pollutant removal. Among the affordable and readily available minerals, it is necessary to examine the influence of their different structures, morphologies and composition on adsorption properties, and environmental applicability in general [11,12].

Various dyes were investigated as model pollutants, such as methylene blue (MB), a cationic thiazine dye, which finds its way into essentially different fields—from pharmacology to the dye/textile industry. This multifunctionality arises from versatile functional groups present in its structure. The fact that it can mimic nitrogen/sulfur-containing pollutants in the environment makes MB an excellent candidate for probe testing for resolving environmental threats.

Not only is it important to find an efficient material [13–15] but the question is also what to do with the saturated adsorbent after the adsorption. Materials after adsorption can be further exploited for the preparation of nanoparticles, but this often requires the addition of metal precursors [16]. An attractive approach was proposed by Tian et al. for hydrothermal treatment of dye-saturated palygorskite/Ag nanoparticles. Obtained carbon composites efficiently adsorbed different dyes and pharmaceuticals [17]. Another proposal for spent MB dye biosorbents is their transformation to biochar, which could be subsequently used for additional dye adsorption [18]. MB dye can be used for the functionalization of cellulose gum sorbent to suit the removal of anionic dye [19]. If the intention is shifted toward applications different from reiterating adsorption cycles after burn-off, incorporation of metal ions is necessary, either as a precursor or as a pollutant adsorbate. For instance, a fuel cell setup may benefit from bio sorbent previously used for platinum adsorption [20]. For metal-free electroactive material production, MB is also a good candidate, as graphene dye functionalization leads to highly active and stable electrode material for charge storage application [21].

Here, we propose natural aluminosilicates selection (bentonite, kaolin, clinoptilolite and diatomites) for the adsorption application of MB as a model pollutant in water tables. MB dye removal was studied in detail in the literature employing natural silicates in raw and modified forms [22–28]. Considering that natural samples have different compositions depending on their origin, we examined different samples to find the optimal adsorbent of the highest capacity that can be further processed. Thus, the universal application of the preferred spent adsorbent may be proposed. After carbonization, the selected MB-saturated silicate sample will be used for the electrocatalytic application. Electrochemical dye degradation on various electrode materials is addressed in the literature as an efficient method [29–31]. To establish a comprehensive environmental approach, this novel electrode material will be employed for the degradation of MB in an electrolytic cell by electrogenerated hydroperoxyl radical to extend the lifecycle of such spent adsorbents.

2. Results and Discussion

2.1. Thermal, Spectral and Structural Properties of Investigated Minerals

Thermal profiles of the investigated silicates are given in Figure 1a. Diatomite samples DK and DP are stable with small differences after a physically adsorbed water loss. Kaoline undergoes a phase transformation to metakaolin at around 500 °C [32], while the BE profile consists of two segments, characteristic of smectite mineral. The first step of mass loss up to 200 °C is related to physisorbed water removal, which is followed by dehydration of Ca^{2+} coordinated water and thermal removal of structural OH groups up to over 600 °C [33]. Clinoptilolite displays a continuous mass decrease up to 700 °C in accordance with previous studies [34].

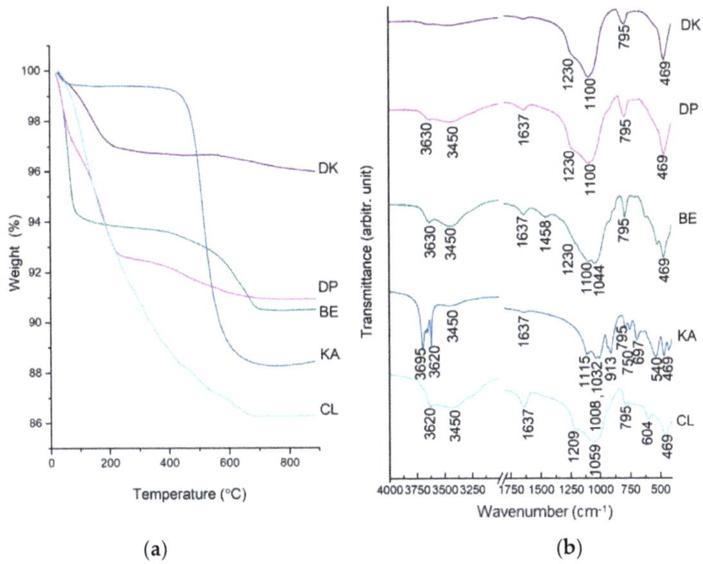

Figure 1. TG profiles (**a**) and FTIR spectra (**b**) of natural silicates—clinoptilolite (CL), Peru diatomite (DP), Kolubara diatomite (DK), kaolin (KA) and bentonite (BE).

Spectral profiles of the tested silicates were analyzed and the FTIR spectra are presented in Figure 1b.

In line with the TG results, the diatomite samples DK and DT express the least resolved band at 3450 cm^{-1} and 1637 cm^{-1} attributed to O–H stretching and H–O–H deformation vibration. These bands are evident in the BE and CL samples, accompanied by a sharp band at about 3630 and 3620 cm^{-1}, respectively (stretching of the structural hydroxyl groups). The most complex spectral features in this region are associated with KA, with four resolved bands of the structural OH groups (3695, 3670, 3650 and 3620 cm^{-1}) [35], characteristic of the kaolinite mineral, indicating its significant structural ordering.

In addition to the absorption bands of structural OH, Si–O groups play an important role in the differentiation of silicate minerals [36].

The FTIR spectra of the diatomaceous earth samples, DK and DP, give a very strong absorption band at 1100 cm^{-1} with a shoulder at 1230 cm^{-1} are attributed to the Si–O–Si stretching vibration. The absorption band at 795 cm^{-1} evolves from the corresponding inter tetrahedral Si–O–Si bending. The absorption peak at around 469 cm^{-1} corresponded to the Si–O–Si bending vibration [37]. The bentonite spectrum reveals strong bands attributed to Si–O elongation in the 1100–1000 cm^{-1} region, characteristic of the layered montmorillonite structure. The bands corresponding to the Al–Al–OH bending vibrations are observed at 915 cm^{-1}, Al–O–Si at 525 cm^{-1} and Si–O–Si bending vibrations at 469 cm^{-1}. The band at 620 cm^{-1} is assigned to coupled Al–O and Si–O out-of-plane vibrations [38]. The band at 1458 cm^{-1} is assigned to the asymmetric stretching vibration of the present carbonate impurity [39] and other, less pronounced bands in the spectrum originate from clay components. Figure S1 gives a bentonite diffractogram to resolve phase content. Reflections at 6.7°, 19.9° and 35.0° 2θ correspond to the montmorillonite (JCPDS 29-1498), while other reflections correspond to low amounts of feldspar in the form of the albite (JCPDS 09-0466) and quartz phase (JCPDS 46-1045) [40].

The 1120–1000 cm^{-1} region resolves several strong absorption bands related to Si–O stretching vibrations in the kaolin sample. Two strong bands at 1032 cm^{-1} and 1008 cm^{-1} belong to the antisymmetric in-plane Si–O–Si stretching. The symmetric Si–O stretching band is located at 1115 cm^{-1} while Al_2OH bending modes are positioned at 935 and 913 cm^{-1} ascribed to inner surface/OH groups. Two low-intensity bands at the 750 and

697 cm^{-1} are associated with perpendicular Si–O modes and possibly surface hydroxyls. The Si–O–Al and Si–O–Si bending are located at 540 and 470 cm^{-1}, in that order [41].

In the clinoptilolite spectrum, an intense 1059 cm^{-1} band is ascribed to the asymmetric stretching vibration modes of internal T–O bonds in TO_4 tetrahedra (T = Si and Al). The 795 cm^{-1} and 469 cm^{-1} bands are attributed to the stretching vibration modes of O–T–O groups and the bending vibration modes of T–O bonds, respectively [42].

2.2. Methylene Blue Adsorption-Quantitative, Spectral and Elemental Analysis

The UV–Vis spectrum of MB shows two characteristic absorption bands at 662 and 607 nm with their intensities dependent on the monomer–dimer structure and the dimer development [43]. Low MB concentration up to 70 ppm gives dominantly monomers/dimmers, while higher concentrations allow aggregation/stacking, thus changing the spectral profile. As the tested concentration range is rather wide, differences in the intensity ratio of the monomer/dimmer band are evident in Figure S2a. Thus, for accurate quantification of MB, the calibration curves are based on the integral band surface instead of the absorbance value.

For MB tested concentrations below 100 ppm, clinoptilolite (CL), diatomite (DP) and bentonite (BE) samples completely and instantly adsorb MB while Kolubara diatomite (DK) and kaolin (KA) enter saturation and therefore, were excluded from further testing. Kaolin's non-expanding structure is fixed without interlayer space, which is the reason for its low adsorption capacity. The CL is getting to its saturation level at 100 ppm while the BE and DP samples retain excellent adsorption performance, (Figure 2a). Furthermore, DP lowered the adsorbed quantity of MB to 54 mg/g, while bentonite demonstrated the best adsorption capacity for the starting MB concentration of 200 ppm. To test the extent of available adsorption sites on bentonite, this sample was further investigated in suspensions with 300, 400 and 500 ppm MB solutions. Complete MB removal was witnessed up to 500 ppm, where 236.6 mg of MB was adsorbed per gram of bentonite adsorbent, approximately 95% of the starting MB quantity. The excellent adsorption performance of bentonite is attributed to its interlayered structure, which is susceptible to expansion and swelling in an aqueous environment, thus forcing the layers apart. To conclude, a mineral from the expending smectite group is a highly superior adsorbent compared to other tested silicate materials. A cationic dye such as MB readily interacts with clay minerals by cation exchange of extra-framework cations in an aluminosilicate structure [44].

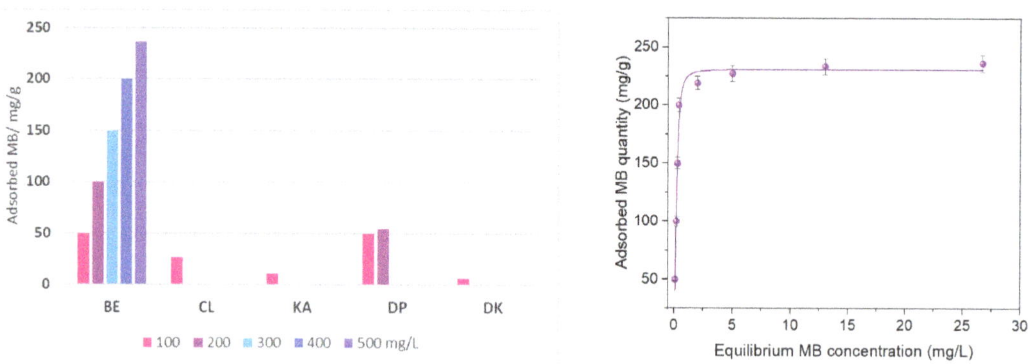

(a) (b)

Figure 2. Retained quantities of MB on investigated minerals, recorded for different initial MB concentrations (**a**); Langmuir–Freundlich adsorption isotherm for MB adsorption on bentonite sample (**b**).

Adsorption isotherm testing (Figure 2b) for several additional starting MB concentrations revealed the Langmuir–Freundlich isotherm model [45] is better suited for the description of the adsorption process (correlation coefficient 0.98) in comparison to the less realistic Langmuir equation [45] Yang (correlation coefficient 0.91). This finding is in accordance with presumed nearly equivalent adsorption centers and the excellent adsorption behavior of the bentonite sample.

The MB post-adsorption spectral investigation (Figure 3) reveals no change in the structure of the tested silicates upon adsorption. The most intense pristine MB bands are located at 1592 and 1386 cm^{-1}. These bands can be observed in the BE spectra at 1604 and 1398 cm^{-1} after MB adsorption. The transition of band positions to higher wavenumbers is a result of band shortening due to MB immobilization on the BE surface.

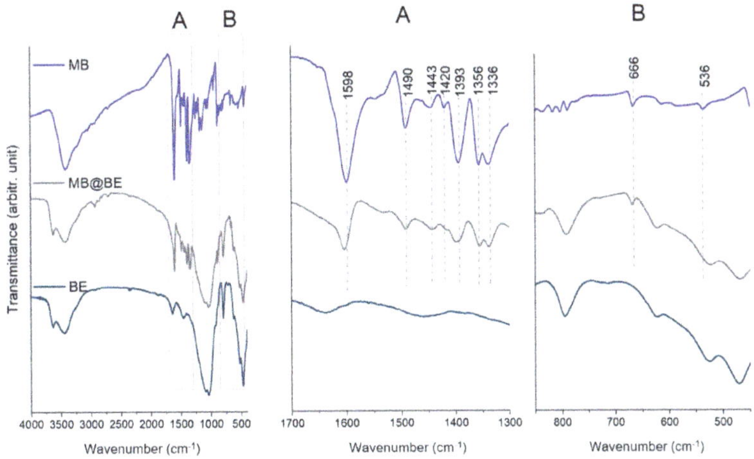

Figure 3. FTIR spectra of MB and BE adsorbent before and after MB adsorption (MB@BE).

As the bentonite sample proved to be the utmost MB adsorbent, carbonization of the saturated adsorbent was the next step in this investigation. In such a way, carbonaceous material comprising heteroatoms was obtained. Raman microanalysis confirmed that the amorphous carbon phase is obtained by thermal treatment of the MB-saturated bentonite sample. The characteristic G (stretching modes, 1585 cm^{-1}) and D (breathing modes, 1373 cm^{-1}) bands are clear indications of MB carbonization with an evident structural disorder [46]. The FTIR spectra give additional information on the preservation of aluminosilicate structure, as witnessed in Figure S3.

To confirm its elemental composition, EDX measurements were performed on the carbonized sample, and the atomic % of dominant elements (>1%) is summarized in Figure 4. The absence of carbon and nitrogen in pristine bentonite is evidenced, while after carbonization, a high portion of the surface is made up of carbon (34 at.%) and nitrogen (7 at.%) atoms. Additionally, the ratio between carbon and nitrogen atoms matches that within the MB molecule. The N-doped carbon phase is evenly present at the aluminosilicate surface in the amount of 25 wt. % carbon and 6 wt. % N, corresponding well to the already-measured highly adsorbed MB quantity.

The favorable elemental composition supports subsequent electroactive testing of this novel material derived from the saturated adsorbent.

(**a**) (**b**)

Figure 4. EDX map (**a**) and elemental composition change for carbonized MB-saturated bentonite (C-MB@BE) in comparison to pristine bentonite (BE) sample (**b**).

2.3. Electro-Applicability of Spent Adsorbent

After carbonization of the MB dye, adsorbed on bentonite, the silicate scaffold was covered with layers of nitrogen-doped carbon, as seen by Raman, FTIR and EDX. Doped carbons are well known for their activity towards ORR in alkaline solution, albeit somewhat lower than those for transition metals and their oxides [47]. Spent biochar adsorbents of heavy metals can also be transferred to electrocatalysts for ORR [48]. Figure 5 contains the CVs for carbonized saturated adsorbent displaying clear dependence of the measured current on the rate of rotation, indicating that the process of oxygen reduction is diffusion-controlled at more negative potential. Koutecký–Levich (K–L) analysis was conducted to assess the number of exchanged electrons per oxygen molecule and to distinguish between the $2e^{-}4e^{-}$ ORR mechanisms [47,49].

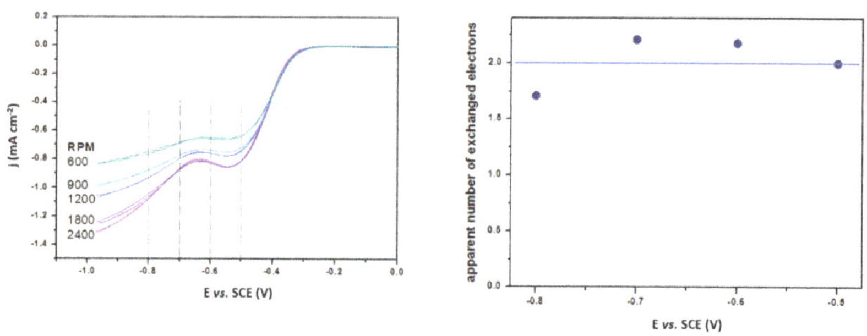

(**a**) (**b**)

Figure 5. Background-corrected CVs for ORR in 0.1 M KOH at different rotation rates (**a**); dependence of the apparent number of exchanged electrons on the given potential, according to K–L analysis (**b**).

Calculated values for the apparent number of exchanged electrons show that $2e^{-}$ are being consumed in the potential region of interest and that, at a moderately negative potential, HO_2^{-} is formed according to reaction $O_{2(g)} + H_2O_{(aq)} + 2e^{-} \leftrightarrow HO_2^{-}{}_{(aq)} + OH^{-}{}_{(aq)}$.

Katafias et al. [50] investigated chemical MB degradation in base and base/peroxide systems and reported that HO_2^{-} nucleophile in alkaline solutions may be an efficient phenothiazine derivative degradation agent. This prompted us to check whether electro-generated HO_2^{-} in the adlayer near or at the electrode surface could be used to reduce MB or degrade it. MB can be involved in a fast, reversible, $2e^{-}$ redox process where MB transforms to a reduced, colorless leuco-MB form [51] or can be demethylated via nucleophile attack [50,52]. Electron transfer enables versatile applications of MB, which can be followed and analyzed by electrochemical methods [53,54].

After preparation of the initial solution of MB standard in KOH, the potential of the electrode was fixed at an HO_2^- generating potential for three consecutive runs with aliquots taken for spectrophotometric measurements. The UV range monitoring was required to differentiate whether the process is only decolorization (due to colorless leuco-MB evolution) [55] or whether we achieved actual MB degradation and if the formed associated colored degradation products [50,52] decreased. Therefore, withdrawn volumes were used to measure the decrease in absorbance characteristics for both MB species, as presented in Figure 6. The decrease in the spectral response of quiescent KOH/MB (without the electrode present) amounted to 14% after 1 h and served as a baseline/guideline for degradation studies, Figure 6a. Interestingly, there is no rise in the UV absorption band due to MB reduction to leuco-MB. Although not resolved due to the stronger MB signal overlap, this characteristic band at 255 nm of leuco-MB form [56] did not contribute to the UV absorption rising. This means that the portion of MB transferred to leuco-MB is moderately low and that a major quantity of MB is degraded in the alkaline environment by HO_2^-.

Figure 6b depicts a slight effect of adsorption upon rotating the electrode due to increased mass flow of the MB solution towards the electrode. Concurrent adsorption and inherent KOH bleaching decreased the signal by 23.1%. Importantly, this decrease is simultaneously seen for both UV and Vis absorption bands, suggesting an actual MB solution concentration decrease without the evolution of the leuco-MB form.

The intended effect of electrogenerated HO_2^- towards MB degradation is seen by a gross 65.9% calculated from a decrease in MB absorption bands in the 450–750 nm region (42.8% when corrected for inherent degradation and effect of rotation) within an hour when potential is held at −0.8 V, Figure 6c. The results can be interpreted in accordance with the literature whereby MB is positively charged [57], suggesting that it will readily adsorb electrostatically on a negative electrode surface. Maintaining the HO_2^- generation at −0.8 V vs. SCE will have the additional effect of attracting MB towards the surface. Once there, MB will simultaneously adsorb at the electrode and be in close proximity to the generated HO_2^-, a known proton donor, with electrons from the electrode enabling successful MB degradation by demethylation/deamination via OH- and HO_2^- intermediates resulting in colored products (Azure A/B/C and Methylene violet) [50]. A rotating disc electrode (RDE) forces the convection of the MB molecules as well as oxygen from the bulk of the solution towards the electrode surface.

Increased convection results in the formation of significantly more HO_2^-, seen as an increase in measured current, and more MB molecules near the surface. By integrating current with respect to time, the amount of charge spent on HO_2^- generation could be assessed and amounted to (0.46, 0.60 and 0.61 C) recalculated to 2.37, 3.10, and 3.14 µmol of HO_2^-, which was sufficient to account for the decrease in initial MB content (0.18 µmol). The combined effect is the significant decrease in MB concentration indicating successful degradation by in situ electrogenerated nucleophile attack.

Using a test reaction with MB, it was demonstrated that the new carbonized material derived from the saturated adsorbent can be successfully used for the generation of HO_2^-. In addition to bleaching, electro-generated HO_2^- could be used for wider environmental applications, such as a non-selective oxidant in water purification systems.

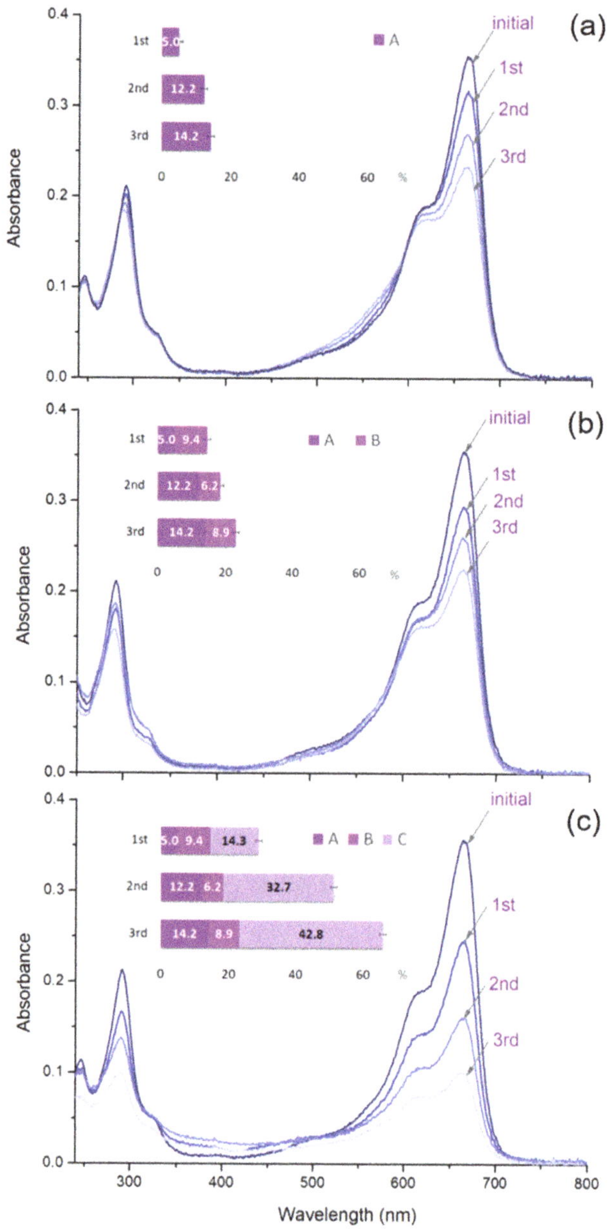

Figure 6. Spectro-electrochemical results on MB degradation in (A) KOH electrolyte (**a**); (A + B) KOH-filled electrolytic cell with rotating electrode (**b**) and (A + B + C) KOH-filled electrolytic cell with rotating electrode and potential applied (**c**).

3. Materials and Methods

3.1. Materials

The natural zeolite, clinoptilolite (CL), used in this study originates from the Zlatokop deposit (Vranje, Serbia). Diatomite samples were examined from two sites, one of domestic origin, the Kolubara basin Serbia (DK), and the other originating from South America,

Peru (DP sample, food-grade diatomaceous earth, freshwater source). The DP sample was neat and used as is, while the DK sample was used after thermal treatment (at 550 °C for 4 h). Bentonite (BE) obtained from Fagron (Greece) and kaolin (KA) from Sigma-Aldrich (St. Louis, MI, USA), were both of the pharmaceutical (Ph. Eur.) grade.

3.2. Characterization of the Natural Silicates

Infrared spectroscopy with Fourier analysis was performed in the range 4000–400 cm^{-1} with 32 scans per spectrum and a 4 cm^{-1} resolution on Nicolet iS20 spectrometer (Thermo Scientific, Waltham, MA, USA) using KBr pellet technique.

X-ray diffraction (XRD) at room temperature was performed on a D8 Advance diffractometer (Bruker, Karlsruhe, Germany) using Cu Kα radiation, at a 40 kV voltage and 40 mA current. Diffraction data were collected in the 2–50° 2θ range with a 0.02° step and 0.1 s retention time per step. Data collection was guided by DIFFRAC Plus Commander v.2.6.1 software (Bruker, Germany).

UV–Vis absorption spectra were recorded on an Evolution 220 spectrometer (Thermo Scientific, Waltham, MA, USA) in a quartz cuvette with an optical path of 1 cm or 0.1 cm, in the 200–800 nm range.

The investigated minerals were subjected to thermogravimetry (TG) on thermobalance SDT 2960 (TA Instruments, New Castle, DE, USA). Measurements were performed up to 900 °C with the heating rate 10 °C min^{-1} in a synthetic air atmosphere with a flow rate of 80 mL min^{-1}.

3.3. Methylene Blue Adsorption, Adsorbents Treatment and Characterization

The natural silicates were tested as adsorbents for MB from aqueous solutions. The batch adsorption suspensions comprised 10 mg of solid adsorbent and 5 mL volume of MB (Reag. Ph. Eur., Merck, Kenilworth, NJ, USA) solution. The suspensions were placed in an ultrasonic bath for 2 × 10 min. For lower MB concentrations (20–80 ppm), 1 mL of the suspension was centrifuged for 5 min at 13,400 rpm and UV–Vis spectra were recorded after 1 h equilibration time. In the case of higher concentrations (100–500 ppm), spectra were recorded after 24 h equilibration time.

All samples were tested for adsorption in 100 ppm and 200 ppm MB concentrations. As bentonite was the only material that completely adsorbed 200 ppm MB, this sample was further investigated for retention of higher concentrations—300, 400 and 500 ppm.

Upon adsorption of Methylene Blue, the selected sample was heated in a tubular furnace within argon atmosphere for 2 h at 700 °C, and 10 °C/min rate. The carbonized sample was denoted C-MB@BE.

In addition to FTIR measurements, a DXR Raman spectrometer with a microscope (Thermo Scientific, Waltham, MA, USA), was employed for spectral analysis of carbonized saturated adsorbent. Laser wavelength was 532 nm and power set at 2.0 mW, with 10 × 10 s exposure time, 900 lines/mm grating and 50 μm aperture.

The elemental composition of the carbonized sample was investigated by an energy-dispersive X-ray spectrometer (EDX) coupled with a Phenom ProX scanning electron microscope (Thermo Scientific, Waltham, MA, USA).

3.4. Electrochemical Measurements

Glassy carbon (GC) electrode coated with a thin film of the carbonized sample was used for electrochemical measurements. Liquid ink suspension was arranged by sonicating 5 mg of C-MB@BE in 0.5 mL of ethanol/water/Nafion mixture. Loading of 250 μg cm^{-2} was achieved by drop-casting 5 μL of the suspension on top of the GC disk (5 mm diameter). Oxygen reduction reaction (ORR) activity and MB degradation were evaluated using cyclic voltammetry (CV) in a conventional three-electrode cell with Pt wire as a counter and a saturated calomel electrode (SCE) as a reference electrode, in 0.1 M KOH aqueous solution as supporting electrolyte. An Ivium VO1107 potentiostat/galvanostat was used to impose the desired potential and measure the current response. Before and during the measurements, slight oxygen flow was kept through the electrolyte to keep the solution saturated with oxygen. The baseline was recorded by purging nitrogen through the cell to eliminate oxygen and concurrently all processes that use oxygen. Measurements were performed at standard pressure and ambient temperature. Koutecký–Levich analysis (K–L) (Equation (1)) was used to define the mechanism of ORR and determine the number of exchanged electrons per oxygen molecule:

$$\frac{1}{i_m} = \frac{1}{i_k} + \frac{1}{B\omega^{1/2}} \quad (1)$$

where i_m is the measured current, i_k is the kinetic current from the electrochemical reaction, B is the Levich constant and ω is the angular rotation rate.

Spectro-electrochemical measurements were performed by dissolving 1 mL of 500 ppm MB aqueous solution in 249 mL 0.1 M KOH. Generation of species was carried out at a constant potential (−0.8 V vs. SCE), at 2400 rpm, with 5 mL aliquots being drawn for UV–Vis measurements every 1200 s, three successive times.

4. Conclusions

Herein, we propose a straightforward selection of the best adsorbent of methylene blue (MB) in aqueous solutions among natural aluminosilicates (bentonite, kaolin, clinoptilolite and diatomites). Nitrogen-rich MB was selected for investigation as a model pollutant in water tables.

Complete MB removal (cca. 95%) was witnessed up to a 500 ppm starting concentration, where 236.6 mg MB was adsorbed per gram of bentonite, while other silicates showed significantly lower capacities. The excellent adsorption performance of bentonite is attributed to its interlayered structure with evenly distributed adsorption sites. The MB interaction with the bentonite sample was confirmed by FTIR analysis along with mineral structure preservation and insight into possible adsorption centers. MB-saturated bentonite was further heated in an inert atmosphere to carbonize the adsorbate. Raman spectral analysis confirmed the amorphous carbon phase on the bentonite sample, and FTIR spectra evidenced the preservation of the aluminosilicate structure. The obtained carbonaceous material comprises N and S as confirmed by the EDX investigation.

The doped carbon material was tested for its activity toward oxygen reduction in an alkaline solution. The Koutecký–Levich analysis confirmed that $2e^-$ are consumed in the selected potential region, therefore, confirming HO_2^- formation. To exploit nucleophile generation, in situ, three complementary experiments were conducted in three 20 min runs. As a baseline, in the absence of rotation, a decrease in MB concentration matches that naturally occurring in an alkaline MB solution in the amount of 14%. In the presence of rotation of the carbonized MB-saturated bentonite electrode, but without the applied potential, MB concentration decreases by 23% due to electrode adsorption in addition to MB alkaline bleaching. Finally, the effect of electrogenerated HO_2^- on MB degradation is seen by a 66% decrease within an hour. For successful MB degradation, the generation of a significant local concentration of HO_2^- is needed in the presence of a reasonable concentration of MB near or at the surface. This local, electro-generated HO_2^- concentration surpasses that found in the solution and successful degradation prefers intimate contact of pollutant molecules and HO_2^- at the electrode surface.

The combined effect of convection and oxygen reduction enables successful degradation of pollutants by in situ electrogenerated nucleophile attack, placing this procedure as a method of choice for spent adsorbent employment.

Supplementary Materials: The following supporting information can be downloaded at: https://www.mdpi.com/article/10.3390/catal12050519/s1, Figure S1: XRD pattern of bentonite sample; Figure S2: Calibration curve for MB in (a) an aqueous solution and (b) KOH, based on integral absorption band surface area; Figure S3: Raman and FTIR spectra of carbonized C-MB@BE sample.

Author Contributions: Conceptualization, D.P., M.M.-R. and D.B.-B.; methodology, D.P., D.K., N.G. and L.I.; formal analysis, D.P.; investigation, D.P., S.M. and N.G.; resources, L.I., S.M. and D.K.; data curation, D.P., D.B.-B. and N.G.; writing—original draft preparation, D.P., D.B.-B. and N.G.; writing—review and editing, M.M.-R., D.B.-B. and N.G.; visualization, M.M.-R.; supervision, D.B.-B. and N.G. All authors have read and agreed to the published version of the manuscript.

Funding: This research was funded by the Ministry of Education, Science and Technological Development of the Republic of Serbia, grant numbers 451-03-68/2022-14/200146 and 451-03-68/2022-14/200161.

Data Availability Statement: Data are contained within the article or Supplementary Materials.

Conflicts of Interest: The authors declare no conflict of interest. The funders had no role in the design of the study; in the collection, analyses, or interpretation of data; in the writing of the manuscript, or in the decision to publish the results.

References

1. Ganguly, S.; Das, P.; Das, T.K.; Ghosh, S.; Das, S.; Bose, M.; Mondal, M.; Das, A.K.; Das, N.C. Acoustic Cavitation Assisted Destratified Clay Tactoid Reinforced in Situ Elastomer-Mimetic Semi-IPN Hydrogel for Catalytic and Bactericidal Application. *Ultrason. Sonochemistry* **2020**, *60*, 104797. [CrossRef]
2. Ganguly, S.; Das, N.C. Water Uptake Kinetics and Control Release of Agrochemical Fertilizers from Nanoclay-Assisted Semi-Interpenetrating Sodium Acrylate-Based Hydrogel. *Polym.-Plast. Technol. Eng.* **2017**, *56*, 744–761. [CrossRef]
3. Krajišnik, D.; Milojević, M.; Malenović, A.A.; Daković, A.; Ibrić, S.; Savić, S.S.; Dondur, V.; Matijašević, S.; Radulović, A.; Daniels, R.; et al. Cationic Surfactants-Modified Natural Zeolites: Improvement of the Excipients Functionality. *Drug Dev. Ind. Pharm.* **2010**, *36*, 1215–1224. [CrossRef] [PubMed]
4. Cerri, G.; Farina, M.; Brundu, A.; Daković, A.; Giunchedi, P.; Gavini, E.; Rassu, G. Natural Zeolites for Pharmaceutical Formulations: Preparation and Evaluation of a Clinoptilolite-Based Material. *Microporous Mesoporous Mater.* **2016**, *223*, 58–67. [CrossRef]
5. Krajišnik, D.; Daković, A.; Malenović, A.; Milojević-Rakić, M.; Dondur, V.; Radulović, Ž.; Milić, J.; Radulović, Z.; Milić, J. Investigation of Adsorption and Release of Diclofenac Sodium by Modified Zeolites Composites. *Appl. Clay Sci.* **2013**, *83–84*, 322–326. [CrossRef]
6. Krajišnik, D.; Daković, A.; Milić, J.; Marković, M. Zeolites as Potential Drug Carriers. In *Modified Clay and Zeolite Nanocomposite Materials*; Elsevier: Amsterdam, The Netherlands, 2019; pp. 27–55.
7. Pagnacco, M.; Maksimović, J.; Mudrinić, T.; Banković, P.; Nedić-Vasiljević, B.; Milutinović-Nikolić, A. Oscillatory Briggs-Rauscher Reaction as "Fingerprint" for Bentonite Identification: The Fine-Tuning of Oscillatory Dynamics with Addition of Clay. *ChemistrySelect* **2020**, *5*, 8137–8141. [CrossRef]
8. Jovic-Jovicic, N.; Bankovic, P.; Mojovic, Z.; Nedic-Vasiljevic, B.; Marinovic, S.; Mudrinic, T.; Milutinovic-Nikolic, A. Ecologically Friendly Chitosan-Montmorillonite Bio-Nanocomposite as Adsorbent for Textile Dyes from Aqueous Solutions. *Sci. Sinter.* **2017**, *49*, 419–429. [CrossRef]
9. Janićijević, J.; Milić, J.; Čalija, B.; Micov, A.; Stepanović-Petrović, R.; Tomić, M.; Daković, A.; Dobričić, V.; Nedić Vasiljević, B.; Krajišnik, D. Potentiation of the Ibuprofen Antihyperalgesic Effect Using Inorganically Functionalized Diatomite. *J. Mater. Chem. B* **2018**, *6*, 5812–5822. [CrossRef] [PubMed]
10. Janićijević, J.; Krajišnik, D.; Čalija, B.; Vasiljević, B.N.; Dobričić, V.; Daković, A.; Antonijević, M.D.; Milić, J. Modified Local Diatomite as Potential Functional Drug Carrier—A Model Study for Diclofenac Sodium. *Int. J. Pharm.* **2015**, *496*, 466–474. [CrossRef]
11. Milojević-Rakić, M.; Popadić, D.; Janošević Ležaić, A.; Jevremović, A.; Nedić Vasiljević, B.; Uskoković-Marković, S.; Bajuk-Bogdanović, D. MFI, BEA and FAU Zeolite Scavenging Role in Neonicotinoids and Radical Species Elimination. *Environ. Sci. Processes Impacts* **2022**, *24*, 265–276. [CrossRef]
12. Barakan, S.; Aghazadeh, V. The Advantages of Clay Mineral Modification Methods for Enhancing Adsorption Efficiency in Wastewater Treatment: A Review. *Environ. Sci. Pollut. Res.* **2021**, *28*, 2572–2599. [CrossRef] [PubMed]
13. Kuang, Y.; Zhang, X.; Zhou, S. Adsorption of Methylene Blue in Water onto Activated Carbon by Surfactant Modification. *Water* **2020**, *12*, 587. [CrossRef]
14. Rosly, N.Z.; Abdullah, A.H.; Ahmad Kamarudin, M.; Ashari, S.E.; Alang Ahmad, S.A. Adsorption of Methylene Blue Dye by Calix [6]Arene-Modified Lead Sulphide (Pbs): Optimisation Using Response Surface Methodology. *Int. J. Environ. Res. Public Health* **2021**, *18*, 397. [CrossRef] [PubMed]
15. Rahmawati, F.; Ridassepri, A.F.; Chairunnisa; Wijayanta, A.T.; Nakabayashi, K.; Miyawaki, J.; Miyazaki, T. Carbon from Bagasse Activated with Water Vapor and Its Adsorption Performance for Methylene Blue. *Appl. Sci.* **2021**, *11*, 678. [CrossRef]
16. Harikishore Kumar Reddy, D.; Vijayaraghavan, K.; Kim, J.A.; Yun, Y.-S. Valorisation of Post-Sorption Materials: Opportunities, Strategies, and Challenges. *Adv. Colloid Interface Sci.* **2017**, *242*, 35–58. [CrossRef]
17. Tian, G.; Wang, W.; Zong, L.; Kang, Y.; Wang, A. From Spent Dye-Loaded Palygorskite to a Multifunctional Palygorskite/Carbon/Ag Nanocomposite. *RSC Adv.* **2016**, *6*, 41696–41706. [CrossRef]

18. Feng, Y.; Dionysiou, D.D.; Wu, Y.; Zhou, H.; Xue, L.; He, S.; Yang, L. Adsorption of Dyestuff from Aqueous Solutions through Oxalic Acid-Modified Swede Rape Straw: Adsorption Process and Disposal Methodology of Depleted Bioadsorbents. *Bioresour. Technol.* **2013**, *138*, 191–197. [CrossRef]
19. Yan, H.; Zhang, W.; Kan, X.; Dong, L.; Jiang, Z.; Li, H.; Yang, H.; Cheng, R. Sorption of Methylene Blue by Carboxymethyl Cellulose and Reuse Process in a Secondary Sorption. *Colloids Surf. A Physicochem. Eng. Asp.* **2011**, *380*, 143–151. [CrossRef]
20. Dimitriadis, S.; Nomikou, N.; McHale, A.P. Pt-Based Electro-Catalytic Materials Derived from Biosorption Processes and Their Exploitation in Fuel Cell Technology. *Biotechnol. Lett.* **2007**, *29*, 545–551. [CrossRef]
21. Zhang, Y.; An, Y.; Wu, L.; Chen, H.; Li, Z.; Dou, H.; Murugadoss, V.; Fan, J.; Zhang, X.; Mai, X.; et al. Metal-Free Energy Storage Systems: Combining Batteries with Capacitors Based on a Methylene Blue Functionalized Graphene Cathode. *J. Mater. Chem. A* **2019**, *7*, 19668–19675. [CrossRef]
22. Badeenezhad, A.; Azhdarpoor, A.; Bahrami, S.; Yousefinejad, S. Removal of Methylene Blue Dye from Aqueous Solutions by Natural Clinoptilolite and Clinoptilolite Modified by Iron Oxide Nanoparticles. *Mol. Simul.* **2019**, *45*, 564–571. [CrossRef]
23. Hong, S.; Wen, C.; He, J.; Gan, F.; Ho, Y.-S. Adsorption Thermodynamics of Methylene Blue onto Bentonite. *J. Hazard. Mater.* **2009**, *167*, 630–633. [CrossRef] [PubMed]
24. Mouni, L.; Belkhiri, L.; Bollinger, J.-C.; Bouzaza, A.; Assadi, A.; Tirri, A.; Dahmoune, F.; Madani, K.; Remini, H. Removal of Methylene Blue from Aqueous Solutions by Adsorption on Kaolin: Kinetic and Equilibrium Studies. *Appl. Clay Sci.* **2018**, *153*, 38–45. [CrossRef]
25. Al-Ghouti, M.A.; Khraisheh, M.A.M.; Ahmad, M.N.M.; Allen, S. Adsorption Behaviour of Methylene Blue onto Jordanian Diatomite: A Kinetic Study. *J. Hazard. Mater.* **2009**, *165*, 589–598. [CrossRef] [PubMed]
26. Yang, B.; Liu, X.; Ma, Z.; Wang, Q.; Yang, J. Synthesis of Nano-ZnO/Diatomite Composite and Research on Photoelectric Application. *Catalysts* **2021**, *11*, 1232. [CrossRef]
27. Liu, Y.; Wang, L.; Xue, N.; Wang, P.; Pei, M.; Guo, W. Ultra-Highly Efficient Removal of Methylene Blue Based on Graphene Oxide/TiO_2/Bentonite Sponge. *Materials* **2020**, *13*, 824. [CrossRef]
28. Kuntubek, A.; Kinayat, N.; Meiramkulova, K.; Poulopoulos, S.G.; Bear, J.C.; Inglezakis, V.J. Catalytic Oxidation of Methylene Blue by Use of Natural Zeolite-Based Silver and Magnetite Nanocomposites. *Processes* **2020**, *8*, 471. [CrossRef]
29. Samarghandi, M.R.; Dargahi, A.; Shabanloo, A.; Nasab, H.Z.; Vaziri, Y.; Ansari, A. Electrochemical Degradation of Methylene Blue Dye Using a Graphite Doped PbO_2 Anode: Optimization of Operational Parameters, Degradation Pathway and Improving the Biodegradability of Textile Wastewater. *Arab. J. Chem.* **2020**, *13*, 6847–6864. [CrossRef]
30. Droguett, T.; Mora-Gómez, J.; García-Gabaldón, M.; Ortega, E.; Mestre, S.; Cifuentes, G.; Pérez-Herranz, V. Electrochemical Degradation of Reactive Black 5 Using Two-Different Reactor Configuration. *Sci. Rep.* **2020**, *10*, 4482. [CrossRef]
31. Łuba, M.; Mikołajczyk, T.; Pierożyński, B.; Smoczyński, L.; Wojtacha, P.; Kuczyński, M. Electrochemical Degradation of Industrial Dyes in Wastewater through the Dissolution of Aluminum Sacrificial Anode of Cu/Al Macro-Corrosion Galvanic Cell. *Molecules* **2020**, *25*, 4108. [CrossRef]
32. Irfan Khan, M.; Khan, H.U.; Azizli, K.; Sufian, S.; Man, Z.; Siyal, A.A.; Muhammad, N.; Faiz ur Rehman, M. The Pyrolysis Kinetics of the Conversion of Malaysian Kaolin to Metakaolin. *Appl. Clay Sci.* **2017**, *146*, 152–161. [CrossRef]
33. Bourliva, A.; Michailidis, K.; Sikalidis, C.; Filippidis, A. Spectroscopic and Thermal Study of Bentonites from Milos Island, Greece. *Bull. Geol. Soc. Greece* **2013**, *47*, 2020. [CrossRef]
34. Krajišnik, D.; Daković, A.; Milojević, M.; Malenović, A.; Kragović, M.; Bajuk-Bogdanović, D.; Dondur, V.; Milić, J. Properties of Diclofenac Sodium Sorption onto Natural Zeolite Modified with Cetylpyridinium Chloride. *Colloids Surf B Biointerfaces* **2011**, *83*, 165–172. [CrossRef] [PubMed]
35. Farmer, V.C. Transverse and Longitudinal Crystal Modes Associated with OH Stretching Vibrations in Single Crystals of Kaolinite and Dickite. *Spectrochim. Acta Part A Mol. Biomol. Spectrosc.* **2000**, *56*, 927–930. [CrossRef]
36. Russell, J.D.; Fraser, A.R. Infrared Methods. In *Clay Mineralogy: Spectroscopic and Chemical Determinative Methods*; Springer: Dordrecht, The Netherlands, 1994; pp. 11–67.
37. El Ouardi, Y.; Branger, C.; Toufik, H.; Laatikainen, K.; Ouammou, A.; Lenoble, V. An Insight of Enhanced Natural Material (Calcined Diatomite) Efficiency in Nickel and Silver Retention: Application to Natural Effluents. *Environ. Technol. Innov.* **2020**, *18*, 100768. [CrossRef]
38. Zaitan, H.; Bianchi, D.; Achak, O.; Chafik, T. A Comparative Study of the Adsorption and Desorption of O-Xylene onto Bentonite Clay and Alumina. *J. Hazard. Mater.* **2008**, *153*, 852–859. [CrossRef]
39. Jović-Jovičić, N.; Milutinović-Nikolić, A.; Banković, P.; Dojčinović, B.; Nedić, B.; Gržetić, I.; Jovanović, D. Synthesis, Characterization and Adsorptive Properties of Organobentonites. *Acta Phys. Pol. A* **2010**, *117*, 849–854. [CrossRef]
40. Zhirong, L.; Azhar Uddin, M.; Zhanxue, S. FT-IR and XRD Analysis of Natural Na-Bentonite and Cu(II)-Loaded Na-Bentonite. *Spectrochim. Acta Part A Mol. Biomol. Spectrosc.* **2011**, *79*, 1013–1016. [CrossRef]
41. Madejová, J.; Gates, W.P.; Petit, S. IR Spectra of Clay Minerals. In *Developments in Clay Science*; Elsevier: Amsterdam, The Netherlands, 2017; pp. 107–149.
42. Faghihian, H.; Kabiri-Tadi, M. Removal of Zirconium from Aqueous Solution by Modified Clinoptilolite. *J. Hazard. Mater.* **2010**, *178*, 66–73. [CrossRef]
43. Fernández-Pérez, A.; Valdés-Solís, T.; Marbán, G. Visible Light Spectroscopic Analysis of Methylene Blue in Water; the Resonance Virtual Equilibrium Hypothesis. *Dye. Pigment.* **2019**, *161*, 448–456. [CrossRef]

44. Kahr, G.; Madsen, F.T. Determination of the Cation Exchange Capacity and the Surface Area of Bentonite, Illite and Kaolinite by Methylene Blue Adsorption. *Appl. Clay Sci.* **1995**, *9*, 327–336. [CrossRef]
45. Yang, R.T. *Adsorption: Fundamentals and Applications*; Wiley-Interscience: Hoboken, NJ, USA, 2003; ISBN 0-471-29741-0.
46. Rupar, J.; Bajuk-Bogdanović, D.; Milojević-Rakić, M.; Krstić, J.; Upadhyay, K.; Gavrilov, N.; Janošević Ležaić, A. Tailored Porosity Development in Carbons via Zn^{2+} Monodispersion: Fitting Supercapacitors. *Microporous Mesoporous Mater.* **2022**, *335*, 111790. [CrossRef]
47. Gavrilov, N.; Pašti, I.A.; Mitrić, M.; Travas-Sejdić, J.; Ćirić-Marjanović, G.; Mentus, S.V. Electrocatalysis of Oxygen Reduction Reaction on Polyaniline-Derived Nitrogen-Doped Carbon Nanoparticle Surfaces in Alkaline Media. *J. Power Sources* **2012**, *220*, 306–316. [CrossRef]
48. Chen, Z.; Zheng, R.; Wei, W.; Wei, W.; Zou, W.; Li, J.; Ni, B.-J.; Chen, H. Recycling Spent Water Treatment Adsorbents for Efficient Electrocatalytic Water Oxidation Reaction. *Resour. Conserv. Recycl.* **2022**, *178*, 106037. [CrossRef]
49. Pérez-Díaz, P.J.; Medina-Ramírez, A.; Esquivel, I.R.G.; Ruiz, G.G.; Ruiz-Camacho, B. Effect of X Zeolite-Carbon Composite Ratio as Support of Pt Nanoparticles for MOR and ORR. *Ionics* **2021**, *27*, 1813–1828. [CrossRef]
50. Katafias, A.; Lipińska, M.; Strutyński, K. Alkaline Hydrogen Peroxide as a Degradation Agent of Methylene Blue—Kinetic and Mechanistic Studies. *React. Kinet. Mech. Catal.* **2010**, *101*, 251–266. [CrossRef]
51. Zhan, R.; Song, S.; Liu, Y.; Dong, S. Mechanisms of Methylene Blue Electrode Processes Studied by in Situ Electron Paramagnetic Resonance and Ultraviolet–Visible Spectroelectrochemistry. *J. Chem. Soc. Faraday Trans.* **1990**, *86*, 3125–3127. [CrossRef]
52. Adamčíková, L.; Pavlíková, K.; Ševčík, P. The Decay of Methylene Blue in Alkaline Solution. *React. Kinet. Catal. Lett.* **2000**, *69*, 91–94. [CrossRef]
53. Salamifar, S.E.; Mehrgardi, M.A.; Kazemi, S.H.; Mousavi, M.F. Cyclic Voltammetry and Scanning Electrochemical Microscopy Studies of Methylene Blue Immobilized on the Self-Assembled Monolayer of n-Dodecanethiol. *Electrochim. Acta* **2010**, *56*, 896–904. [CrossRef]
54. Kelley, S.O.; Barton, J.K.; Jackson, N.M.; Hill, M.G. Electrochemistry of Methylene Blue Bound to a DNA-Modified Electrode. *Bioconjugate Chem.* **1997**, *8*, 31–37. [CrossRef]
55. Begum, R.; Najeeb, J.; Sattar, A.; Naseem, K.; Irfan, A.; Al-Sehemi, A.G.; Farooqi, Z.H. Chemical Reduction of Methylene Blue in the Presence of Nanocatalysts: A Critical Review. *Rev. Chem. Eng.* **2020**, *36*, 749–770. [CrossRef]
56. Pegu, R.; Majumdar, K.J.; Talukdar, D.J.; Pratihar, S. Oxalate Capped Iron Nanomaterial: From Methylene Blue Degradation to Bis(Indolyl)Methane Synthesis. *RSC Adv.* **2014**, *4*, 33446–33456. [CrossRef]
57. Wei, X.; Wang, Y.; Feng, Y.; Xie, X.; Li, X.; Yang, S. Different Adsorption-Degradation Behavior of Methylene Blue and Congo Red in Nanoceria/H_2O_2 System under Alkaline Conditions. *Sci. Rep.* **2019**, *9*, 4964. [CrossRef] [PubMed]

catalysts

MDPI

Article

Influence of the Type and the Amount of Surfactant in Phillipsite on Adsorption of Diclofenac Sodium

Danijela Smiljanić [1], Aleksandra Daković [1,*], Milena Obradović [1], Milica Ožegović [1], Marija Marković [1], George E. Rottinghaus [2] and Bruno de Gennaro [3]

1 Institute for Technology of Nuclear and Other Mineral Raw Materials, Franche D' Epere 86, 11000 Belgrade, Serbia
2 Veterinary Medical Diagnostic Laboratory, College of Veterinary Medicine, University of Missouri, Columbia, MO 65211, USA
3 ACLabs—Applied Chemistry Labs, Department of Chemical, Materials Engineering and Industrial Production, University of Naples Federico II, P.le V. Tecchio 80, 80125 Naples, Italy
* Correspondence: a.dakovic@itnms.ac.rs

Abstract: Modified phillipsite samples were prepared with two different amounts (monolayer and bilayer coverage) of surfactants octadecyldimethylbenzylammonium chloride (O) and dodecylamine (D). Composites were characterized by Fourier transform infrared spectroscopy with attenuated total reflectance (FTIR–ATR), thermal analysis and determination of zeta potential, and subsequently tested for removal of diclofenac sodium (DCF). Drug adsorption experiments were performed under different initial DCF concentrations and different contact times. In order to investigate the influence of the chemical structure of surfactants used for modification of phillipsite on the preparation and properties of composites and DCF adsorption, experimental data were compared with previously published results on DCF adsorption by composites containing phillipsite and the same amounts of surfactants cetylpyridinium chloride (C) and Arquad®2HT-75 (A). DCF adsorption isotherms for O and D composites showed a better fit with the Langmuir model with maximum adsorption capacities between 12.3 and 38.4 mg/g and are similar to those for C and A composites, while kinetics run followed a pseudo-second-order model. Composites containing either benzyl or pyridine functional groups showed higher adsorption of DCF, implying that surfactant structure has a significant impact on drug adsorption. Drug adsorption onto O, D, C and A composites was also confirmed by FTIR–ATR spectroscopy and zeta potential measurements.

Keywords: zeolite-rich tuff; phillipsite; surfactant modification; drug adsorption; diclofenac

Citation: Smiljanić, D.; Daković, A.; Obradović, M.; Ožegović, M.; Marković, M.; Rottinghaus, G.E.; de Gennaro, B. Influence of the Type and the Amount of Surfactant in Phillipsite on Adsorption of Diclofenac Sodium. *Catalysts* **2023**, *13*, 71. https://doi.org/10.3390/catal13010071

Academic Editor: Aurora Santos

Received: 30 November 2022
Revised: 23 December 2022
Accepted: 26 December 2022
Published: 30 December 2022

1. Introduction

Zeolites are aluminosilicate minerals with a three-dimensional framework consisting of SiO_4 and AlO_4 tetrahedra linked through shared oxygen atoms in a regular way creating cavities and channels. Due to the isomorphic substitution of Si^{4+} with Al^{3+}, the zeolitic structure has a net negative charge that is compensated by exchangeable cations located in channels and at the zeolitic surface together with water molecules [1,2]. Besides being abundant and eco-friendly, natural zeolites are also reasonably priced and have a high cation exchange capacity, physicochemical stability, and good hydraulic properties [3–5]. Due to the cation exchange property, these minerals have received substantial attention for the removal of heavy metal cations and ammonium from various systems, making them a potential solution in environmental cleaning processes [5–11]. However, because of the net negative charge and the presence of inorganic cations in the zeolite structure, these minerals are hydrophilic and have no affinity for the adsorption of anions and/or hydrophobic molecules [12,13]. By using large organic cations, such as surfactant, to modify the zeolitic surface, this restriction can be overcome. Surfactant molecules are too large to enter into zeolite channels or access internal cation exchange sites, so their adsorption is limited only to the external zeolitic surface. The amount of inorganic cations at the external surface that can be replaced with long-chain organic cations is expressed as the external cation exchange capacity (ECEC) of zeolite. Depending on the amount of surfactant used for modification of the zeolitic surface, their adsorption onto a negatively-charged surface involves both cation exchange and hydrophobic bonding. The surface charge of the composite will vary depending on how much surfactant is employed for the functionalization. If the amount of surfactant corresponds to 100% of the zeolite's ECEC value, a monolayer is created, providing hydrophobic sites at zeolite surface, thus surface charge changes from negative to neutral. If the amount of surfactant is above 100% of the ECEC of zeolite, tail-to-tail interaction of surfactant molecules will lead to formation of surfactant bilayer. In the bilayer case, cationic surfactant heads in the outer surfactant layer make zeolitic surface positive and will provide adsorption sites for anionic species [13–15].

Surfactant-modified zeolites have been widely researched for the removal of various pollutants from the aquatic environment, for example, anionic species of metals (such as chromates and arsenates), organic compounds (benzene, ethylbenzene, toluene, xylene, chlorophenol, perchloroethylene, dyes, pesticides, bisphenol A, etc.), pathogens [5,7,12,16–27] and more recently nonsteroidal anti-inflammatory drugs (NSAIDs) [13,28,29]. Diclofenac (DCF) is an over-the-counter drug from the class of NSAIDs frequently used, in both human and animal medicine, to treat various pain states and fevers [30,31]. As a result, increased consumption causes environmental contamination due to release in water systems [32–36]. Diclofenac caught worldwide attention as a contaminant in 2004 when a large number of vulture deaths were reported in the Indian subcontinent. Vultures died from kidney failure after being fed with carcasses of cows that were treated with diclofenac before death. In 2006, countries on the Indian subcontinent began banning diclofenac in veterinary medicine, and after this measure, the vulture population increased substantially [33,37,38]. In 2013, in Europe, DCF was added to the first Watch List of contaminants of emerging concern (Directive 2013/39/EU (Article 8b)) in order to gather the data needed to assess this drug's potential for pollution of the environment [35]. Recently, in the review of Bonnefille et al. (2018), it was reported that prolonged exposure of different marine organisms to DCF (even at concentrations in ng/L) causes a variety of effects, including the modification of vital biological functions [39]. This suggests that chronic exposure should not be lightly disregarded. Thus, it is necessary to find an economical way to remove DCF from wastewater before discharge into the environment. The literature data reported a high potential of minerals treated with surfactants for DCF adsorption and that increase in surfactant concentration at the mineral's surface (zeolites) or in minerals (clays) resulted in greater drug adsorption [13,29,40–43].

In a previous paper [29], results on the adsorption of DCF and ketoprofen onto composites prepared by modification of two zeolites (clinoptilolite and phillipsite) with two different amounts (monolayer and bilayer coverage) of surfactants—cetylpyridinium chloride (C) and Arquad®2HT-75 (A) were presented. Results showed that adsorption of both drugs increased with increasing the amount of each surfactant at zeolite surfaces and with increasing the initial concentration of each drug. It was determined that depending on the zeolitic content in the zeolitic tuff, the type and the number of surfactants and the type of the drug, the characteristic adsorption capacity of investigated composites varied from 1 to 35 mg/g. Moreover, composites of clinoptilolite and phillipsite with C gave higher adsorption of DCF and ketoprofen.

The present study aims to examine if composites of phillipsite-rich tuff with octadecyldimethylbenzylammonium chloride (O) and dodecylamine (D) (in amounts of 100% and 200% of ECEC values) are able to adsorb the nonsteroidal anti-inflammatory drug DCF. The prepared composites were used for the first time for DCF adsorption, and the adsorption of the drug onto composites containing aliphatic primary amine—D and quaternary ammonium salt with aromatic ring—O was compared. Additionally, since generally, surfactant ions in composites are responsible for the adsorption of drugs and their adsorption is dependent on the type of surfactant used for modification of the zeolitic surface, results were discussed together with previous results on adsorption of DCF on composites of phillipsite with C (quaternary ammonium salt with pyridine ring) and A (quaternary ammonium salt with two alkyl chains) [29]. The main aim of this research is to investigate the influence of the chemical structure of surfactants used for modification of the phillipsite surface on the preparation and properties of composites and subsequent DCF adsorption.

The DCF adsorption experiments on O and D composites were performed in batch conditions, testing the influence of different initial drug concentrations and contact time. The mechanism of DCF adsorption by composites was suggested based on the drug adsorption experiments. Detailed characterization of surfactant-modified zeolites was performed by several characterization methods: FTIR–ATR analysis, thermal analysis, and zeta potential measurements. Additionally, to confirm the presence of the DCF at the modified zeolitic surfaces, composites containing the adsorbed drug were characterized by using FTIR–ATR spectroscopy and zeta potential measurements.

2. Results and Discussion

2.1. Characterization of O and D Composites

FTIR–ATR spectra of starting material—P, surfactants—O and D, and composites PO100, PO200, PD100 and PD200 are given in Figure 1A,B. Infrared analysis of P has displayed typical adsorption bands of zeolite-rich tuff associated with internal and external framework vibration of primary and secondary building units [15,29,44,45].

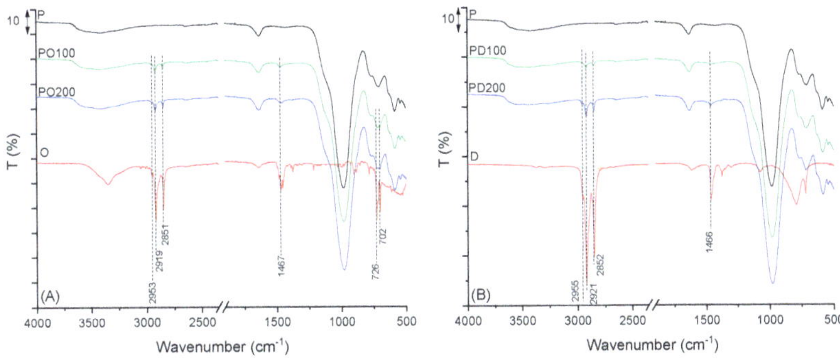

Figure 1. FTIR–ATR spectra of (**A**) starting material—P, surfactant octadecyldimethylbenzylammonium chloride—O, and composites PO100 and PO200; (**B**) starting material—P, surfactant dodecyl amine—D, and composites PD100 and PD200.

A broad peak at around 3500 cm^{-1} and a smaller one at 1635 cm^{-1} belong to O-H stretching and bending vibrations in water molecules, respectively. A high-intensity peak at 985 cm^{-1} originates from asymmetric stretching vibrations of the T-O group, where T stands for Al or Si and O is an oxygen atom. Additionally, T-O-T stretching and bending vibrations are confirmed with peaks below 800 cm^{-1} [44]. All bands characteristic for P are also evident in the spectra of composites, proving that the functionalization process and presence of surfactant molecules in composites did not cause any changes in the structure of the tuff. After modification with O, in the spectra of composites (PO100 and PO200), new peaks have appeared (Figure 1A), proving the presence of surfactant molecules. The peak at 2953 cm^{-1} belongs to the CH_3 group, whereas peaks at 2919 cm^{-1} and 2851 cm^{-1} are due to asymmetric and symmetric C-H stretching vibrations of CH_2 groups in the surfactant alkyl chain, respectively. Additionally, the low-intensity peak at 1467 cm^{-1} is from C-H bending vibrations. Peaks at 726 and 702 cm^{-1} are associated with vibrations of C–H groups in the benzene ring of O molecules [46]. When D was used for phillipsite's surface modification (Figure 1B), peaks associated with surfactant were visible at 2955 cm^{-1}, 2921 cm^{-1}, and 2852 cm^{-1} (C-H stretching vibration), as well as at 1466 cm^{-1} (bending C-H vibration). It is observed that intensities of peaks indicative of the surfactant presence increased with increasing the surfactant content at the zeolitic surface, thus confirming that FTIR spectroscopy served to validate successful modification of the zeolite's surface with no changes of the zeolitic structure. Similar peaks were reported for composites of phillipsite with C and A (PCM, PCB, PAM, and PAB) when bands originating from C-H stretching and bending vibrations confirmed successful modification of phillipsite-rich tuff with C or A in amounts of 100 and 200% of ECEC [29].

Thermal analysis (DSC/TGA) was used to study the thermal properties of the starting material and prepared composites, as well as to supply information on the bonding between organic species and P. The differential scanning calorimetry (DSC) curves of P and composites (PO100, PO200, PD100, and PD200) are presented in Figure 2, while the mass loss from thermogravimetric (TGA) curves for all samples in different temperature regions is given in Table 1.

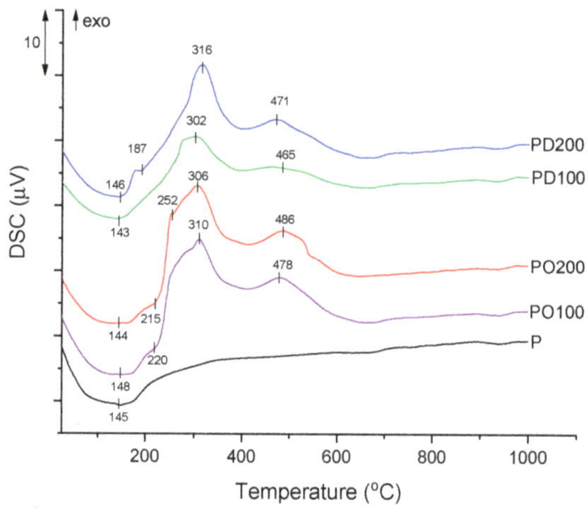

Figure 2. DSC analysis of starting material—P and composites PO100, PO200, PD100 and PD200.

Table 1. Mass loss (%) from TGA curves for starting material-P and composites.

Sample	Mass Loss (%)			
	25–200 °C	200–700 °C	700–1000 °C	∑25–1000 °C
P	7.44	3.77	0.08	11.29
PO100	6.46	8.02	0.07	14.55
PO200	6.35	11.35	0.03	17.73
PD100	6.47	5.94	0.04	12.45
PD200	7.49	6.98	0.05	14.52

Based on the results given in Table 1 and TGA curves (not shown), for P, a continuous mass loss in the temperature range of 25–1000 °C is observed, and the total mass loss was 11.29%. The presence of both surfactants led to an increase in the mass loss and was proportional to the amount of surfactant molecules at P surface, meaning that mass loss for composites increased with increasing the amount of each surfactant at the P surface. Thus, in the case of PO100 and PO200, total mass losses were 14.55% and 17.73%, respectively, while in the case of D composites PD100 and PD200, mass losses were 12.45% for PD100 and 14.52% for PD200. These results were in agreement with the literature and confirmed the presence of surfactants at the P surface.

From Table 1, it can also be seen that for P, most of the mass loss occurred up to 200 °C (7.44%) and is associated with physically adsorbed water. Usually, in this region, for composites of zeolite containing surfactants in amounts up to 100% of ECEC, due to the increased hydrophobicity of the zeolitic surface, the mass loss should be lower. When the amount of surfactant at the zeolitic surface is above 100% ECEC, besides dehydration from physically adsorbed water, mass loss is also associated with the desorption of non-adsorbed organic compound or elimination of weakly bound surfactant. After modification with surfactants, mass loss in the first temperature region for PO100, PO200 and PD100 was slightly lower (around 6.40%) compared to the mass loss for P (7.44%), confirming hydrophobicity of composites surfaces. Composite PD200 showed a mass loss (7.49%) similar to the mass loss of P, suggesting the desorption of a small fraction of unbound D molecules [47,48]. In the second temperature range from 200–700 °C, P continued to gradually lose water molecules coordinated to the cations in zeolite channels (mass loss of 3.77%), while for composites, mass loss mainly originates from the oxidation of organic phase present at the zeolitic surface. It is observed that for composites, mass loss increased with increasing the amount of O and D in composites. Thus, PO100 had a mass loss of 8.02%, and PO200 had a mass loss of 11.35%, while for D composites, PD100 and PD200, mass loss was 5.94% and 6.98%, respectively. Additionally, higher mass losses were reported in the case of composites with O (PO100 and PO200) due to the higher molecular weight of the O molecule.

The DSC curve of P showed only one visible wide endothermic peak (with a maximum of around 145 °C) originating from the desorption of water molecules. Besides the peak around 145 °C, composites PO100, PO200 and PD200 have additional low-intensity shoulder, around 220 °C, 215 °C, and 187 °C respectively, that may be connected to the loss of non-adsorbed or loosely bound surfactant molecules. In the case of PD200, this shoulder appeared below 200 °C (187 °C), suggesting that this composite desorption of weakly bound D occurred at a lower temperature and is confirmed by the slightly higher mass loss in the first temperature region. DSC analysis has shown that in composites, between 200–700 °C, exothermic peaks occurred due to the oxidation of the organic phase present at the P surface. When O was used for modification in the amount equal to 100% of ECEC (PO100), exothermic peaks at 310 °C and 478 °C were observed, while with an increase of the amount of surfactant (PO200), a small peak at 252 °C and more intensive exothermic peaks at 306 °C and 486 °C were noticed, suggesting the higher energy requirements to burn higher surfactant amount [48]. Similar behavior was noticed for composites with D: DSC curves showed exothermic peaks at 302 °C and 465 °C for PD100, while for PD200, these peaks were at 316 °C and 471 °C. Again, the more intensive peak was in the case of composite with a higher surfactant amount (PD200). DSC curves of O and D composites showed that intensities of the exothermic peaks for PD100 and PD200 were noticeably lower than for PO100 and PO200, being a consequence of the higher molecular weight of the O molecule.

Zeta potential is a reflection of surface potential, and it changes depending on surfactant loading at the zeolite surface [13]. When surfactants are used in an amount equal to 100% of ECEC, zeta potential should be close to zero indicating increased surface hydrophobicity, while when the amount of surfactants is equal to 200% of ECEC, a bilayer of surfactant should be formed at the zeolite surface making the composite surface more positive. This means that surfactant molecules are covering negative active sites at the zeolite surface and, in this way, are reducing the negative zeta potential value of unmodified zeolite.

Zeta potential data for P and O and D composites (see Section 2.3) showed that the initial negative surface of P (−34.0 mV) after modification with O turned to positive (+2.3 mV for PO100 and +20.2 mV for PO200) while after modification with D, the negative charge of P was changed to less negative as the amount of D increased in D composites (−14.1 mV for PD100 and to −12.6 mV for PD200). For both types of composites, zeta potential values confirmed successful modification and the presence of O and D at the surface of P. It is also observed that changes in zeta potential value are more pronounced in the case of quaternary ammonium ion—O, than in the case of modification with primary amine—D.

In the case of O composites, when the amount of O was equal to ECEC value (PO100), zeta potential approached zero, confirming nearly monolayer formation and almost complete hydrophobicity of the P surface. At O amount equal to 200%, ECEC zeta potential became positive, pointing charge reversal and bilayer formation at the P surface. Similar changes in zeta potential values were noticed when P was modified with the same amounts of quaternary ammonium surfactants C and A. When the amount of these surfactants was equal to 100% ECEC, zeta potential became slightly positive (+2.8 mV for PCM, and +2.1 mV for PAM), while when the amount of C and A was 200% of ECEC reversed charge was reported (+30.6 mV for PCB, and +33.5 mV for PAB) [29]. These results indicated that although these quaternary ammonium salts have different functional groups and different chain lengths: O (one alkyl chain with C_{18}), C (one alkyl chain with C_{16}) and A (two alkyl chains with $C_{14–18}$), they have the same effect on zeta potential meaning that they are similarly positioned at the P surfaces. However, when primary amine—D was used, zeta potential after modification remained negative for composites PD100 and PD200, and these results may be explained by its different chemical structure, much shorter chain length (one alkyl chain with C_{12}) and smaller cationic head. The negative values of zeta potential mean that despite primary amine being adsorbed at P surfaces, due to the much smaller chain length, the P surface is still partially uncovered. Results of zeta potential measurements for O and D composites suggest that although equivalent amounts of the organic phase (O and D) were used for modification, differences in surfactant molecular structures will lead to different arrangements of the surfactant molecules at composite surfaces [49]. Similar differences in zeta potential values were reported by Orta et al. (2019) [50] when montmorillonite (Mt) was modified with quaternary ammonium ion octadecyltrimethylammonium (ODTMA) and primary amine octadecylamine (ODA) in two different amounts 50 and 100% of ECEC. Namely, the authors noticed that zeta potentially changes from negative (for Mt) to positive for modification with ODTMA, while when ODA was used for modification, the surface of Mt remained negative. The behavior of composite with primary amine was explained with the smaller size of the ammonium head groups in the ODA ($-NH_3^+$) compared to the size of ODTMA ($-N(CH_3)_3^+$). They also indicated that primary ammonium head groups might form additional hydrogen bonds with water molecules and chloride anions onto the clay surface, and consequently, if a positive surface charge is generated, it would be almost neutralized [50].

2.2. Adsorption of Diclofenac Sodium

Adsorption isotherms are a crucial step in figuring out the tested materials' adsorptive capacities and determining the DCF adsorption mechanism. In Figure 3, adsorption isotherms of DCF from buffer solution on O and D composites are presented. For all composites, adsorption of DCF increased with increasing the initial drug concentration and with increasing the amount of organic phase in both types of composites. This means that either quaternary ammonium ion or primary amine at the surface of P is responsible for the adsorption of DCF by O and D composites since unmodified P has no affinity for the adsorption of this drug. It is noticed that drug adsorption increased in the following order: PO200 > PD200> PO100 > PD100, making it simpler to see that bilayer composites had a higher adsorption capacity than monolayer composites. Also, comparing the adsorption of DCF by O and D composites containing the same amount of organic phase, composites containing quaternary ammonium ion are more efficient for the adsorption of DCF than composites with primary amine. Under the applied experimental conditions, the maximum adsorbed amounts of DCF by O and D composites with an amount of O or D equal to 100% ECEC were 11.1 mg/g for PD100 and 19.9 mg/g for PO100. For composites with an organic phase equal to 200% of ECEC, the adsorption of DCF was 26.8 mg/g and 37.9 mg/g for PD200 and PO200, respectively.

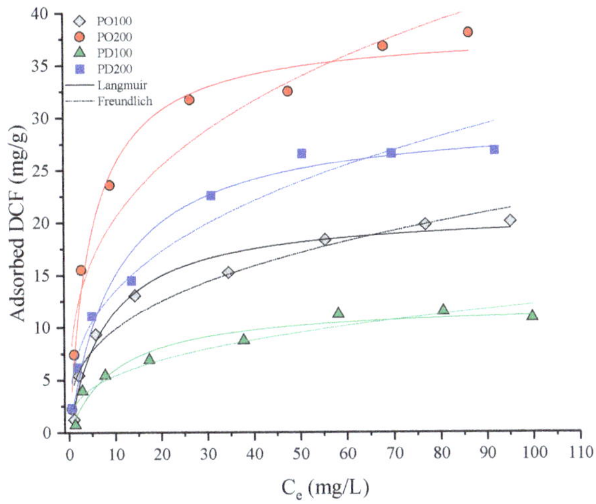

Figure 3. Equilibrium isotherms of DCF adsorption on composites of P.

Since the adsorption of DCF followed nonlinear isotherms, Langmuir and Freundlich isotherm models (Equations (1) and (2), respectively) were applied to fit the experimental data:

$$Q_e = \frac{Q_{max} K_L C_e}{1 + K_L C_e} \quad (1)$$

$$Q_e = K_F C_e^{1/n} \quad (2)$$

where Q_e (mg/g) is the amount of loaded DCF at equilibrium, C_e (mg/L) is the concentration of DCF in solution at equilibrium, Q_{max} (mg/g) is the highest concentration of the drug, K_L (L/mg) and K_F (L/mg) are Langmuir and Freundlich constants, respectively, and 1/n is a constant related to the strength of adsorption [51].

Due to the Langmuir isotherm's (Equation (1)) presumption that adsorbate molecules can only form one layer, the curve reaches a plateau [28,29]. Freundlich isotherm (Equation (2)) takes into consideration the heterogeneity of adsorbents represented by the constant n in the Freundlich equation [18]. According to the n constant, the favorability of adsorption can be estimated: values above 2 represent good, between 1–2 medium, and below 1 poor adsorbate-adsorbent interactions [18,52]. The determination coefficients (R^2) and Akaike weight parameter (AICw) were used to evaluate the selected mathematical models [53,54]. For all composites, obtained n values were higher than 2, suggesting good affinity of O and D composites for DCF adsorption. The computed adsorption isotherm parameters are listed in Table 2. Both mathematical models, Langmuir and Freundlich, had R^2 above 0.9, suggesting good agreement with experimental data. However, according to slightly higher R^2 and based on AICw values, the Langmuir model consistently provided a better fit.

The adsorption of DCF on several surfactant-modified zeolites was also found to exhibit nonlinear isotherms that are in good agreement with the Langmuir isotherm model [29,43,47,48]. The results of this study (Figure 3) and a review of the literature both point to nonlinear isotherms for the adsorption of DCF. It was previously mentioned that the adsorption of DCF may be dependent on the type of organic phase used for the modification of the P surface. In a previous study, the adsorption of DCF by composites containing the same amounts of C and A (100 and 200% ECEC) was studied under the same experimental conditions. Composites with C and A amounts equal to 100% of ECEC were denoted as PCM and PAM, while composites with surfactants amounting equal to 200% ECEC were denoted as PCB and PAB. It was observed that the adsorption of DCF increased with increasing the amount of each surfactant in composites as well as with increasing the initial concentration of DCF. In all cases, nonlinear adsorption isotherms described the adsorption of DCF by C and A composites. From Langmuir isotherms (Table 3), maximum adsorption capacities for monolayer composites were 17.2 mg/g for PCM and 7.4 for PAM, while for bilayer composites, adsorption of the drug was 30.0 mg/g and 19.9 mg/g for PCB and PAB, respectively. Comparing the results of DCF adsorption by composites containing C and A with those for adsorption of the drug by O and D composites, for monolayer composites, adsorption of DCF increased, depending on the type of organic phase, in the following order A < D < C < O. The highest adsorption of DCF by monolayer composites were obtained with O (one alkyl chain with C_{18})—21.1 mg/g and C (one alkyl chain with C_{16})—17.2 mg/g, suggesting that longer chain but also the presence of aromatic rings (benzyl in O and pyridine in C) in their structure contribute to the higher adsorption of DCF that also has aromatic ring [41,43]. Lower adsorption of DCF by PD100 (12.3 mg/g) can be explained by the shorter alkyl chain in the D structure. However, although A has two alkyl chains with relatively high lengths ($C_{14–18}$), its monolayer resulted in the lowest DCF adsorption (7.4 mg/g). Composites containing C, O and D in the amount of 200% ECEC showed similar adsorption of DCF (30.0 mg/g for PCB, 38.8 mg/g for PO200 and 30.8 mg/g for PD200), while only composite PAB showed much lower adsorption of DCF (19.9 mg/g). It is possible that the presence of two alkyl chains in A hindered DCF adsorption by composites PAM and PAB.

Table 2. Mathematical models for the adsorption isotherms: parameters and goodness-of-fit.

Drug	Sample	Model	Parameters			Goodness-of-Fit	
			K (L/mg)	n	Q_{max} (mg/g)	R^2	AICw
DCF	PO100	Langmuir	0.12 ± 0.03		21.1 ± 0.9	0.974	0.974
		Freundlich	4.5 ± 0.8	2.9 ± 0.4		0.937	0.026
	PO200	Langmuir	0.20 ± 0.04		38.4 ± 0.9	0.984	0.998
		Freundlich	9.9 ± 0.9	3.7 ± 0.3		0.925	0.002
	PD100	Langmuir	0.10 ± 0.03		12.3 ± 0.8	0.947	0.852
		Freundlich	2.4 ± 0.5	2.9 ± 0.4		0.919	0.148
	PD200	Langmuir	0.14 ± 0.05		30.2± 0.7	0.974	0.918
		Freundlich	0.10 ± 0.03	2.8 ± 0.4		0.952	0.082

Table 3. DCF adsorption on P composites with different surfactants.

Surfactant Amount	Surfactant Type	* Adsorbed DCF mg/g	Reference
100% of ECEC	C	17.2	[29]
	A	7.4	[29]
	O	21.1	This study
	D	12.3	This study
200% of ECEC	C	30.0	[29]
	A	19.9	[29]
	O	38.8	This study
	D	30.8	This study

* Langmuir model values.

DCF is a hydrophobic molecule that exists in anionic form at pH levels above 3. Thus, nonlinear isotherms obtained for O, D, C and A composites suggest a complex adsorption mechanism that involves ion exchange and hydrophobic partitioning for monolayer composites as well as additional anion exchange in bilayer composites. Therefore, nonlinear isotherms for DCF adsorption on monolayer composites suggest an adsorption mechanism via both hydrophobic interactions between the alkyl chains of the O, D, C or A and the hydrophobic part of DCF, electrostatic interactions with the cationic heads from organic phase attached directly to the P surface as well as anion exchange of the counter anions in organic molecules from local bilayer with the anionic DCF. The improved DCF adsorption capacities by bilayer composites demonstrated that anion exchange plays an important role throughout the entire adsorption process. The lowest adsorption of DCF by A composites (PAM and especially PAB) indicates that two alkyl chains in A are more densely packed and make the P surface more hydrophobic and prevent adsorption of an ionic form of DCF on the local bilayer in PAM and in complete bilayer in PAB. Although it was expected that composites with primary amine show lower adsorption of DCF, results indicated that these composites are good adsorbents for DCF. Overall results showed that the adsorption of DCF by composites containing different quaternary ammonium ions was dependent on the type of surfactant used for the modification of the P surface.

Figure 4 illustrates the influence of contact time on DCF adsorption onto O and D composites. According to the results, it can be seen that when the amount of organic phase utilized in the modification of P was 200% of ECEC values (PO200 and PD200), the adsorption rate increased rapidly (adsorption equilibrium was established in around 10–15 min). Adsorption of DCF was likewise fast for the PO100 sample, while for the PD100, adsorption of the drug was slower, and equilibrium was reached after approximately 45 min. For the initial concentration of DCF of 20 mg/L, maximum adsorbed amounts of the drug were 23.6 for PO200 and 19.1 mg/g for PD200, respectively, whereas maximum adsorbed values of DCF in composites with lower concentrations of organic phase were 13.0 mg/g and 8.5 mg/g, for PO100 and PD100, respectively.

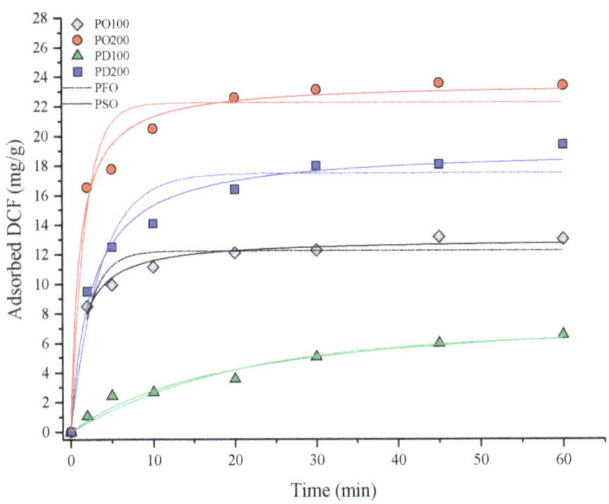

Figure 4. Kinetics run for DCF adsorption on composites of P. PFO = pseudo-first-order; PSO = pseudo-second-order.

The pseudo-first-order (PFO) and pseudo-second-order (PSO) models, which are denoted by Equations (3) and (4), are frequently used to analyze the kinetic runs for surfactant-modified minerals

$$Q_t = Q_{max}\left(1 - e^{-K_1 t}\right) \tag{3}$$

$$Q_t = \frac{K_2 Q_{max}^2 t}{K_2 Q_{max} t} \tag{4}$$

where Q_{max} is the adsorbate concentration at equilibrium, Q_t is the amount of drug adsorbed by the adsorbent as a function of time t (min), and K_1 and K_2 are the pseudo-first and second-order constants, respectively.

According to statistical criteria (Table 4), PSO had a higher goodness-of-fit (correlation coefficients R^2). The PSO kinetics indicates that the chemical adsorption is the rate-determining step controlling DCF adsorption by O and D composites and is in agreement with the results reported by other researchers [45,55,56] as well as with the kinetic results for the adsorption of DCF by C and A composites [29].

Table 4. Mathematical models of kinetic runs: parameters and goodness-of-fit.

Drug	Sample	Model	Parameters			Goodness-of-Fit	
			K_1 (min^{-1})	K_2 (g mg^{-1} min^{-1})	Q_{max} (mg/g)	R^2	AICw
DCF	PO100	PFO	0.5 ± 0.1		12.2 ± 0.4	0.676	0.005
		PSO		0.06 ± 0.01	13.0 ± 0.3	0.929	0.995
	PO200	PFO	0.6 ± 0.1		22.3 ± 0.8	0.950	0.006
		PSO		0.040 ± 0.008	23.6 ± 0.6	0.986	0.994
	PD100	PFO	0.05 ± 0.01		6.7 ± 0.7	0.952	0.225
		PSO		0.006 ± 0.002	8.5 ± 0.9	0.965	0.775
	PD200	PFO	0.27 ± 0.06		17.5 ± 0.8	0.937	0.005
		PSO		0.021 ± 0.004	19.1 ± 0.6	0.983	0.995

Composites prepared by modifying P with C had similar Q_{max} as composites with O (12.0 mg/g for PCM, 13.0 mg/g for PO100, 21.9 mg/g for PCB, and 23.6 mg/g for PO200). This additionally confirmed that the presence of the aromatic ring in the structure of both O and C as well similar chain length of these two surfactants, contribute to the higher adsorption of DCF. Among investigated surfactants, a composite with A had lower DCF capacities well as a composite with the primary amine—D. Thus, in the case of monolayer composites, DCF adsorbed amounts were 2.48 for PAM and 8.5 mg/g for PD100. However, adsorption kinetic was faster for PAM when equilibrium was reached after 10 min. For composites with an organic phase equal to 200% of ECEC, DCF adsorption was 13.1 and 19.1 mg/g for PAB and PD200, respectively.

2.3. *Characterization after Adsorption: FTIR–ATR and Zeta Potential*

In order to confirm the presence of DCF on O, D, C and A composites, the characterization of composite/drug complexes was performed by using FTIR spectroscopy and zeta potential measurements. Analyses were performed on samples collected after kinetic runs from this study (PO100, PO200, PD100, PD200) and also on samples from previous kinetic runs on DCF adsorption by PCM, PCB, PAM, and PAB [29].

Figure 5 shows the FTIR–ATR spectra for DCF and composites both before and after DCF adsorption. Similar behavior was noticed for all samples. According to the data, peaks at around 2920 cm^{-1} and 2850 cm^{-1} (symmetric and asymmetric C-H stretching vibrations) that originate from surfactant on the zeolitic surface were reduced after the adsorption of DCF. This most likely occurred due to the small desorption of the surfactant molecules that were weakly bonded at the zeolitic surface [47,57]. Additionally, low-intensity peaks in the spectra of O, D, C and A composites between 1590–1360 cm^{-1} have appeared after drug adsorption, providing further evidence of DCF presence at the surfactants modified P surface. Namely, peaks assigned to asymmetric (1573 cm^{-1}) and symmetric stretching (1396 cm^{-1}) vibration of the carboxylate group in the DCF molecule (Figure 5A) were shifted after drug adsorption onto P composites. For composites with O and D, after adsorption of DCF, these vibrations were shifted to 1576 cm^{-1} and 1376 cm^{-1} (Figure 5B), and for composites containing C and A to 1576 cm^{-1} and 1373 cm^{-1}(Figure 5C). Similar behavior was reported for surfactant-modified bentonite [41] and kaolinite [58] after the adsorption of DCF, suggesting that the COO^- group is involved in drug adsorption onto composites. After drug incorporation, no appreciable changes in the wavenumbers of the bands ascribed to the P were seen, indicating that the zeolitic framework was unaffected by the drug adsorption. Similar behavior was reported in the case when salicylic acid was adsorbed on P composites (PCM, PCB, PAM, and PAB) [57], when IBU and DCF were adsorbed on composites of different minerals with O [13], and when IBU was adsorbed on composites of clinoptilolite-rich tuff with benzalkonium chloride and C [49].

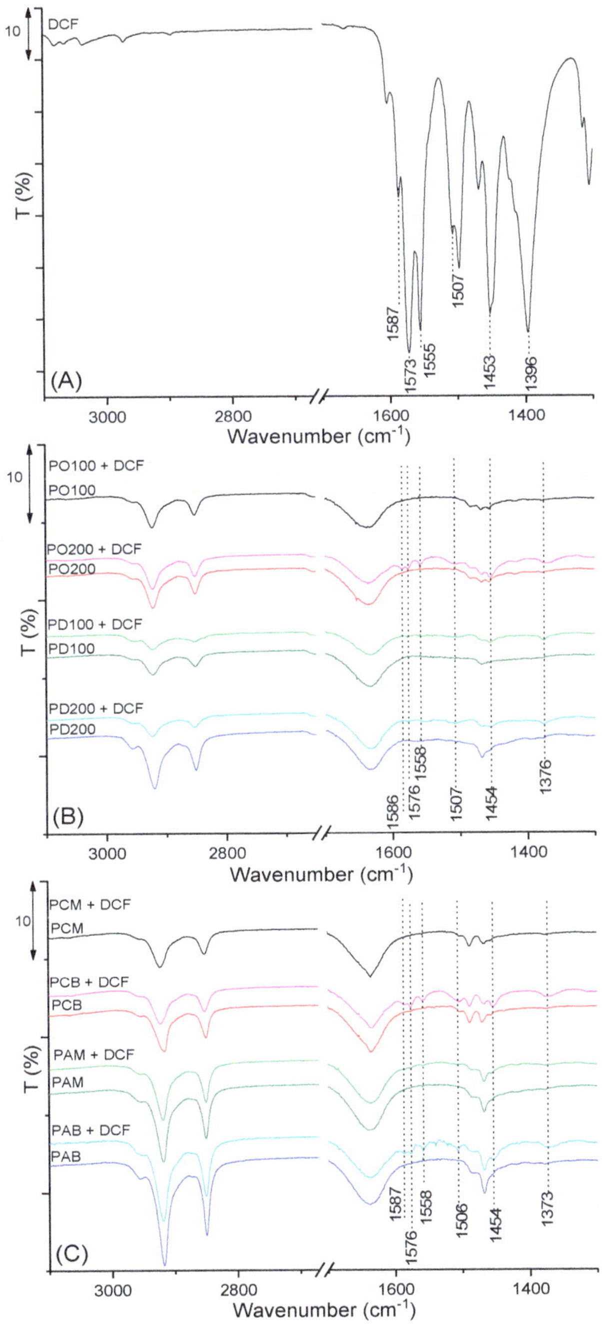

Figure 5. FTIR–ATR spectra of: DCF (**A**), spectra of P composites with dodecylamine and octadecyldimethylbenzylammonium chloride (PO100, PO200, PD100, and PD200) before and after adsorption of DCF (**B**), and spectra of P composites with cetylpyridinium chloride and Arquad®2HT-75 (PCM, PCB, PAM, and PAB) before and after adsorption of DCF [29] (**C**).

Results for the determination of the zeta potential of O, D, C and A composites before and after adsorption of DCF are presented in Figure 6. It can be seen that, after DCF adsorption, the zeta potential of the composite containing O altered from +2.3 mV to −3.4 mV for sample PO100 and from +20.2 mV to +15.5 mV in the case of PO200. When DCF was adsorbed on composites with D, zeta potential changed from −11.4 mV to −25.2 mV for sample PD100 and from −12.6 mV to −21.2 mV for sample PD200. Thus, when the amount of O, C, and A was equal to ECEC value (PO100, PCM, and PAM), the zeta potential was close to zero, confirming nearly monolayer formation and almost complete hydrophobicity of zeolitic surfaces. When the amount of both surfactants was 200% of the ECEC value, zeta potential became positive in the case of PO200, PCB, and PAB, pointing to charge reversal and bilayer formation at phillipsite surfaces. When D was used for modification potential of the phillipsite sample (PD100 and PD200) remained negative. After adsorption of DCF, zeta potential for all composites decreased: changed from positive to negative (PO100, PCM, PCB, and PAM), became more negative (PD100 and PD200), or became less positive (PO200 and PAB). The findings imply that the adsorption of DCF led to a change in the surface charge of composites, providing evidence that DCF molecules are present at the surface of O, D, C, and A composites. Differences in values of zeta potential after adsorption of DCF on P composites may be caused by the different arrangements of surfactant ions at their surfaces and are the consequence of the different structures of these molecules.

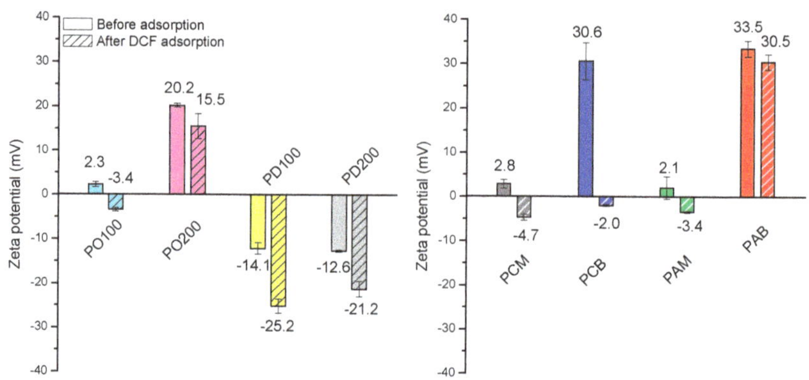

Figure 6. Zeta potential measured for P composites before and after adsorption of DCF. Measured values were compared with the zeta potential of the composites before drug adsorption previously reported in Smiljanic et al. (2020) [29]. Data were used with permission from Microporous and Mesoporous Materials, Elsevier, license number: 5436600349228.

3. Materials and Methods

In this study, natural zeolite (P) from Italy (trade name PHIL75) with 58% by weight of zeolite content was used as the starting material. The main zeolitic component of the tuff was phillipsite (44 wt%), whereas smaller amounts of analcime (10 wt%) and chabazite (4 wt%) were reported [29,47,59–61]. The ECEC value of the starting material was 0.13 mEq/g [62].

The surface of the zeolite-rich tuff was modified with the surfactants: quaternary ammonium salt—octadecyldimethylbenzylammonium chloride (O) (Hoechst AG, Frankfurt, Germany) and primary amine—dodecylamine (D) (Fluka, Buchs, Switzerland) (Figure 7). Two amounts of each surfactant equivalent to 100% or 200% of the external cation exchange capacity (ECEC) of the zeolite were used for the modification process. The composites were prepared according to the following procedure: 10 g of zeolite-rich tuff was added to 200 mL of distilled water containing surfactant equivalent to 100 and 200% ECEC zeolite.

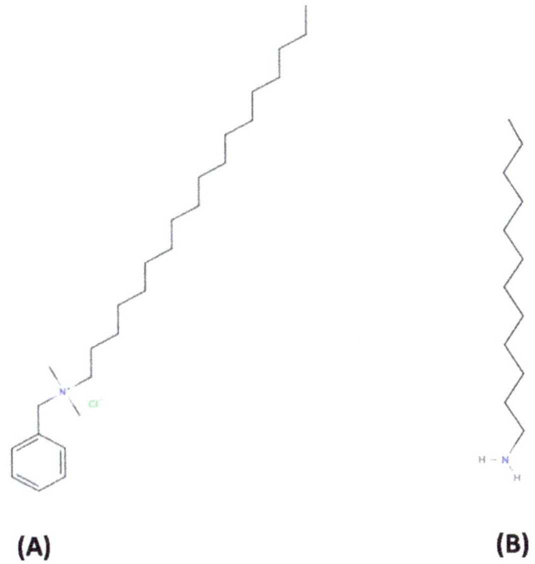

Figure 7. The structural formula of (A) octadecyldimethylbenzylammonium chloride (O) and (B) dodecylamine (D).

Modifications were made using a mixer at 6000 rotations per minute with a short activation time (about half an hour). When D was used to modify the zeolite surface to protonate the amino groups of the surfactant, the pH of the D solution was neutralized by the addition of 1M HCl. Composites of P with O are designated as PO, and the corresponding numbers 100 or 200 were added depending on the organic concentration, resulting in the composite names: PO100 and PO200. For composites of P and D, names were derived in the same manner: PD100 and PD200.

Several methods [13,29,62–64] were used to characterize the properties of the P and composites. FTIR spectra of starting material and both surfactants and composites before and after adsorption of the drug were collected using Nicolet iS50 spectrophotometer (Thermo Fisher Scientific, Waltham, MA, USA) with a diamond ATR accessory in the range 4000–400 cm^{-1} and with the resolution of 2 cm^{-1} and 32 scans. Starting material and prepared composites were characterized by simultaneous thermal analysis—STA (TGA/DSC). Prior to thermal analysis, all samples were dried for 2 h at 60 °C and then placed in a desiccator for 24 h at a relative humidity of 75%. The analysis was carried out in a synthetic air atmosphere (flow 70 mL/min), using PtRh crucibles, in the temperature range 25 °C–1000 °C with a heating rate of 10 °C/min (NETZSCH STA 449 F5 Jupiter, Selb, Germany). Additionally, the zeta potentials of starting material and composites before and after adsorption of the drug were measured using a Zetasizer Nano ZS90 (Malvern Instruments, Malvern, UK) that was calibrated using latex dispersion. Aqueous suspensions of 0.1 mg/mL for each material were prepared, and the mean value was calculated from 5 measurements.

The effect of contact time and initial drug concentration on DCF adsorption was studied in batch adsorption experiments following the procedures given by Smiljanic et al. (2020) [29]. The diclofenac sodium salt of the analytical grade (≥98%) used in experiments was purchased from Sigma Aldrich (Schnelldorf, Germany). The stock solution of DCF (1000 mg/L) was prepared in methanol and then used to prepare 2–100 mg/L working solutions of the DCF in 0.01 M potassium phosphate buffer (pH 6.5). Suspensions were prepared by adding 10 mL of DCF working solutions (in various initial concentrations) to 5 mg of each composite. Suspensions were shaken for 60 min (time found sufficient to achieve equilibrium conditions) at room temperature using a laboratory shaker (Heidolph Unimax 1010, Schwabach, Germany) with speed adjusted at 300 rpm. After adsorption, suspensions were filtered, and the concentration of DCF in filtrates was measured using UV–VIS spectroscopy (Spectrophotometer UV—1800, Shimadzu, Kyoto, Japan) at a wavelength of 276 nm. In order to carry out kinetic runs of DCF adsorption onto composites, 500 mg of each composite was added to 1000 mL of DCF water solution (C_0 = 20 mg/L, pH~6) and stirred for 60 min. At fixed time intervals, 5 mL aliquots were taken, filtered, and DCF concentration in filtrates was measured on UV–VIS.

The starting material-P used in this study was also used in the work of Smiljanic et al. (2020) [29] when cationic surfactants C and A served for modification of the mineral surface. Surfactant concentration was equivalent to 100% or 200% of zeolites' ECEC value, necessary for monolayer and bilayer formation, respectively, resulting in four composites with the following names [abbreviation used in the original work]: P+100% C [PCM], P+200% C [PCB], P+100% A [PAM], and P+200% A [PAB] [29]. Solid samples collected after kinetic runs in this study using composites (PO100, PO200, PD100, and PD200) and after kinetic runs performed in previous work (PCM, PCB, PAM, and PAB) were dried at room temperature and stored for further characterization. The zeta potential of all eight composites before and after DCF adsorption was evaluated using water suspension (0.1 g/L). Additionally, pure DCF and composites before and after DCF adsorption were characterized by FTIR spectroscopy.

4. Conclusions

Natural zeolite—phillipsite was modified with two different amounts of quaternary ammonium salt—O and primary amine—D, and composites were tested as adsorbents of the anti-inflammatory drug DCF. Results were compared with the adsorption of DCF by composites of phillipsite and quaternary ammonium salts—C and A. The effects of the type of surfactant, the initial concentration of the drug and the contact time on the adsorption of DCF were investigated. For O and D composites, adsorption of DCF increased with increasing the amount of each surfactant at the P surface as well as with increasing the initial concentration of the drug. The highest adsorption of DCF was achieved with PO200, which contains the highest amount of O. Comparison of results reported in this paper with previous results showed that adsorption of DCF by composites containing different quaternary ammonium ions (O, C and A) was dependent on the type of the surfactant used for the modification of the P surface. The presence of the aromatic ring in the structure of both O and C as well similar chain length of these two surfactants, contribute to the higher adsorption of DCF. Although it was expected that composites with primary amine—D would show lower adsorption of DCF, results indicated that these composites are good adsorbents for DCF and showed even higher adsorption of the drug than composites of P and surfactant-containing two alkyl chains (A). Overall results indicated that both composites of P with quaternary ammonium salt O or primary amine—D have the potential for removal of nonsteroidal anti-inflammatory drug DCF from contaminated water systems.

Author Contributions: Conceptualization, A.D.; formal analysis, M.O. (Milica Ožegović); investigation, D.S. and M.O. (Milena Obradović); resources, A.D.; data curation, M.M. and M.O. (Milena Obradović); writing—original draft preparation, D.S. and A.D.; writing—review and editing, A.D., M.M., G.E.R. and B.d.G.; visualization, M.O. (Milica Ožegović) and D.S.; supervision, A.D., B.d.G. and G.E.R. All authors have read and agreed to the published version of the manuscript.

Funding: This research was funded by the Ministry of Education, Science, and Technological Development of the Republic of Serbia, grant number 451-03-68/2022-14/200023.

Data Availability Statement: The data presented in this study are available on request from the corresponding author.

Conflicts of Interest: The authors declare no conflict of interest.

References

1. Inglezakis, V.J.; Zorpas, A.A. *Handbook of Natural Zeolites*; Bentham Science Publishers: Sharjah, United Arab Emirates, 2012. [CrossRef]
2. Smith, J.V. Definition of a Zeolite. *Zeolites* **1984**, *4*, 309–310. [CrossRef]
3. Delkash, M.; Ebrazi Bakhshayesh, B.; Kazemian, H. Using Zeolitic Adsorbents to Cleanup Special Wastewater Streams: A Review. *Microporous Mesoporous Mater.* **2015**, *214*, 224–241. [CrossRef]
4. Lemić, J.; Tomašević-Čanović, M.; Adamović, M.; Kovačević, D.; Milićević, S. Competitive Adsorption of Polycyclic Aromatic Hydrocarbons on Organo-Zeolites. *Microporous Mesoporous Mater.* **2007**, *105*, 317–323. [CrossRef]
5. Misaelides, P. Application of Natural Zeolites in Environmental Remediation: A Short Review. *Microporous Mesoporous Mater.* **2011**, *144*, 15–18. [CrossRef]
6. Erdem, E.; Karapinar, N.; Donat, R. The Removal of Heavy Metal Cations by Natural Zeolites. *J. Colloid Interface Sci.* **2004**, *280*, 309–314. [CrossRef] [PubMed]
7. Wang, S.; Peng, Y. Natural Zeolites as Effective Adsorbents in Water and Wastewater Treatment. *Chem. Eng. J.* **2010**, *156*, 11–24. [CrossRef]
8. Albino, V.; Cioffi, R.; Pansini, M.; Colella, C. Disposal of Lead-Containing Zeolite Sludges in Cement Matrix. *Environ. Technol.* **1995**, *16*, 147–156. [CrossRef]
9. Cioffi, R.; Pansini, M.; Caputo, D.; Colella, C. Evaluation of Mechanical and Leaching Properties of Cement-Based Solidified Materials Encapsulating Cd-Exchanged Natural Zeolites. *Environ. Technol.* **1996**, *17*, 1215–1224. [CrossRef]
10. Colella, C.; de' Gennaro, M.; Langella, A.; Pansini, M. Evaluation of Natural Phillipsite and Chabazite as Cation Exchangers for Copper and Zinc. *Sep. Sci. Technol.* **1998**, *33*, 467–481. [CrossRef]
11. Pansini, M.; Colella, C. Dynamic Data on Lead Uptake from Water by Chabazite. *Desalination* **1990**, *78*, 287–295. [CrossRef]
12. Haggerty, G.M.; Bowman, R.S. Sorption of Chromate and Other Inorganic Anions by Organo-Zeolite. *Environ. Sci. Technol.* **1994**, *28*, 452–458. [CrossRef] [PubMed]
13. Obradović, M.; Daković, A.; Smiljanić, D.; Ožegović, M.; Marković, M.; Rottinghaus, G.E.; Krstić, J. Ibuprofen and Diclofenac Sodium Adsorption onto Functionalized Minerals: Equilibrium, Kinetic and Thermodynamic Studies. *Microporous Mesoporous Mater.* **2022**, *335*, 111795. [CrossRef]
14. De Gennaro, B.; Mercurio, M.; Cappelletti, P.; Catalanotti, L.; Daković, A.; De Bonis, A.; Grifa, C.; Izzo, F.; Kraković, M.; Monetti, V.; et al. Use of Surface Modified Natural Zeolite (SMNZ) in Pharmaceutical Preparations. Part 2. A New Approach for a Fast Functionalization of Zeolite-Rich Carriers. *Microporous Mesoporous Mater.* **2016**, *235*, 42–49. [CrossRef]
15. Cappelletti, P.; Colella, A.; Langella, A.; Mercurio, M.; Catalanotti, L.; Monetti, V.; de Gennaro, B. Use of Surface Modified Natural Zeolite (SMNZ) in Pharmaceutical Preparations Part 1. Mineralogical and Technological Characterization of Some Industrial Zeolite-Rich Rocks. *Microporous Mesoporous Mater.* **2017**, *250*, 232–244. [CrossRef]
16. Reeve, P.J.; Fallowfield, H.J. The Toxicity of Cationic Surfactant HDTMA-Br, Desorbed from Surfactant Modified Zeolite, towards Faecal Indicator and Environmental Microorganisms. *J. Hazard. Mater.* **2017**, *339*, 208–215. [CrossRef] [PubMed]
17. Li, Z.; Bowman, R.S. Sorption of Perchloroethylene by Surfactant-Modified Zeolite as Controlled by Surfactant Loading. *Environ. Sci. Technol.* **1998**, *32*, 2278–2282. [CrossRef]
18. Mirzaei, N.; Hadi, M.; Gholami, M.; Fard, R.F.; Aminabad, M.S. Sorption of Acid Dye by Surfactant Modificated Natural Zeolites. *J. Taiwan Inst. Chem. Eng.* **2016**, *59*, 186–194. [CrossRef]
19. Schulze-Makuch, D.; Pillai, S.D.; Guan, H.; Bowman, R.; Couroux, E.; Hielscher, F.; Totten, J.; Espinosa, I.Y.; Kretzschmar, T. Surfactant-Modified Zeolite Can Protect Drinking Water Wells from Viruses and Bacteria. *EOS* **2002**, *83*, 193–201. [CrossRef]
20. Zhan, Y.; Lin, J.; Qiu, Y.; Gao, N.; Zhu, Z. Adsorption of Humic Acid from Aqueous Solution on Bilayer Hexadecyltrimethyl Ammonium Bromide-Modified Zeolite. *Front. Environ. Sci. Eng. China* **2011**, *5*, 65–75. [CrossRef]
21. Bowman, R.S. Applications of Surfactant-Modified Zeolites to Environmental Remediation. *Microporous Mesoporous Mater.* **2003**, *61*, 43–56. [CrossRef]
22. De Gennaro, B. Surface modification of zeolites for environmental applications. In *Modified Clay and Zeolite Nanocomposite Materials: Environmental and Pharmaceutical Applications*; Elsevier Inc.: Amsterdam, The Netherlands, 2018; pp. 57–85. [CrossRef]
23. De Gennaro, B.; Catalanotti, L.; Bowman, R.S.; Mercurio, M. Anion Exchange Selectivity of Surfactant Modified Clinoptilolite-Rich Tuff for Environmental Remediation. *J. Colloid Interface Sci.* **2014**, *430*, 178–183. [CrossRef] [PubMed]
24. De Gennaro, B.; Aprea, P.; Liguori, B.; Galzerano, B.; Peluso, A.; Caputo, D. Zeolite-Rich Composite Materials for Environmental Remediation: Arsenic Removal from Water. *Appl. Sci.* **2020**, *10*, 6939. [CrossRef]
25. Dimas Rivera, G.L.; Martínez Hernández, A.; Pérez Cabello, A.F.; Rivas Barragán, E.L.; Liñán Montes, A.; Flores Escamilla, G.A.; Sandoval Rangel, L.; Suarez Vazquez, S.I.; De Haro Del Río, D.A. Removal of Chromate Anions and Immobilization Using Surfactant-Modified Zeolites. *J. Water Process Eng.* **2020**, *39*, 101717. [CrossRef]
26. Apreutesei, R.E.; Catrinescu, C.; Teodosiu, C. Surfactant-Modified Natural Zeolites for Environmental Applications in Water Purification. *Environ. Eng. Manag. J.* **2008**, *7*, 149–161. [CrossRef]
27. Hrenovic, J.; Rozic, M.; Sekovanic, L.; Anic-Vucinic, A. Interaction of Surfactant-Modified Zeolites and Phosphate Accumulating Bacteria. *J. Hazard. Mater.* **2008**, *156*, 576–582. [CrossRef]

28. Smiljanić, D.; de Gennaro, B.; Daković, A.; Galzerano, B.; Germinario, C.; Izzo, F.; Rottinghaus, G.E.; Langella, A. Removal of Non-Steroidal Anti-Inflammatory Drugs from Water by Zeolite-Rich Composites: The Interference of Inorganic Anions on the Ibuprofen and Naproxen Adsorption. *J. Environ. Manage.* **2021**, *286*, 112168. [CrossRef]
29. Smiljanić, D.; de Gennaro, B.; Izzo, F.; Langella, A.; Daković, A.; Germinario, C.; Rottinghaus, G.E.; Spasojević, M. Removal of Emerging Contaminants from Water by Zeolite-Rich Composites: A First Approach Aiming at Diclofenac and Ketoprofen. *Microporous Mesoporous Mater.* **2020**, *289*, 110057. [CrossRef]
30. Burian, M.; Geisslinger, G. COX-Dependent Mechanisms Involved in the Antinociceptive Action of NSAIDs at Central and Peripheral Sites. *Pharmacol. Ther.* **2005**, *107*, 139–154. [CrossRef]
31. Jung, C.; Boateng, L.K.; Flora, J.R.V.; Oh, J.; Braswell, M.C.; Son, A.; Yoon, Y. Competitive Adsorption of Selected Non-Steroidal Anti-Inflammatory Drugs on Activated Biochars: Experimental and Molecular Modeling Study. *Chem. Eng. J.* **2015**, *264*, 1–9. [CrossRef]
32. Gros, M.; Petrović, M.; Ginebreda, A.; Barceló, D. Removal of Pharmaceuticals during Wastewater Treatment and Environmental Risk Assessment Using Hazard Indexes. *Environ. Int.* **2010**, *36*, 15–26. [CrossRef]
33. Lonappan, L.; Brar, S.K.; Das, R.K.; Verma, M.; Surampalli, R.Y. Diclofenac and Its Transformation Products: Environmental Occurrence and Toxicity—A Review. *Environ. Int.* **2016**, *96*, 127–138. [CrossRef] [PubMed]
34. Loos, R.; Carvalho, R.; António, D.C.; Comero, S.; Locoro, G.; Tavazzi, S.; Paracchini, B.; Ghiani, M.; Lettieri, T.; Blaha, L.; et al. EU-Wide Monitoring Survey on Emerging Polar Organic Contaminants in Wastewater Treatment Plant Effluents. *Water Res.* **2013**, *47*, 6475–6487. [CrossRef]
35. Sousa, J.C.G.; Ribeiro, A.R.; Barbosa, M.O.; Pereira, M.F.R.; Silva, A.M.T. A Review on Environmental Monitoring of Water Organic Pollutants Identified by EU Guidelines. *J. Hazard. Mater.* **2018**, *344*, 146–162. [CrossRef] [PubMed]
36. Yang, Y.; Ok, Y.S.; Kim, K.H.; Kwon, E.E.; Tsang, Y.F. Occurrences and Removal of Pharmaceuticals and Personal Care Products (PPCPs) in Drinking Water and Water/Sewage Treatment Plants: A Review. *Sci. Total Environ.* **2017**, *596–597*, 303–320. [CrossRef]
37. Oaks, J.L.; Meteyer, C.U.; Rideout, B.A.; Shivaprasad, H.L.; Gilbert, M.; Virani, M.; Watson, R.T.; Khan, A.A. Diagnostic Investigation of Vulture Mortality: The Anti-Inflammatory Drug Diclofenac Is Associated with Visceral Gout. In Proceedings of the Raptor Worldwide, 6th World Conference on Birds of Prey and Owls, Budapest, Hungary, 18–23 May 2003; pp. 241–243.
38. Balmford, A. Pollution, Politics, and Vultures. *Science* **2013**, *339*, 653–655. [CrossRef]
39. Bonnefille, B.; Gomez, E.; Courant, F.; Escande, A.; Fenet, H. Diclofenac in the Marine Environment: A Review of Its Occurrence and Effects. *Mar. Pollut. Bull.* **2018**, *131*, 496–506. [CrossRef]
40. De Oliveira, T.; Guégan, R.; Thiebault, T.; Le Milbeau, C.; Muller, F.; Teixeira, V.; Giovanela, M.; Boussafir, M. Adsorption of Diclofenac onto Organoclays: Effects of Surfactant and Environmental (PH and Temperature) Conditions. *J. Hazard. Mater.* **2017**, *323*, 558–566. [CrossRef]
41. França, D.B.; Trigueiro, P.; Silva Filho, E.C.; Fonseca, M.G.; Jaber, M. Monitoring Diclofenac Adsorption by Organophilic Alkylpyridinium Bentonites. *Chemosphere* **2020**, *242*, 125109. [CrossRef]
42. Maia, G.S.; de Andrade, J.R.; da Silva, M.G.C.; Vieira, M.G.A. Adsorption of Diclofenac Sodium onto Commercial Organoclay: Kinetic, Equilibrium and Thermodynamic Study. *Powder Technol.* **2019**, *345*, 140–150. [CrossRef]
43. Sun, K.; Shi, Y.; Wang, X.; Li, Z. Sorption and Retention of Diclofenac on Zeolite in the Presence of Cationic Surfactant. *J. Hazard. Mater.* **2017**, *323*, 584–592. [CrossRef]
44. Izzo, F.; Mercurio, M.; de Gennaro, B.; Aprea, P.; Cappelletti, P.; Daković, A.; Germinario, C.; Grifa, C.; Smiljanic, D.; Langella, A. Surface Modified Natural Zeolites (SMNZs) as Nanocomposite Versatile Materials for Health and Environment. *Colloids Surfaces B Biointerfaces* **2019**, *182*, 110380. [CrossRef]
45. Mercurio, M.; Izzo, F.; Langella, A.; Grifa, C.; Germinario, C.; Daković, A.; Aprea, P.; Pasquino, R.; Cappelletti, P.; Graziano, F.S.; et al. Surface-Modified Phillipsite-Rich Tuff from the Campania Region (Southern Italy) as a Promising Drug Carrier: An Ibuprofen Sodium Salt Trial. *Am. Mineral.* **2018**, *103*, 700–710. [CrossRef]
46. Govindasamy, P.; Gunasekaran, S. Quantum Mechanical Calculations and Spectroscopic (FT-IR, FT-Raman and UV) Investigations, Molecular Orbital, NLO, NBO, NLMO and MESP Analysis of 4-[5-(4-Methylphenyl)-3-(Trifluoromethyl)-1H-Pyrazol-1-Yl] Benzene-1-Sulfonamide. *J. Mol. Struct.* **2015**, *1081*, 96–109. [CrossRef]
47. Marković, M.; Daković, A.; Krajišnik, D.; Kragović, M.; Milić, J.; Langella, A.; de Gennaro, B.; Cappelletti, P.; Mercurio, M. Evaluation of the Surfactant/Phillipsite Composites as Carriers for Diclofenac Sodium. *J. Mol. Liq.* **2016**, *222*, 711–716. [CrossRef]
48. Krajišnik, D.; Daković, A.; Milojević, M.; Malenović, A.; Kragović, M.; Bogdanović, D.B.; Dondur, V.; Milić, J. Properties of Diclofenac Sodium Sorption onto Natural Zeolite Modified with Cetylpyridinium Chloride. *Colloids Surfaces B Biointerfaces* **2011**, *83*, 165–172. [CrossRef]
49. Krajišnik, D.; Daković, A.; Malenović, A.; Kragović, M.; Milić, J. Ibuprofen Sorption and Release by Modified Natural Zeolites as Prospective Drug Carriers. *Clay Miner.* **2015**, *50*, 11–22. [CrossRef]
50. Del Orta, M.M.; Flores, F.M.; Morantes, C.F.; Curutchet, G.; Torres Sánchez, R.M. Interrelations of Structure, Electric Surface Charge, and Hydrophobicity of Organo-Mica and -Montmorillonite, Tailored with Quaternary or Primary Amine Cations. Preliminary Study of Pyrimethanil Adsorption. *Mater. Chem. Phys.* **2019**, *223*, 325–335. [CrossRef]
51. Foo, K.Y.; Hameed, B.H. Insights into the Modeling of Adsorption Isotherm Systems. *Chem. Eng. J.* **2010**, *156*, 2–10. [CrossRef]

52. Santhana Krishna Kumar, A.; Ramachandran, R.; Kalidhasan, S.; Rajesh, V.; Rajesh, N. Potential Application of Dodecylamine Modified Sodium Montmorillonite as an Effective Adsorbent for Hexavalent Chromium. *Chem. Eng. J.* **2012**, *211–212*, 396–405. [CrossRef]
53. Spiess, A.-N.; Neumeyer, N. An Evaluation of R2 as an Inadequate Measure for Nonlinear Models in Pharmacological and Biochemical Research: A Monte Carlo Approach. *BMC Pharmacol.* **2010**, *10*, 6. [CrossRef]
54. Glatting, G.; Kletting, P.; Reske, S.N.; Hohl, K.; Ring, C. Choosing the Optimal Fit Function: Comparison of the Akaike Information Criterion and the F-Test. *Med. Phys.* **2007**, *34*, 4285–4292. [CrossRef]
55. Malvar, J.L.; Martín, J.; del Orta, M.M.; Medina-Carrasco, S.; Santos, J.L.; Aparicio, I.; Alonso, E. Simultaneous and Individual Adsorption of Ibuprofen Metabolites by a Modified Montmorillonite. *Appl. Clay Sci.* **2020**, *189*, 105529. [CrossRef]
56. Ghemit, R.; Makhloufi, A.; Djebri, N.; Flilissa, A.; Zerroual, L.; Boutahala, M. Adsorptive Removal of Diclofenac and Ibuprofen from Aqueous Solution by Organobentonites: Study in Single and Binary Systems. *Groundw. Sustain. Dev.* **2019**, *8*, 520–529. [CrossRef]
57. Smiljanić, D.; Daković, A.; Obradović, M.; Ožegović, M.; Izzo, F.; Germinario, C.; de Gennaro, B. Application of Surfactant Modified Natural Zeolites for the Removal of Salicylic Acid—A Contaminant of Emerging Concern. *Materials* **2021**, *14*, 7728. [CrossRef]
58. Sun, K.; Shi, Y.; Wang, X.; Rasmussen, J.; Li, Z.; Zhu, J. Organokaolin for the Uptake of Pharmaceuticals Diclofenac and Chloramphenicol from Water. *Chem. Eng. J.* **2017**, *330*, 1128–1136. [CrossRef]
59. De Gennaro, B.; Aprea, P.; Pepe, F.; Colella, C. Cation Selectivity of a Ca2+ Pre-Exchanged Clinoptilolite Tuff. *Stud. Surf. Sci. Catal.* **2007**, *170*, 2128–2133. [CrossRef]
60. de Gennaro, B.; Aprea, P.; Colella, C.; Buondonno, A. Comparative Ion-Exchange Characterization of Zeolitic and Clayey Materials for Pedotechnical Applications-Part 2: Interaction with Nutrient Cations. *J. Porous Mater.* **2007**, *16*, 667–673. [CrossRef]
61. De Gennaro, B.; Colella, A.; Cappelletti, P.; Pansini, M.; de' Gennaro, M.; Colella, C. Effectiveness of Clinoptilolite in Removing Toxic Cations from Water: A Comparative Study. *Stud. Surf. Sci. Catal.* **2005**, *158B*, 1153–1160. [CrossRef]
62. Marković, M.; Daković, A.; Rottinghaus, G.E.; Kragović, M.; Petković, A.; Krajišnik, D.; Milić, J.; Mercurio, M.; de Gennaro, B. Adsorption of the Mycotoxin Zearalenone by Clinoptilolite and Phillipsite Zeolites Treated with Cetylpyridinium Surfactant. *Colloids Surf. B Biointerfaces* **2017**, *151*, 324–332. [CrossRef]
63. Zhu, S.; Khan, M.A.; Kameda, T.; Xu, H.; Wang, F.; Xia, M.; Yoshioka, T. New Insights into the Capture Performance and Mechanism of Hazardous Metals Cr^{3+} and Cd^{2+} onto an Effective Layered Double Hydroxide Based Material. *J. Hazard. Mater.* **2022**, *426*, 128062. [CrossRef]
64. Zhu, S.; Xia, M.; Chu, Y.; Khan, M.A.; Lei, W.; Wang, F.; Muhmood, T.; Wang, A. Adsorption and Desorption of Pb(II) on L-Lysine Modified Montmorillonite and the Simulation of Interlayer Structure. *Appl. Clay Sci.* **2019**, *169*, 40–47. [CrossRef]

catalysts

Review

Environmental and Pharmacokinetic Aspects of Zeolite/Pharmaceuticals Systems—Two Facets of Adsorption Ability

Nataša R. Mijailović [1], Bojana Nedić Vasiljević [2], Maja Ranković [2], Vladimir Milanović [3] and Snežana Usković-Marković [4,*]

1 Department of Pharmacy, Faculty of Medical Sciences, University of Kragujevac, Svetozara Markovića 69, 34000 Kragujevac, Serbia; nacakg@gmail.com
2 Faculty of Physical Chemistry, University of Belgrade, Studentski Trg 12-16, 11000 Belgrade, Serbia; bojana@ffh.bg.ac.rs (B.N.V.); majarankovic998@gmail.com (M.R.)
3 Zemun Community Health Center, Rada Končara 46, 11080 Belgrade, Serbia; vladamilanovic86@gmail.com
4 Faculty of Pharmacy, University of Belgrade, Vojvode Stepe 450, 11221 Belgrade, Serbia
* Correspondence: snezaum@pharmacy.bg.ac.rs

Abstract: Zeolites belong to aluminosilicate microporous solids, with strong and diverse catalytic activity, which makes them applicable in almost every kind of industrial process, particularly thanks to their eco-friendly profile. Another crucial characteristic of zeolites is their tremendous adsorption capability. Therefore, it is self-evident that the widespread use of zeolites is in environmental protection, based primarily on the adsorption capacity of substances potentially harmful to the environment, such as pharmaceuticals, pesticides, or other industry pollutants. On the other hand, zeolites are also recognized as drug delivery systems (DDS) carriers for numerous pharmacologically active agents. The enhanced bioactive ability of DDS zeolite as a drug carrying nanoplatform is confirmed, making this system more specific and efficient, compared to the drug itself. These two applications of zeolite, in fact, illustrate the importance of (ir)reversibility of the adsorption process. This review gives deep insight into the balance and dynamics that are established during that process, i.e., the interaction between zeolites and pharmaceuticals, helping scientists to expand their knowledge necessarily for a more effective application of the adsorption phenomenon of zeolites.

Keywords: zeolites; drugs; adsorption; drug delivery systems; theoretical approach

Citation: Mijailović, N.R.; Nedić Vasiljević, B.; Ranković, M.; Milanović, V.; Usković-Marković, S. Environmental and Pharmacokinetic Aspects of Zeolite/Pharmaceuticals Systems—Two Facets of Adsorption Ability. *Catalysts* **2022**, *12*, 837. https://doi.org/10.3390/catal12080837

Academic Editor: Narendra Kumar

Received: 7 July 2022
Accepted: 27 July 2022
Published: 30 July 2022

1. Introduction

Zeolites are porous, hydrated aluminosilicate minerals with a three-dimensional structure and loosely bounded cations of alkali or alkali earth metals [1–3]. Due to the progress of mineralogy, the traditional definition of zeolite as aluminosilicate frameworks was found to be too inflexible. In the definition of a zeolite mineral recommended by the International Mineralogical Association, Commission on New Minerals and Mineral Names, structures containing an interrupted framework of tetrahedra are accepted where other zeolitic properties predominate, and complete substitution by elements other than Si and Al is allowed [4]. These include the divalent cations Be^{2+} and Zn^{2+}, other trivalent cations such as B^{3+}, Ga^{3+}, and Fe^{3+}, as well as tetravalent cations such as Ti^{4+} and Ge^{4+}.

According to their origin, they can be natural and synthetic [5]. Zeolites are formed in nature in the reaction of volcanic rocks and ash with water of high pH value and a high concentration of salt [1,2,5,6]. Natural zeolites possess high selectivity for heavy metal ions and ammonium ions and are therefore important for environmental protection [6]. There are around 50 known natural zeolites, and the most important ones are clinoptilolite, mordenite, and chabazite [2]. Natural zeolites usually contain impurities of other minerals—for example, feldspar and quartz, metals, etc. [2], and, as a result, their application is limited when it is required to use zeolites of high purity and uniformity [1]. It is important to note that zeolite deposits are a non-renewable resource. Laboratory synthesis of zeolites is based on the application of natural and synthetic silicates as carriers. Zeolites synthesised using natural substrates are never 100% pure. However, their price is significantly lower compared to zeolites obtained from synthetic substrates [6]. Studies have shown that synthetic zeolites have numerous advantages over natural zeolites. The pore size of synthetic zeolites is significantly larger than natural zeolite, which allows the adsorption of larger molecules—for example, diesel fuels and used engine oil. Also, the efficiency of removing radioactive waste and heavy metal ions from the environment is higher for synthetic zeolites [2,6].

Zeolites can be classified on the basis of their crystal structure, chemical composition, pore size, etc. [1]. The Si/Al ratio is an important characteristic of zeolites that determines their ion-exchange abilities. Increasing this ratio changes the surface selectivity from hydrophilic to hydrophobic [1,2,7]. For this reason, silicon-rich zeolites are stabilized in the synthesis process by adding various organic species to the reaction mixture [8].

Zeolites can be classified according to the Si/Al ratio as follows:

1. Low Silica Zeolites: $Si/Al \leq 2$
2. Intermediate Silica Zeolites: $2 < Si/Al \leq 5$
3. High Silica Zeolites: $Si/Al > 5$ [9].

Another classification of zeolites is based on the diameters of their pores:

1. Small-pore zeolites (8-member rings) with pore diameter of 0.3–0.45 nm
2. Medium-pore zeolites (10-member rings) with pore diameter of 0.45–0.6 nm
3. Large-pore zeolites (12-member rings) with pore diameter of 0.6–0.8 nm
4. Extremely large-pore zeolites (14-member rings) with pore diameter of 0.8–1.0 nm [1–3].

Zeolites can be represented by the structural formula based on a crystallographic unit cell:

$$M_{x/n}[(AlO_2)_x(SiO_2)_y]\cdot wH_2O$$

where n represents the valence of a cation M, w is the number of water molecules per unit cell, x and y is the number of tetrahedra per unit cell, and the y/x ratio is most often in the range of 1 to 5 [3,10].

The zeolite structure is composed of SiO_4 and AlO_4 tetrahedra interconnected through oxygen atoms [5,10]. In the zeolite structure, silicon is tetravalent and forms an electroneutral SiO_4 tetrahedron, while aluminium is trivalent and each AlO_4 tetrahedra carries a negative charge. This charge is balanced by the presence of non-framework easily exchangeable cations [10]. The connection of the SiO_4 and AlO_4 tetrahedra (both commonly marked as TO_4) leads to the formation of cavities and channels with an internal surface area that reaches several hundred square meters per gram of zeolite. This characteristic makes zeolites extraordinarily effective ion exchangers. The pore diameter is usually in the range of 0.3 nm to 1.0 nm, in the group of aluminosilicates, and extends to about 1.4 nm in the respective phosphates, and they contain water molecules and cations [11].

The primary building blocks of zeolites TO_4 tetrahedra can be assembled into secondary building units by linking through oxygen atoms, resulting in the final structures of zeolites with a regular distribution of pores and cavities [3,5,10].

These secondary building units (SBU) or blocks have various different compositions and can have up to 16 T-atoms. SBUs can be single 4-, 5-, 6-, 8-member rings or double 4-, 6-, 8-member rings, or some other complicated structures like two connected 5-member rings (5–3) or 6-member rings followed by 4-member one (6–2), etc. The space combination of these SBUs sometimes gives a rise to specific cages in structure. For example, a combination of single 6- and 4-member rings creates a well-known sodalite cage. Using different parts of this cage to connect cages, different zeolite frameworks arise. Illustrations of double 4- and 6-membered rings are presented in Figure 1. Sodalite cages directly connected through 4-member rings create a sodalite structure (Figure 2 left), while cages on some distance create a double 4-member ring giving a rise of Linde Type A structure type

(Figure 2 middle). Similarly, creating a double 6-member ring makes a Faujasite structure type (Figure 2 right). Each zeolite framework is marked using a three-letter type code assigned by the International Zeolite Association [12]. The above-mentioned zeolites have codes SOD, LTA and FAU, respectively. Silicon and aluminium could be substituted by phosphorus, gallium or germanium ions, creating aluminophosphates, galophosphates etc. In the zeolite structure, cations can be reversibly replaced with other cations from solution in contact with the zeolite in the ion exchange process, and water can be reversibly removed at temperatures usually below 400 °C, leaving the crystal structure undamaged [1,3,4,13].

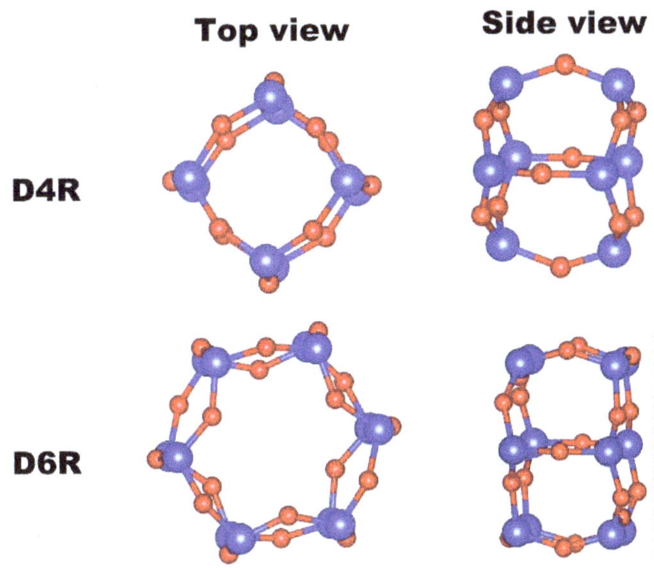

Figure 1. A double 4-members ring (D4R) and double 6-members ring (D6R) as examples of SBUs. Red atoms are oxygen and blue atoms are silicon or aluminium atoms.

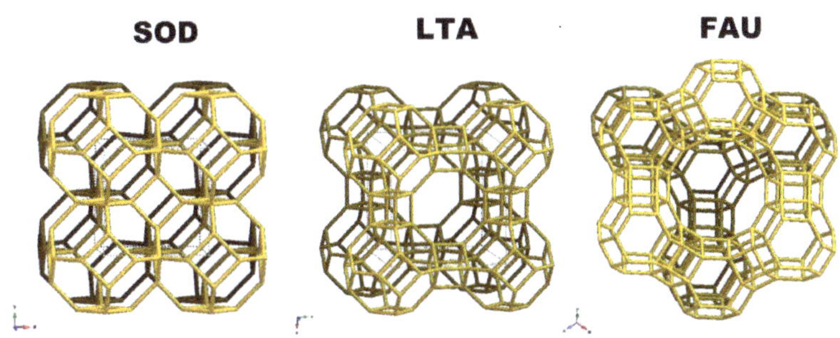

Figure 2. The framework structure for SOD (**left**), LTA (**middle**) and FAU (**right**) zeolitic network.

The wide and still not completely finished list of applications of zeolites is related mainly with their porous structure, high adsorption capacity, and ion exchange properties [14]. Zeolites are used in various technological applications, such as catalysts and as molecular sieves, including separating various molecules, as detergents, etc. [15,16].

Biotechnology and medicine are other promising fields for applications of zeolites [17]. These include detoxification of animal and human organisms, improvement of the nutrition status and immunity of farm animals, separation of various biomolecules and cells, construction of biosensors and detection of biomarkers of various diseases, controlled drug delivery systems (DDS), radical scavenging, and tissue and bone engineering. As components of hemostatics, as gastroprotective drugs, or as antioxidative agents, zeolites can also be applied.

Zeolites can be widely applied in nowadays extremely important sustainable chemistry, including biomass conversion, exhaust post-treatment, radionuclide removal, fuel cells, thermal energy storage, CO_2 capture and conversion, air-pollution remediation, water and air purification, etc. [18]. Water purification with zeolites includes removal of radioactive contaminants, and their great efficiency in pesticide removal [19–23].

Many of these applications are based on the extraordinary adsorption capacity of zeolites. Isotherm modelling and kinetic investigation are applied in order to get insight into the main mechanisms behind the adsorption processes.

2. Zeolites in the Removal of Pharmaceuticals from the Environment

2.1. Where to Start?

Wide applications of pharmaceuticals, which are defined as substances used in the diagnosis, treatment, or prevention of disease and for restoring, correcting, or modifying organic functions, have contributed to the incredible improvement of human health and raised the quality and length of life to a great extent [24]. The most recent example is the explosion of the use of pharmaceuticals during the SARS-CoV-2 pandemic, in the treatment, but also in the prevention and diagnosis of diseases on a global scale. Despite the many benefits of modern medicine, its usage is a double-edged sword, because of the fact that a significant percentage of drugs used will end their journey in the environment.

In addition to frequently administered drugs like antidepressants, lipid-lowering agents, β-blockers, etc., very often investigated the occurrence of antibiotics, and non-steroidal anti-inflammatory drugs-NSAIDs and their residues in wastewater [25,26]. Although antibiotics levels are ng/L to μg/L, they are considered toxic compounds which contribute to the evolution of antibiotic-resistant bacteria and genes [27,28].

Ciprofloxacin, azithromycin and cephalexin, as frequently administered drugs, are suitable as markers of antibiotic pollution of the aquatic environment [25]. Interestingly, the effects on aquatic systems of birth control pills or better defined, endocrine-disrupting chemicals, are the subject of numerous reported studies for more than 40 years [29].

Excretion is the dominant way that drugs reach the environment, sometimes in their parent form or as metabolites. The environment's physicochemical and bioconversion of excreted pharmaco-active compounds, which occurred after their biotransformation pathways, give as a result various and sometimes unpredictable products [30]. The fate of pharmaceuticals in the aquatic environment is therefore a very attractive field of study [31]. Although pharmaceuticals are generally considered as susceptible to diverse transformation reactions, the resulting products are often very stable. Such transformed hydrophilic compounds easily pass-through sewage treatment plants [32]. Anyway, the drugs of even the same pharmacology active groups possess a variety of non-predictable excretion rates, with considerable impact to the aquatic environment [33].

The importance of inadequate disposal of discarded drugs should not be neglected, although this area is officially regulated in most countries [34].

Luckily, this topic is nowadays under the watchful eye of scientists, and water is an environment of interest for monitoring pharmaceutical levels, as well as the resulting impacts it leads to. Focus is placed on surface water samples, sewage treatment plants, and drinking water [35–37].

2.2. Examples of Effective Removal of Pharmaceuticals with Zeolites

The procedures for removal of pharmaceuticals from the aquatic environment are numerous and more and more innovative. For example, pharmaceuticals can be removed from wastewater by biological processes, hydrodynamic cavitation and ultraviolet light treatment [38]. Besides the general approach, it is usually necessary to create a specific action toward targeted compounds.

Although the conventional procedures, like an advanced oxidation, hydrolysis, and photo-degradation, are usually an efficient way for removal of pollutants, these strategies applied to pharmaceutical rich waters, such as hospital waste effluents or pharmaceutical industries are, can potentially cause environmental and health harmful products [39,40], and must be considered in detail [41].

Among numerous attempts, the use of zeolite for water purification from pollutants was highlighted as a simple, efficient, low-cost, and eco-friendly procedure. Zeolites (either in pristine or modified forms), are well-known materials for the adsorption of pesticides from wastewaters [19–23].

The most important characteristic of zeolite for this application is its adsorption ability. The efforts of scientists in this field are focused on finding materials with improved adsorption capacities, appropriate kinetics, and examining the reversibility of the adsorption process, as well as the possibility of reuse.

First, investigations often start with neat structures, to elucidate key parameters, such as Si/Al ratio, surface area and extra-framework ions. The experimental design relies on sample loading optimization; however, the practical approach imposes the use of low loadings to boost adsorption capacity. To participate in this quest, researchers sometimes report loading that exceeds the starting adsorbent amount several times, although this has no physical meaning. Another challenge is to detect important parameters for efficient removal of environmentally significant concentrations, which are often for pharmaceuticals at the ng/mL level. Such a low concentration masks real adsorbent performance, and lab research usually employs mg/L concentration to boost experimental sensitivity [42].

A search of the literature can reveal the published results of testing a large number of zeolite structures with the aim of testing their ability to remove drugs, i.e., their adsorption from wastewater. Table 1. gives the most explored types of zeolites for this purpose.

Table 1. Some types of zeolites.

Synthetic		Natural	
Structure	**Zeolite Type**	**Structure**	**Zeolite Type**
LTL	Zeolite L	HEU	Clinoptilolite
LTA	Zeolite A	MOR	Mordenite
MFI	ZSM-5 Zeolite	CHA	Chabazite
FAU	Zeolite X		
	Zeolite Y		
BEA	Beta Zeolite		

Among different zeolites, the FAU framework is especially present for wastewater treatment due to its substantial adsorption capacity toward different molecules. Additionally, in some cases there is no need for functionalization, and commercially available zeolite can be readily employed. Following its beneficial adsorption behaviour, FAU zeolite has been so far employed for the removal of sulfonamide [43,44] and fluoroquinolone antibiotics [45,46]. A special study was dedicated to the macrolide class representative, azithromycin removal by FAU zeolite [47]. Another reason for FAU selection lies in its fast kinetics, which takes up to several minutes for half an hour, which is of utmost importance for environmental application. Antibiotics are often oxygen and nitrogen-rich compounds whose main mechanism of interaction with the zeolite surface is hydrogen bonding. If this is the case, it is necessary to examine pH effects on the adsorption process [47]. Scheme 1 illustrates the drug removal by zeolites.

Zeolites sometimes require functionalization/composite preparation to establish new and improved features. MFI zeolite, for example, suffers from low adsorption capacity, and combination with excellent adsorbents enabled the removal technique to be extended from adsorption to catalytic degradation in the presence of suitable oxidants. For instance, composite prepared of MFI zeolite and carbon adsorbent is proposed for 150 ppm ciprofloxacin adsorption [48].

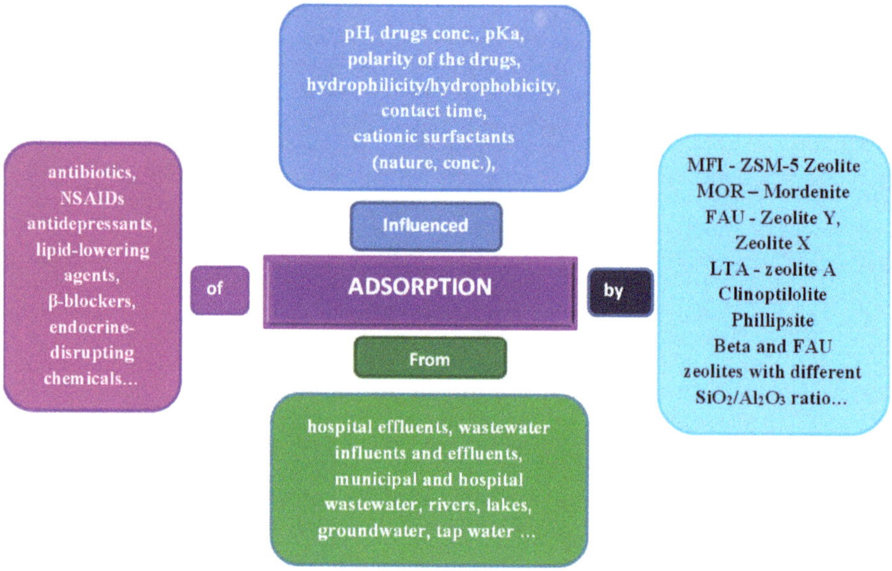

Scheme 1. Pharmaceutical adsorption by zeolites from aquatic media.

These procedures, however, sometimes tend to be costly and complicated or even pose a bigger environmental threat, due to hazardous solvents or higher toxicity, leaving of constituents, etc.

The adsorption conditions of antibiotics on zeolite are often examined, like in the study of chlortetracycline, oxytetracycline (OTC), ofloxacin, and enrofloxacin adsorption on natural zeolite [49]. The Langmuir-Freundlich sorption model was employed to estimate the maximum sorption capacity, and it was found that the capacity increased if the solution pH decreased. The presence of natural organic matter reduced the sorption of OTC but improved the sorption of the remaining antibiotics. An ofloxacin removal was additionally tested on LTA zeolite prepared from red mud/fly ash/spinel iron oxide nanoparticles [50]. Tested adsorption was investigated in different matrices—tap and river water—by spiking the samples with 10 μg/L antibiotics.

As another example, erythromycin (ERY) and levofloxacin (FLX) were almost completely adsorbed by Y zeolite (as an example of organophilic zeolite) from water samples collected at the outlet of a wastewater treatment plant [45].

The "2-in-1" idea to lower the overall costs and to use environmentally undesirable materials is reported in the study where coal fly ash (CFA) driven zeolites were applied for the adsorptive removal of ceftazidime, a broad-spectrum antibiotic [51], where CFA is a leftover product of burned coal, very harmful for the environment.

Sometimes reviews about zeolites and mesoporous silica materials as effective adsorbents of drugs, besides antibiotics, are focused on the removal of NSAIDs, as another most frequently used pharmaceuticals. The review of Grela et al. summarized available literature data and concluded that the highest concentrations of diclofenac, ibuprofen, and ketoprofen were found in wastewater influents, municipal wastewater, and hospital effluents, and gave the order of NSAIDs and antibiotics concentrations in different types of water samples: hospital effluents > wastewater influents > municipal wastewater > secondary wastewater > river water > wastewater effluents > groundwater > surface water > seawater > tap water > hospital wastewater > surface water (lakes) [52].

High concentrations of pollutants in simulations of real water samples are necessary to elucidate key parameters for adsorption and to select the best adsorbent. But, directing further research towards lower concentrations in environmental real water samples (real effluents in environmentally relevant concentrations) and information on the practical implementation of these materials in real-life wastewater treatment is of extreme importance. Recently, Ajo et al. showed that the actual removal rates of ibuprofen cannot be accurately estimated in the context of real wastewaters without negative bias from simultaneous reformation [53]. The review by Shearer et al. focused on metformin and macrolides and assesses isotherm, kinetic and thermodynamic studies, as well as the adsorption mechanisms, with discussion on some identified mistakes and inconsistencies. The review also sought to identify gaps in knowledge, particularly real-world applications, which should be priorities for future investigations [54]. The removal of pharmaceuticals from wastewater can be challenging due to its complex matrix. The examination of removal efficiency of fluoroquinolone antibiotics, norfloxacin and ofloxacin, by using nanoscale zero-valent iron loaded zeolites modified with polyethylene glycol surfactant in samples taken from the Yellow River is directed towards such applications [55]. A similar investigation, considering real river samples, was conducted for the Photo-Fenton treatment of water sample from the Meurthe River in France for the removal of 21 different pollutants, including 17 pharmaceutical compounds such as diclofenac, erythromycin, ibuprofen, ketoprofen, and lidocaine using Faujasite Y zeolite containing iron as catalyst [56]. In an interesting study, De Sousa et al. studied wastewater effluent samples from Girona wastewater treatment plants in Spain, which included wastewater from hospitals, homes, and urban areas. Two FAU zeolites with different SiO2/Al2O3 ratios were used as adsorbents for determining the concentration of azithromycin, ofloxacin, and sulfamethoxazole [47]. However, a larger number of studies refer to the analysis of samples that do not actually contain pollutants, but real matrix (for example, river water) spiked with pollutants.

Natural zeolite, like Jordanian zeolite (Intermediate silica), was successfully applied as an adsorbent for the removal of several frequently used pharmaceuticals such as ibuprofen, diclofenac sodium, indomethacin, chlorpheniramine maleate, and paracetamol from water [57]. The study showed the optimal was pH 2 for the removal of all tested compounds, except for diclofenac sodium it was pH 6, with 80 min as the optimum adsorption time. After optimization, the highest removal was found to be 88.3% for NSAID ibuprofen, and 85.8% for antihistaminic chlorpheniramine maleate. The adsorption efficiencies were evaluated, and it turned out that Freundlich isotherm fits the experimental data for both ibuprofen and chlorpheniramine maleate. The results of a continuous flow experiment performed on ibuprofen under constant influent concentration and fixed flow rate indicated that the percentage removal of ibuprofen on zeolite was the highest after fraction 9 with 78% removal.

The introduction of cationic surfactants (cetylpyridinium chloride and Arquad® 2HT-75) into natural zeolites, such as clinoptilolite (CLI) and phillipsite, leads to a better effect of the obtained composites in the removal of ibuprofen and naproxen [58]. The zeolitic surfaces were prepared as monolayer and bilayer surfactant coverage. The Langmuir model gives the conclusion that the highest adsorption capacity for the composite characterized by a bilayered surfactant at the clinoptilolite surface was 19.7 mg/g for ibuprofen and 16.1 mg/g for naproxen. The influence of the initial drug concentrations and contact time on adsorption of ibuprofen and naproxen and zeolite clinoptilolite and phillipsite, are surely very important factors for the process of drug adsorption from buffer solutions [58].

Simply discovering that a certain zeolite has the ability to adsorb a pharmaceutical, so it can be used for wastewater treatment, can in no way satisfy the scientific public. It is extremely important to shed light on the adsorption process itself, e.g., to determine the factors that affect the capacity, dynamics and reversibility of this process. For this purpose, a multidisciplinary approach and a number of modern instrumental methods are used. This approach will be illustrated by citing some examples from the literature.

Adsorption isotherms and thermogravimetric analysis show that ERY, FLX and carbamazepine (CBZ) are adsorbed in remarkable amounts by Y zeolite. X-ray structure analyses carried out on zeolite after adsorption revealed the selected drugs inside the Y cage. The study indicates that the adsorption properties of zeolitic materials do not only depend on micropore size, and that zeolite shape selectivity also depends on structural features [45].

The beta zeolites with different SiO_2/Al_2O_3 ratio (i.e., 25, 38 and 360) were tested for adsorption of ketoprofen, hydrochlorothiazide and atenolol from diluted aqueous solutions, with changing the ionic strength and the pH, before and after thermal treatment of the adsorbents [59]. The processes were followed by thermogravimetry and X-ray diffraction. The study confirmed that the adsorption capacity was dependent on both the solution pH and the alumina content of the beta zeolites. The noticed difference was explained as a function of the interactions between drug molecules and zeolite surface functional groups. Atenolol was adsorbed on the less hydrophobic zeolite, under pH conditions in which electrostatic interactions were predominant, while ketoprofen adsorption was mainly determined by hydrophobic interactions. The adsorption capability for undissociated molecules increased with the increase of hydrophobicity.

Certain studies were conducted with the aim to better understand the interaction between the natural zeolite clinoptilolite and antibiotics that caused gastric side effects, such as metronidazole and sulfamethoxazole [60]. Beside the considerable importance of pH on the adsorption process, the study reported that interaction of metronidazole and sulfamethoxazole with the clinoptilolite and its forms is fundamentally related with the polarity of the molecules and the nature of the zeolitic material.

Other agents that can modify the properties of the solid surface of zeolites and improve the adsorption of some pharmaceutics are surfactants. The study of Lam et al. reported the results of semiempirical calculations applied on the systems formed by surfactants, drugs, water and a clinoptilolite channel model [61]. Special attention was paid to the interaction of each drug molecule with the external surface of the clinoptilolite model. The cationic surfactant seems to be well adsorbed on the clinoptilolite model, contrary to the anionic surfactant. The polarity of the drugs plays a very important role in the adsorption process from the solution: the most polar studied drug, metronidazole, was best adsorbed on the zeolite model, followed by acetylsalicylic acid and sulfamethoxazole. If the same system contains the cationic surfactant, the order of the drug adsorption is opposite: the adsorption of sulfamethoxazole as a hydrophobic molecule is more pronounced. Those conclusions can help in adjusting the adsorption of certain drugs on clinoptilolite in the desired direction by the presence of surfactant on the zeolite external surface.

Zeolitic imidazolate metal organic framework group of compounds, such as ZIF-8, belongs to metal-organic frameworks (MOFs), and possess similar characteristics to zeolite, including the high adsorption capability, and therefore can be a great candidate for the testing of removal of the pharmaceuticals from wastewater. As an example, ZIF-8 exhibits ultra-high adsorption capacity to tetracycline from aquatic media [62].

2.3. *Use of Theoretical Calculation for Predicting Interactions*

In a well-designed experiment, theoretical calculations should precede empirical research. Afterwards, when optimizing interactions of interest is concluded, and adsorbates and adsorbents are selected, it makes sense to start an experiment. The fact is, however, that predictions do not always produce ideal results, thus the reverse order of research is resorted to. Namely, most methods include theoretical support that follows after preceedingly obtained experimental results as a form of deeper explanation of adsorption phenomena.

There is great power in using theoretical calculations like density-functional theory (DFT), semi-empirical methods or molecular dynamic simulations. These methods could be used for predicting the stability and reactivity of molecules or predicting adsorption energies.

Drugs are usually organic molecules, so HOMO and LUMO orbitals and associated energies can be calculated. HOMO energy is connected to the ability of a molecule to donate electrons, while LUMO energy is connected to the ability to accept electrons. The gap between these two orbitals helps in the description of chemical behaviour. Lower energies are correlated with higher reactivity and lower stability of molecules [63].

Modern DFT calculations can be used to predict material properties and it is a tool that is indispensable. Based on the first principal energies of the particular systems can be calculated, followed by the prediction of adsorption energies [64].

Zeolite structures are complex, so calculations of energies in systems of interest could be complicated and some approximations had to be used. Mainly, the structure of zeolite is approximated by a cluster which represents the main cage or pore, in the particular zeolite. This was done in the work of Brachi et al., where the structure of FAU zeolite was represented by the central cage [65]. Different sulfonamide antibiotic molecules were initially optimised using DFT and minimal energy conformers were found. These steps were followed by calculations of interaction energies, bond lengths and optimal dimer conformations in the FAU cage.

In their work, Hessou et al. used the primitive crystal lattice of FAU zeolite and modeled the adsorption of dibenzyl disulfide, which can be considered as a theoretical model of pollutant molecules [66]. A theoretical prediction of the interaction energy was modeled in the presence of various monovalent extra framework cations. They predicted that CsY, AgY and CuY had high adsorption energy for dibenzyl disulfide, while not favoring production of dissociated species.

While molecular mechanics/molecular dynamics calculations are widely used to investigate the adsorption of various organic molecules with different complexity on full silica zeolites, application of these calculations on drug adsorption is quite low but emerging lately.

Fatouros and his co-workers applied molecular dynamic calculations to predict interactions and adsorption of theophylline and salbutamol on beta zeolite. Theophylline and salbutamol are molecules with similar dimensions, but salbutamol is a more flexible one. Calculations predicted that salbutamol could be adsorbed into the channels and pores of beta zeolite, while theophylline could not. These theoretical findings are supported by experimental data showing the significantly less adsorbed amount of theophylline on beta zeolite, compared with salbutamol [67].

The same research group applied molecular dynamics to predict diffusion rate and release the possibility of 5-fluorouracil from beta and FAU zeolite [68]. This drug had significantly different diffusion rate coefficients in FAU and beta systems. The slower release rate from beta zeolite is a consequence of a smaller pore system, and increased van der Waals interactions with drug molecules, compared with FAU zeolite.

In the final example of using theoretical models for predicting interactions of drugs with zeolite adsorbents, the author employed molecular dynamic simulations, as well as DFT calculations. Simulations were performed for over 20 widely used drugs and two zeolites as adsorbents, mordenite and faujasite. The author showed that the interaction energy of the zeolite-drug system from minimal energy configuration could be useful for predicting the effective removal of drugs from water. The zeolite framework was kept rigid in the simulation process in order to lower system complexity. For mordenite, it was calculated that 8 out of 21 drugs (i.e., triclosan, ibuprofen, oxybenzone...) had interaction energies between 200 and 270 kJ/mol and experimentally it was found that mordenite removes quantitatively these drugs. Nine drugs (i.e., diazepam and hydrocodone) not adsorbed by mordenite had interaction energies close to zero.

The calculations take considerable time and computational resources, but this type of screening should precede experiments whenever possible [69].

3. Zeolite-Based Biomaterials for Biomedical Application

The biocompatibility and mechanical strengths of zeolite make them suitable for biomedical application as pharmaco active compounds or as biomaterial used for dental fillers, bone grafts, implant coating, or as a drug carrier agent [70].

An illustration of the positive example of application of zeolites themselves in pharmacotherapy is clinoptilolite which can stand out as a potent detoxifying, antioxidant, and anti-inflammatory agent [71,72]. Zeolites are currently regarded as dietary super-materials [71], and this, sometimes, is an overstatement. Every drugstore is selling zeolite-based nutritional formulations enhanced with vitamins, enzymes, etc., often without any scientific basis to support the claims of high nutritional benefits. Antioxidant, antimicrobial, detoxifying, and anticancer activity is often attributed to both synthetic and natural zeolites. However, a strict survey of the available literature gives somewhat controversial findings. For instance, zeolite's ability for trapping radical species in order for them to be safely removed from the body is a cornerstone of many studies [73]. A part of this is actually true, zeolite structure does enable radical species removal [74,75], although they shouldn't be a part of a human diet in spite of being used for cattle because tightly bind mycotoxins from animal feed in the gastrointestinal tract and thereby decrease their bioavailability [76]. Concerning this, Ipek et al. assessed the effect of natural zeolite, clinoptilolite supplementation on the oxidative status in cows and concluded that it did not support cow's systems against oxidative stress [77].

3.1. Zeolites for Dental Applications

In addition to their wide application in medicine and other sciences, zeolites have also displayed a significant potential for use in the field of dentistry. Zeolites may be applied in root canal therapy, periodontics, implant and restorative dentistry, and tissue engineering [78]. Zeolites were introduced into dental practice mainly as root filler materials, based on their hemolytic and cytotoxic properties [79], or their antimicrobial and mechanical characteristics [78]. Remineralizing the ability of calcium-rich zeolite makes it a promising candidate as a dental composite filler [80]. We can say that a wide application of zeolite in dentistry is not primarily based on adsorption as a phenomenon, but certain functions of zeolite are significantly improved thanks to the adsorption effect, which will be illustrated by examples.

In general, zeolites alone have little or no effect on antimicrobial properties, unless the zeolites are doped with ions, such as silver or zinc. The ability of zeolites to uptake and release ions, combined with their exceptional biocompatibility and long-lasting effects, has been used for the antimicrobial treatment against pathogenic oral microorganisms. Regarding dental restorative materials, zeolites are generally combined with glass ionomer cements (GIC), resin cements, or bonding agents. The antibacterial effects of GIC containing silver-zeolite (AgZ) were demonstrated on *Streptococcus mutans* in vitro [81]. Similar antimicrobial results can also be found in a zinc-doped zeolite (ZnZ) against *Escherichia coli*, *Staphylococcus aureus*, *Pseudomonas aeruginosa*, *Bacillus subtilis*, and *Candida albicans* [82]. A functional dental restorative platform consisting of zeolite nanoparticles as a drug delivery carrier loaded with chlorhexidine (CHX), was incorporated into commercial dental GIC, demonstrating a stronger inhibitory effect on *S. mutans* compared to GIC alone [83]. AgZ combined with GIC sealer showed a stronger antimicrobial effect towards *E. faecalis* compared to GIC sealer alone, which was concentration- and time-independent [84].

The root canal treatment has been related to persistent periradicular lesions, including primary endodontic infections and persistent infectious progression with the *E. faecalis* as the main etiological factor in these diseases. Zeolites in endodontics were generally added to calcium hydroxide and mineral trioxide aggregate (MTA) and used as root canal irrigants to enhance their antimicrobial properties. Ghatole et al. showed that adding AgZ to calcium hydroxide enhanced the antimicrobial effect against *E. faecalis* compared to the control or when chlorhexidine is added [85]. Among the root end filling materials, MTA is regarded as biocompatible and is most commonly used in clinical applications, but with limited antimicrobial activity. The addition of AgZ to MTA enhanced antimicrobial effects toward selected oral microflora such as *E. faecalis*, *S. aureus*, and *Candida albicans* in a concentration-dependent manner throughout 72 h [86].

Two types of zeolites, Zeolite A and ZSM-5 were investigated for their potential to adsorb volatile sulphide compounds (VSC), which are produced in the oral environment by Gram-negative bacteria and cause periodontal disease and mouth odour. The amount of H_2S adsorbed on zeolite A was found to be larger than that on ZSM-5, suggesting that the adsorptive property of zeolites depends on their Si/Al ratio. By optimising Si/Al ratio, it is expected to develop an adsorbent material, which highly adsorbs VSC, and may contribute to oral health [87].

Zeolite in prosthesis can be added to both acrylic resin and non-acrylic materials, such as ceramic. The non-acrylic materials that were tested with zeolite demonstrated that adding AgZ to soft liners enhanced its antimicrobial properties against *C. albicans* and gram-negative bacteria while also maintaining its viscoelastic properties [88]. Sodalite zeolite is a subtype of zeolite that can easily infiltrate other materials due to its selectivity and strong catalytic activity and has been often applied to ceramic prostheses to improve their mechanical properties [89]. Incorporation of Ag-Zn zeolite with acrylic resin materials showed beneficial effects by improving their surface finish and resistance to surface damage by increasing hardness [90].

Application of zeolites in dentistry can also extend to antibacterial coatings on implants since coating titanium implants with AgZ was effective in inhibiting methicillin-resistant *S. aureus* growth [91].

Zeolites are known for their applicability in different composite materials, including dental materials. Zeolite fillers associated with phenol−formaldehyde resins and poly-(vinylidene fluoride) are often used [92]. An important fact is that zeolites have a positive impact on the mechanical properties of composites, which is combined with their ability to deliver the calcium ions (Ca^{2+}) to the tooth surface, showing the remineralizing potential of the hydroxyapatite structure of dentin and enamel [93]. FAU zeolites (X and Y type) have the highest remineralisation potential [94] and can be also used as scaffolds because they do not affect cell viability [80]. Since Gram-negative bacteria do not adhere to the type X zeolite, composite materials with zeolite X will not be sensitive to secondary caries [80].

Dental composites consist of inorganic fillers and organic resin matrix with some amounts of additives [95]. Due to the low adhesion of inorganic fillers to resins, it is often necessary to modify their surface before mixing these components. One of the main types of modifications that is used in dental fillings is silanization. This modification empowers bringing into contact the groups that are involved in cross-linking of the composite to a filler's surface [96]. Silane molecules that ensure the binding between the organic matrix and inorganic filler make the composite have a more rigid structure and improve the mechanical features of the composite, such as compressive and flexural strength, to obtain similar mechanical characteristics of the tooth structure [97]. Application of calcium-rich 13X zeolites as active fillers improved the remineralizing effect of examined composites by providing sustained release of calcium ions in conditions simulating a natural oral environment. Furthermore, the silanization of these composites significantly improved flexural strength and compressive strength values. The beneficial effects of silanization are the consequence of a stronger bond between fillers particles and resins from the organic matrix due to the introduction of methacrylic groups to the fillers' surface, which form covalent bonds between the resin and the filler [80].

Modification of the zeolite surface by a 4-(dimethylamino)benzenediazonium cation to acquire an active filler in methacrylic-resin-based composites was performed to verify some mechanical properties, as well as crosslinking ability [98]. All conducted tests proved that the addition of modified zeolite improved the compressive and flexural strength of the composite. The modification process, as well, has a crucial impact on these values. The results also show that it affected the crosslinking properties of the resin.

Soft and hard tissue engineering of the oral cavity represents the therapeutic approach with great potential. Zeolites, due to their favourable properties, have appeared promising as scaffolds in bone- and tooth-tissue engineering. The clinoptilolite-composite scaffolds enhanced mechanical, physical, and biological properties of polymer-based scaffolds, increased in vitro protein adsorption capacity of the scaffold and led to higher osteoinductivity and intracellular calcium deposition [99].

It is expected that the further research effort should focus on zeolite-based materials and concentrate on zeolite effects in response to microbial challenges in vivo, but also to determine the proper concentration of zeolite that may be incorporated into various dental materials and establish zeolite's impact on their mechanical properties. The possibilities and beneficial effects of various surface modification methods should also be involved. Zeolite's favourable properties and wide range of the composition and hierarchical pore structure makes the zeolitic materials convenient for tissue engineering purposes.

3.2. *Zeolites as Drug Carriers in Pharmacotherapy*

The next logical step in the medical application of zeolites was their usage as drug carriers. That idea was surely provoked by two main groups of reasons. The first one is the reversibility of the adsorption of pharmaceuticals by zeolites, and which requires thorough studying of many important factors which influence this process. The focus, for sure, must be put on the ability to control and prolong drug release.

At the same time, scientists have made great efforts to exploit the possible synergistic bioactivities of zeolites with applied drugs. The resulting effects are often unpredictable and require detailed study, especially when cytotoxic effects on fibroblasts and tumour cells are performed in parallel.

The application of zeolites in biomedicine, as drug carriers, is based on their large specific surface area, high adsorption capacity [100], biocompactness, low toxicity [101], and microporous structure, which allows drug encapsulation within the zeolite and the ability to control and prolong drug release. Drug release is monitored in liquids with different pH values that correspond to the pH of the regions through which the drug-loaded zeolite carrier passes [102]. Drug encapsulation in drug delivery systems (DDS) eliminates side effects while maintaining treatment effectiveness [103], reduces drug concentration, and provides targeted delivery [104].

Zeolite nanocarriers enable drug delivery to a specific target and drug release without affecting surrounding healthy cells [105], and they display enhanced permeability, controlled drug distribution, and prolonged life in the blood system [106]. Synergistic effects of zeolites, such as gastroprotective influence, especially in the case of clinoptilolite DDS, could be ascribed to zeolites ability to attach to hydrogen ions and biologically active amines and nitrates [107].

The differences between the zeolite pore size and the targeted drug, and in hydrophilicity between zeolites and drugs, which are usually considered as limitation in their loading capacity, can be improved by surface modification of the zeolite [108], with the aim to adjust the surface of a zeolite depending on the delivered drug characteristics.

The efficiency of zeolites as DDS can be improved by the adsorption of cationic surfactants on the surface of zeolites through ion exchange and hydrophobic interactions, leading to the formation of monolayers or bilayers, depending on their concentration [109]. It looks that very important role here concerns the concentration of the cationic surfactant in relation to external cation exchange capacity—ECEC value, confirmed by the study performed on the natural zeolite with high clinoptilolite content with different levels of cetylpyridinium chloride (CPC) [110]. If the surfactant loading level was equal to ECEC, a monolayer of the organic phase is present at the zeolitic surface. When the amount of the CPC was above the ECEC value, a less extended bilayer is formed, while the sample with the highest surfactant content ordered bilayer or admicelles exist at the zeolitic surface. Similar results, obtained for the cationic surfactant-hexadecyltrimethylammonium bromide (HB) showed proportional increase of drug adsorption by increasing the amount of surfactant used for zeolite modification, leading to interactions between DS and HB at the zeolitic surface [111].

3.2.1. Zeolites as Carriers of Anti-inflammatory Drugs

Diclofenac and ibuprofen are the most commonly used nonsteroidal anti-inflammatory drugs (NSAIDs) with low toxicity and thermal and chemical stability [100,112]. Due to the short half-life and the possibility of side effects [109,112], the subject of many studies is the encapsulation of drugs into systems that allow prolonged drug release, improving therapeutic activity and reducing side effects [112]. The ability of natural zeolites as carriers of anti-inflammatory drugs can be improved by modifying their surface with different cationic surfactants [100,109].

For the sorption of diclofenac sodium (DS) as a model drug by zeolites modified with cationic surfactant cetylpyridinium chloride (CPC), the best fit to the equilibrium data over the entire tested concentration range provides the Langmuir equation [110]. The phase resulting from adsorbed CPC was the primary sorption phase for the DS, enabling the adsorption and partitioning processes at the same time.

Results of DS release in vitro of the hexadecyltrimethylammonium bromide modified zeolite composites showed that the prolonged DS release throughout 8 h was achieved, making them promise as a functional drug formulation excipient [111]. Because of the strong affinity of hexadecyltrimethylammonium bromide for DS, even corresponding physical mixtures showed prolonged DS release.

Serri et al. tested the DS delivery system using natural zeolites clinoptilolite (CLI), chabazite (CHA), and phillipsite (PHI) modified with cetylpyridinium chloride. The drug loading process took less than five minutes, with roughly equal amounts of DS in clinoptilolite and phillipsite and loading capacities greater than anion exchange capacity (AEC), while chabazite had the lowest loading capacity. The DS release process was sustained for five hours for all three zeolites, through particle diffusion for CLI and PHI, and a combination of film diffusion and particle diffusion for CHA, with a release of about 40% in the first hour. Despite the higher loading efficiency, CLI and PHI achieved a lower percentage of release compared to CHA (100%), indicating stronger interaction with the surfactant [112]. When DS was adsorbed in granulated clinoptilolite modified with cetylpyridinium chloride, the release process was extended up to 9 h, with approximately 22% of the drug released in the first 2 h in the solution at pH 1.2. The release of DS from the zeolite followed quasi-zero order kinetics for the first 8 h [108].

Considering the zeolite capability as a carrier for sustained drug release, a clinoptilolite was superficially modified with CLC and loaded with DS, to enhance its adsorption. The release profile was found to be reversible with the DS being gradually released in the presence of an ionic medium, such as simulated intestinal fluid (pH~6.8). The results pointed out the ability of superficially modified nanozeolites to prolong DS release for 5 h. This release was predominantly governed by very fast ion exchange, while the drug diffusion through the boundary layer was the rate-controlling step of the process [100].

The modification of clinoptilolite and chabazite with cetyltrimethylammonium bromide results in higher DS adsorption capacities of 21.9% and 33.7%, respectively [113]. Diclofenac sodium was released rapidly from chabazite, with 97.7% of the drug released in the first hour and 100% released after only 3 h. In the case of clinoptilolite, an initial release of 57% was observed after one hour, followed by a prolonged release of 84% after six hours.

3.2.2. Zeolites as Carriers of Anticancer Drugs

Cancer is one of the most common diseases whose treatment includes surgery, chemotherapy, and radiotherapy [102,114]. The most frequently used anticancer drugs include doxorubicin, 5-fluorouracil, cisplatin, and cyclophosphamide, among others, which can be used alone or in combination with other drugs [114]. However, many of the drugs used in clinical practice cause undesired mutagenic and cytotoxic changes in normal cells, leading to numerous side effects. Zeolites have emerged as suitable carriers for anticancer drugs which can be encapsulated within them and hence may reduce the side effects of anticancer treatment on healthy cells, but also may provide a prolonged exposure of the drug to the cancer cell, resulting in an enhanced anticancer effect [101,106]. As the environment of tumour tissue is more acidic (5.5–6.0) compared to normal tissue (7.4), the focus of many studies is the development of DDS with pH-controlled drug release, which increases the effectiveness of therapy and reduces side effects [115]. This pH-dependent drug release can be especially important when analysing zeolite as a drug carrier for gastrointestinal cancer treatment because pH values significantly vary through different parts of the digestive tract.

Polyphenol curcumin exhibits a wide range of biomedicine applications based on its anticancer, antioxidant, antifungal, and anti-inflammatory properties [105,116–119]. Encapsulation of curcumin within zeolites improves its solubility and stability in water and increases bioavailability [105]. Karimi et al. investigated the effects of surface modification with polyethylene glycol PEG on drug encapsulation and release efficiency using two synthetic zeolites, Y and ZSM-5 (PEG/Y and PEG/ZSM-5). The results of BET analysis revealed that the efficiency of curcumin encapsulation in zeolite Y (60.06%) was higher compared to ZSM-5 zeolite, while in the case of modified zeolite, the pores were partially coated with PEG, resulting in a lower encapsulation efficiency. Nitrogen adsorption-desorption analysis indicated that zeolite pore volume and surface area decreased after the encapsulation of the drug in the zeolite. In vitro release of curcumin in this study showed that more drug was released in solution with a lower pH value, with a higher amount of drug released from modified zeolites due to the increase in solubility of curcumin in the presence of PEG as well as the weakening of zeolites-curcumin interactions. A maximum drug release efficiency of 67% was achieved for PEG/Y over a period of 120 h [117]. Investigation of cyclodextrin-modified ZSM-5 zeolite for hydrothermal delivery of curcumin showed that higher efficiency was achieved in a more acidic medium (pH 5.5) compared to buffer solution at pH 7.4 [105].

Magnetite-zeolite nanocomposites (MZNC) were tested as carriers for the anticancer drug 5-fluorouracil (5-FU), with a loading capacity that increased with the concentration of 5-FU. Measuring the amount of drug released in buffer solutions showed that 90% of 5-fluorouracil was released from MZNC at pH 5.0 over 360 min. 5-FU encapsulated in MZNC showed concentration-dependent inhibitory effects on the proliferation of gastric carcinoma cells [114].

In another study, three types of micronized zeolites (ZSM-5, Zeolite A, and Faujasite NaX) were investigated and loaded with 5-FU as delivery systems to establish the drug release behaviour in a simulated gastric fluid environment, but also to reveal the cytotoxic effect of zeolites without/with 5-FU on colon cancer cells. The aluminosilicate structure was easier to be altered and decompose in a more acidic solution (pH 1.6) than in a mild acidic solution (pH 5) and the drug was released easier. Furthermore, all applied zeolites had a safe behaviour towards colon carcinoma cells, while the cytotoxic effect on these cell lines was confirmed in their 5-FU-loaded versions, showing the most potency for ZSM-5, followed by ZA and ZX [102]. Therefore, these zeolites can be used as good carriers for anticancer drugs which release in a controlled way.

When compared to Linde Type L (LTL) zeolite, which has monodimensional nanochannels, drug delivery systems based on NaY zeolite showed higher loading efficiency for 5-FU due to the three-dimensional structure. Both NaY and LTL were nontoxic to three cell lines analysed (breast cancer, colon cancer, and melanoma cell lines), even when applied in the highest concentrations. On the other hand, both host zeolites combined with 5-FU exerted toxic effects on all three cell lines and these effects were concentration- and cell type-dependent, with 5-FU/LTL showing a stronger inhibitory effect on cell viability compared to 5-FU/NaY. The toxic effect was more prominent on breast cancer and colon carcinoma cells. In vivo results also showed cell-specificity and led to a higher tumour reduction in breast cancer cells [120]. Encapsulation of 5-FU in LTL zeolite modified with positive amino groups leads to increased DDS internalisation by breast cancer cells and MCF-10 non-cancer epithelial mammary cells, with a more prominent effect in cancer cells compared to normal, which can be attributed to the higher metabolic and growth rates present in malignant cells compared to healthy ones. The results of this study showed that the surface nanoparticle functionalization with positive charges enhanced their internalisation and improved the zeolite efficacy as DDS, by improving their electrostatic interactions with the negatively charged cell membrane of the chosen cell lines. Furthermore, this effect was achieved with the application of a lower drug amount. Overall, these findings emphasise the importance of zeolite structures as DDS and the significance of surface modification in enhancing their efficacy as anti-cancer agents.

Linde type A zeolites and their magnetite nanocomposites can be used for efficient loading and slow-release applications. While the zeolites and their nanocomposites were nontoxic to breast cancer cell lines, the loading of zeolites with doxorubicin (DOX) enhanced cell growth inhibition along with the increase in concentration compared to non-encapsulated DOX, with the highest efficiency for DOX-4A and magnetic DOX-4A [121].

Doxorubicin loaded on ZSM-5/chitosan core-shell (ZSM-5/CS) nanodisks was examined as pH-sensitive DDS for the treatment of osteosarcoma. Its mesoporous structure contributed to the high loading efficiency of 97.7% while chitosan core-shell layers on the surface improved drug delivery efficiency and controlled DOX release. After 7 days of incubation, the maximum cumulative release ratio of DOX was 71% in solutions at pH 5.5. The cytotoxic effect on osteoblastic human osteosarcoma-derived cell line (MG63) cells showed that ZSM-5/CS/DOX toxicity was higher at lower pH (5.5), causing inhibition of cell growth more than free DOX [122].

In research by Abasian et al., chitosan/PLA/NaX/Fe_3O_4/DOX nanofibers obtained by an electrospinning method were tested as a carrier for doxorubicin. The loading efficiency of the NaX-DOX system reached a maximum of 97%, while the loading efficiency of the DOX-loaded chitosan/PLA/NaX/Fe_3O_4 nanofiber was reduced to 92% due to DOX release during the electrospinning process. The prolonged release of doxorubicin in two steps consisted of release from NaX, followed by release from nanofibers. The presence of magnetic nanoparticles in the fibers contributed to slowing down the process of drug release, while the presence of a magnetic field showed the opposite effect. The main drug-release mechanism was Fickian diffusion of DOX into buffer solution, as described by Korsmeyer-Peppas kinetics. The maximum death of the human lung epidermoid carcinoma cell line after 7 days of treatment was 82% for DOX-loaded composite nanofibers in the presence of an external magnetic field [123].

Cisplatin (CIS) has been used for the treatment of numerous human cancers, with limited application due to its numerous and serious side effects, extensive resistance, and toxicity [124,125]. Furthermore, because of the drug's non-specific delivery, high doses are required for treatment, which are very toxic to healthy cells [123–126]. Investigating the delivery of CIS on synthesized zeolite Y nanoparticles revealed that this DDS was pH-, concentration- and time-dependent. In vitro drug release assessment showed that CIS release from ZC-NPs was faster in an acidic environment (pH 5.4) with a maximum drug release of 93.8%. The zeolites alone increased the viability of MG63 cells, while the CIS incorporated within zeolite decreased cell viability, but to a less extent compared to CIS alone [124].

Similar results were obtained in a study that investigated a polycaprolactone-zeolite (PCL-Z) nanocomposite Y-scaffold as a drug delivery system for CIS in the treatment of osteosarcoma. The release rate of a CIS from the zeolite carrier was pH-dependent, showing that 87.6% of the drug was released at pH 5.4 during 28 days of incubation. The presence of the zeolite in the PCL-Z scaffold increased MG63 cell viability, while the release of CIS from this nanocomposite showed a cytotoxic effect on these cells [125].

Two synthetic zeolites FAU and LTA in their sodium forms (NaY and NaA) were used for the encapsulation and controlled release of an anticancer agent, α-Cyano-4-hydroxycinnamic acid (CHC). The zeolite NaA adsorbed 67–76% of the CHC in solution, while the adsorption capacity of the NaY zeolite was 85%. Both applied zeolites exerted no toxic effect on cancer cells. On the other hand, a significant increase in the drug effect on the human colon cancer cells was observed when applying CHC in the zeolite system compared to CHC alone, and this effect was concentration-dependent and more prominent in zeolite combination with NaY compared to NaA. CHC-zeolite combination led to an inhibition of cell viability up to 146-fold (CHC/NaA) and even 585-fold (CHC/NaY) when compared to the non-encapsulated drug. These results could be attributed to the more open structure of NaY that allows drug diffusion, making it a more efficient drug delivery system [103].

The current challenge in treating cancers is to replace conventional chemotherapy with an in-situ drug delivery system to enhance drug efficiency. Currently, zeolites as pH-sensitive delivery systems with their exceptional physiological stability are very promising contestants in cancer treatment procedures and the development of a new delivery system with multifunctional activity. In experimental models in vivo and in vitro, the use of zeolites as scaffolds for anticancer drugs, can provide effective and sustained drug release and may aggravate the inhibitory effect on various cancer cells compared to the anticancer drug applied alone. Zeolites are denoted as a promising system for targeted delivery of chemotherapeutic agents by promoting their therapeutic efficiency, reducing undesired side effects, and diminishing their toxicity on the healthy surrounding tissue.

Besides zeolites, metal-organic frameworks (MOFs) possess certain advantages compared to other DDSs such as definite crystalline structure and flexibility in creating them from scratch [127]. Zeolitic imidazolate frameworks (ZIFs) seem to be nowadays the most frequently studied MOFs as DDSs, especially because of their biocompatibility, and simple synthesis procedures [128]. ZIFs have especially harvested interest as pH-sensitive drug carriers with high drug loading capacities and biodegradability, which remains stable in neutral and basic media, but quickly collapses in the strong acidic aqueous solution [129]. ZIFs were particularly successfully confirmed as anticancer DDS [130].

Hao et al. reported the evaluation of the possible beneficial applications of zeolites and ZIFs as DDS for anticancer drugs, including doxorubicin (DOX), 5-fluorouracil (5-FU), curcumin, cisplatin, and others [131]. Following PRISMA guidelines, after screening the full texts published till August 2021, 53 articles remained and were included in the analysis. Clinoptilolite is currently the only zeolite registered in the EU as a medical device, and can be used in oral treatment, as drug carriers and delivery systems. Despite that fact, CLI was included as anticancer DDS in only three studies. The study claimed that 35 studies included ZIFs based supports for DDSs, of which in 31 considered particularly ZIF-8. FAU types of zeolites follow in the number of reported results as DDS carriers, i.e., 19 studies included NaX, NaY, or nano NaY zeolites.

What is the secret of such an interest in ZIF-8 as a carrier in anticancer DDS? Computer simulations and molecular modelling helped to clarify that the diffusion of 5-fluorouracil (5-FU) and caffeine (CAF), model drugs, between neighbouring pores in ZIF-8 is strictly restricted due to large energy barriers. The study of Proenza and Longo shows that the inner pores of ZIF-8 surface model were inaccessible to the 5-FU and CAF, but accessible to the solvents (methanol or water) [132]. The outstanding reliability of the ZIF-8 surface model relies in its appropriate explanation of the surface and by exposing adsorption sites such as undercoordinated zinc ions to interactions with large molecules, achieved by changing the periodic conditions from the atomistic level to a higher molecular level, such as a ZIF-8 nanocrystal. Perhaps future considerations of zeolites as drug carriers should take into account the comparative advantages demonstrated by ZIF, modifying them to be more suitable, reliable and effective.

Current research trends in the adsorption phenomenon using porous materials [133,134] along with innovative adsorbent characterization [135] reveal the utmost importance of the subject in both medical and environmental areas and offer novel and comprehensive ideas and solutions which may further assist drug/carrier design.

4. Conclusions

As a conclusion, here are listed some of the perspectives that may be addressed in future zeolite use as a removal tool of pharmaceuticals and/or other pollutants from the aquatic environment:

- main interactions in the pollutant-zeolite system assisted by spectroscopic methods, especially in post-adsorption studies;

- targeted interactions lead to a comprehensive understanding of the adsorption mechanism. Once we know the mechanism in detail, we can elucidate a number of target pollutants. If a designed zeolite adsorbent shows substantial adsorption capacity for one species, can it be applied for the other or their occurring mixtures?
- sometimes zeolites are designated as costly materials, and novel routes for synthesis, from waste materials, are beneficial;
- what to do with the spent adsorbent, does this impose a significant shortcoming of mainly physical removal techniques? Some innovative solutions are offered, mostly in the pyrolysis of the spent adsorbents and subsequent employment as electrode materials;
- environmentally relevant concentrations and/or flow techniques may be employed in the second step of the adsorption test aimed at pharmaceuticals removal. This requires HPLC/UPLC techniques, preferably with sensitive detection such as mass spectrometry. For volatile pharmaceuticals, GC/MS methods are also available.
- a focus needs to be shifted to real effluents, with a range of concurrent adsorbing ions, mostly metals, and organic matter;
- test whether the adsorption, as a removal technique, leaves the environment more toxic than the pollutant itself;
- apply a range of quantum mechanical calculations to guide future adsorbent design as this state-of-the-art calculation can point out exactly what to expect from your adsorption system and enable future predictions.

If we talk about zeolites as drug carriers in pharmacotherapy, the above listed considerations are more or less appropriate for this application of interaction between zeolites and pharmaceuticals. Future directions of investigations must cover several important issues:

- first, there is a need to study and ensure the lowest possible level of toxicity;
- expand the therapeutic range of DDS, improving more benefits and less side effects of pharmacological active compounds;
- raise the specificity of targeted sensitive sites of DDS action;
- work to achieve appropriate kinetics of release of pharmacologically active components. Parallel development of reliable, sensitive and specific analytical methods for following such a low concentration level of drugs in situ are more than desired.

This is certainly a topic within which researchers will be motivated to shed more light in the future, because of its important potential benefits, and as to a final conclusion—adsorption does matter.

Author Contributions: Conceptualization, S.U.-M. and B.N.V.; data curation, N.R.M., M.R. and S.U.-M.; writing—original draft preparation, M.R., V.M., B.N.V.; visualization, B.N.V.; writing—review and editing S.U.-M., N.R.M. All authors have read and agreed to the published version of the manuscript.

Funding: This research was funded by the Ministry of Education, Science and Technological Development of the Republic of Serbia, grant numbers 451-03-68/2022-14/200146 and 451-03-68/2022-14/200161 and grant for collaboration with the Joint Institute for Nuclear Research, Dubna, Russia (JINR-Serbia_P12).

Conflicts of Interest: The authors declare no conflict of interest. The funders had no role in the design of the study; in the collection, analyses, or interpretation of data; in the writing of the manuscript, or in the decision to publish the results.

References

1. Ramesh, K.; Reddy, D.D. Zeolites and Their Potential Uses in Agriculture. *Adv. Agron.* **2011**, *113*, 219–241. [CrossRef]
2. Cataldo, E.; Salvi, L.; Paoli, F.; Fucile, M.; Masciandaro, G.; Manzi, D.; Massini, C.M.; Mattii, G.B. Application of Zeolites in Agriculture and Other Potential Uses: A Review. *Agronomy* **2021**, *11*, 1547. [CrossRef]
3. Jacobs, P.A.; Flanigen, E.M.; Jansen, J.C.; van Bekkum, H. *Introduction to Zeolite Science and Practice*; Elsevier: Amsterdam, The Netherlands, 2001.

4. Coombs, D.S.; Alberti, A.; Armbruster, T.; Artioli, G.; Colella, C.; Galli, E.; Grice, J.D.; Liebau, F.; Mandarino, J.A.; Minato, H.; et al. Recommended Nomenclature for Zeolite Minerals: Report of the Subcommittee on Zeolites of the International Mineralogical Association, Commission on New Minerals and Mineral Names. *Can. Mineral.* **1997**, *35*, 1571–1606.
5. Derbe, T.; Temesgen, S.; Bitew, M.A. Short Review on Synthesis, Characterization, and Applications of Zeolites. *Adv. Mater. Sci. Eng.* **2021**, *2021*, 6637898. [CrossRef]
6. Król, M. Natural vs. Synthetic Zeolites. *Crystals* **2020**, *10*, 622. [CrossRef]
7. Yazdi, M.K.; Zarrintaj, P.; Hosseiniamoli, H.; Mashhadzadeh, A.H.; Saeb, M.R.; Ramsey, J.D.; Ganjali, R.M.; Mozafari, M. Zeolites for theranostic applications. *J. Mater. Chem. B* **2020**, *8*, 5992–6012. [CrossRef] [PubMed]
8. Barrer, R.M. Porous crystals: A perspective. *Pure Appl. Chem.* **1986**, *58*, 1317. [CrossRef]
9. Kianfar, E. Zeolites: Properties, Applications, Modification and Selectivity. In *Zeolites: Advances in Research and Application*; Mahler, A., Ed.; Nova Science Publishers, Inc.: Hauppauge, NY, USA, 2020.
10. Bacakova, L.; Vandrovcova, M.; Kopova, I.; Jirka, I. Applications of zeolites in biotechnology and medicine—A review. *Biomater. Sci.* **2018**, *6*, 974. [CrossRef]
11. Möller, K.; Bein, T. Mesoporosity—A new dimension for zeolites. *Chem. Soc. Rev.* **2013**, *42*, 3689. [CrossRef]
12. Database of Zeolite Structures. Available online: http://www.iza-structure.org/databases/ (accessed on 5 July 2022).
13. Coombs, D.S.; Alberti, A.; Armbruster, T.; Artioli, G.; Colella, C.; Galli, E.; Grice, J.D.; Liebau, F.; Mandarino, J.A.; Minato, H.; et al. Recommended nomenclature for zeolite minerals: Report of the subcommittee on zeolites of the International Mineralogical Association, Commission on New Minerals and Mineral Names. *Mineral. Mag.* **1998**, *62*, 533–571. [CrossRef]
14. Rhodes, C.J. Properties and applications of zeolites. *Sci. Prog.* **2010**, *93*, 223–284. [CrossRef]
15. Li, Y.; Yu, J. Emerging applications of zeolites in catalysis, separation and host–guest assembly. *Nat. Rev. Mater.* **2021**, *6*, 1156–1174. [CrossRef]
16. Zhang, Q.; Yu, J.; Corma, A. Applications of Zeolites to C1 Chemistry: Recent Advances, Challenges, and Opportunities. *Adv. Mater.* **2020**, *32*, 2002927. [CrossRef] [PubMed]
17. Serati-Nouri, H.; Jafari, A.; Roshangar, L.; Dadashpour, M.; Pilehvar-Soltanahmadi, Y.; Zarghami, N. Biomedical applications of zeolite-based materials: A review. *Mater. Sci. Eng. C* **2020**, *116*, 111225. [CrossRef]
18. Li, Y.; Li, L.; Yu, J. Applications of Zeolites in Sustainable Chemistry. *Chem* **2017**, *3*, 928–949. [CrossRef]
19. Vasiljević, B.N.; Obradović, M.; Bajuk-Bogdanović, D.; Milojević-Rakić, M.; Jovanović, Z.; Gavrilov, N.; Holclajtner-Antunović, I. In Situ synthesis of potassium tungstophosphate supported on BEA zeolite and perspective application for pesticide removal. *J. Environ. Sci.* **2019**, *81*, 136–147. [CrossRef]
20. Janićijević, D.; Uskoković-Marković, S.; Ranković, D.; Milenković, M.; Jevremović, A.; Nedić Vasiljević, B.; Milojević-Rakić, M.; Bajuk-Bogdanović, D. Double active BEA zeolite/silver tungstophosphates—Antimicrobial effects and pesticide removal. *Sci. Total Environ.* **2020**, *735*, 139530. [CrossRef]
21. Jevremović, A.; Vasiljević, B.N.; Popa, A.; Uskoković-Marković, A.; Ignjatović, L.; Bajuk-Bogdanović, D.; Milojević-Rakić, M. The environmental impact of potassium tungstophosphate/ZSM-5 zeolite: Insight into catalysis and adsorption processes. *Microporous Mesoporous Mater.* **2021**, *315*, 110925. [CrossRef]
22. Jevremović, A.; Božinović, N.; Đorđević, D.; Marmakov, S.; Vasiljević, B.N.; Uskoković-Marković, S.; Bajuk-Bogdanovic, D.; Milojević-Rakić, M. Modulation of cytotoxicity by consecutive adsorption of tannic acid and pesticide on surfactant functionalized zeolites. *Environ. Sci. Process. Impacts* **2020**, *22*, 2199. [CrossRef]
23. Milojević-Rakić, M.; Bajuk-Bogdanović, D.; Vasiljević, B.N.; Rakić, A.; Škrivanj, S.; Ignjatović, L.; Dondur, V.; Mentus, S.; Ćirić-Marjanović, G. Polyaniline/FeZSM-5 composites—Synthesis, characterization and their high catalytic activity for the oxidative degradation of herbicide glyphosate. *Microporous Mesoporous Mater.* **2018**, *267*, 68–79. [CrossRef]
24. Lichtenberg, F.R. How many life-years have new drugs saved? A three-way fixed-effects analysis of 66 diseases in 27 countries, 2000–2013. *Int. Health* **2019**, *11*, 403. [CrossRef] [PubMed]
25. Rodriguez-Mozaz, S.; Vaz-Moreira, I.; Giustina, S.V.D.; Llorca, M.; Barceló, D.; Schubert, S.; Berendonk, T.; Michael-Kordatou, I.; Fatta-Kassinos, D.; Martinez, J.L.; et al. Antibiotic residues in final effluents of European wastewater treatment plants and their impact on the aquatic environment. *Environ. Int.* **2020**, *140*, 105733. [CrossRef] [PubMed]
26. Thalla, A.K.; Vannarath, A.S. Occurrence and environmental risks of nonsteroidal anti-inflammatory drugs in urban wastewater in the southwest monsoon region of India. *Environ. Monit. Assess.* **2020**, *192*, 193. [CrossRef] [PubMed]
27. European Commission. Communication from the Commission to the European Parliament, the Council, and the European Economic and Social Committee: European Union Strategic Approach to Pharmaceuticals in the Environment. *Communication* **2019**, *128*, 1–12. Available online: https://www.eumonitor.eu/9353000/1/j9vvik7m1c3gyxp/vkwpg8sb9wx8 (accessed on 5 July 2022).
28. Merlin, C. Reducing the Consumption of Antibiotics: Would That Be Enough to Slow Down the Dissemination of Resistances in the Downstream Environment? *Front. Microbiol.* **2020**, *11*, 33. [CrossRef]
29. Kidd, K.A.; Paterson, M.J.; Rennie, M.D.; Podemski, C.L.; Findlay, D.L.; Blanchfield, P.J.; Liber, K. Direct and indirect responses of a freshwater food web to a potent synthetic oestrogen. *Phil. Trans. R. Soc. B* **2014**, *369*, 20130578. [CrossRef]
30. Heberer, T. Occurrence, fate, and removal of pharmaceutical residues in the aquatic environment: A review of recent research data. *Toxicol. Lett.* **2002**, *131*, 5–17. [CrossRef]

31. Khetan, S.K.; Collins, T.J. Human pharmaceuticals in the aquatic environment: A challenge to green chemistry. *Chem. Rev.* **2007**, *107*, 2319–2364. [CrossRef]
32. Halling-Sørensen, B.; Nielsen, S.N.; Lanzky, P.; Ingerslev, F.; Lützhøft, H.H.; Jørgensen, S. Occurrence, fate and effects of pharmaceutical substances in the environment—A review. *Chemosphere* **1998**, *36*, 357–393. [CrossRef]
33. Jelic, A.; Gros, M.; Ginebreda, A.; Cespedes-Sánchez, R.; Ventura, F.; Petrovic, M.; Barceló, D. Occurrence, partition and removal of pharmaceuticals in sewage water and sludge during wastewater treatment. *Water Res.* **2011**, *45*, 1165–1176. [CrossRef]
34. Safe Disposal of Medicines. Available online: https://www.fda.gov/drugs/ensuring-safe-use-medicine/safe-disposal-medicines (accessed on 5 July 2022).
35. Nikolaou, A.; Meric, S.; Fatta, D. Occurrence patterns of pharmaceuticals in water and wastewater environments. *Anal. Bioanal. Chem.* **2007**, *387*, 1225. [CrossRef] [PubMed]
36. der Beek, T.A.; Weber, F.-A.; Bergmann, A.; Hickmann, S.; Ebert, I.; Hein, A.; Küster, A. Pharmaceuticals in the environment-Global occurrences and perspectives. *Environ. Toxicol. Chem.* **2016**, *35*, 823. [CrossRef] [PubMed]
37. Heberer, T. Tracking persistent pharmaceutical residues from municipal sewage to drinking water. *J. Hydrol.* **2002**, *266*, 175. [CrossRef]
38. Zupanc, M.; Kosjek, T.; Petkovšek, M.; Dular, M.; Kompare, B.; Širok, B.; Blažeka, Ž.; Heath, E. Removal of pharmaceuticals from wastewater by biological processes, hydrodynamic cavitation and UV treatment. *Ultrason. Sonochem.* **2013**, *20*, 1104. [CrossRef] [PubMed]
39. Li, W.C. Occurrence, sources, and fate of pharmaceuticals in aquatic environment and soil. *Environ. Pollut.* **2014**, *187*, 193. [CrossRef] [PubMed]
40. Jaseem, M.; Kumar, P.; John, R.M. An overview of waste management in pharmaceutical industry. *Pharma Innov.* **2017**, *6 Pt C*, 158–161.
41. Rojas, M.R.; Leung, C.; Bonk, F.; Zhu, Y.; Edwards, L.; Arnold, R.G.; Sáez, A.E.; Klečka, G. Assessment of the effectiveness of secondary wastewater treatment technologies to remove trace chemicals of emerging concern. *Crit. Rev. Environ. Sci. Technol.* **2013**, *43*, 1281. [CrossRef]
42. Baresel, C.; Cousins, A.P.; Hörsing, M.; Ek, M.; Ejhed, H.; Allard, A.-S.; Magnér, J.; Westling, K.; Wahlberg, C.; Fortkamp, U. Pharmaceutical Residues and Other Emerging Substances in the Effluent of Sewage Treatment Plants. IVL Svenska Miljöinstitutet. 2015. (B-Rapport). Available online: http://urn.kb.se/resolve?urn=urn:nbn:se:ivl:diva-2969 (accessed on 5 July 2022).
43. Fukahori, S.; Fujiwara, T.; Ito, R.; Funamizu, N. pH-Dependent adsorption of sulfa drugs on high silica zeolite: Modeling and kinetic study. *Desalination* **2011**, *275*, 237–242. [CrossRef]
44. Blasioli, S.; Martucci, A.; Paul, G.; Gigli, L.; Cossi, M.; Johnston, C.T.; Marchese, L.; Braschi, I. Removal of sulfamethoxazole sulfonamide antibiotic from water by high silica zeolites: A study of the involved host–guest interactions by a combined structural, spectroscopic, and computational approach. *J. Colloid Interface Sci.* **2014**, *419*, 148–159. [CrossRef]
45. Martucci, A.; Pasti, L.; Marchetti, N.; Cavazzini, A.; Dond, F.; Alberti, A. Adsorption of pharmaceuticals from aqueous solutions on synthetic zeolites. *Microporous Mesoporous Mater.* **2012**, *148*, 174–183. [CrossRef]
46. Genç, N.; Dogan, E.C. Adsorption kinetics of the antibiotic ciprofloxacin on bentonite, activated carbon, zeolite, and pumice. *Desalin. Water Treat.* **2013**, *53*, 785–793. [CrossRef]
47. de Sousa, D.N.R.; Insa, S.; Mozeto, A.A.; Petrovic, M.; Chaves, T.F.; Fadini, P.S. Equilibrium and kinetic studies of the adsorption of antibiotics from aqueous solutions onto powdered zeolites. *Chemosphere* **2018**, *205*, 137–146. [CrossRef]
48. Al-Jubouri, S.M.; Al-Jendeel, H.A.; Rashid, S.A.; Al-Batty, S. Antibiotics adsorption from contaminated water by composites of ZSM-5 zeolite nanocrystals coated carbon. *J. Water Process Eng.* **2022**, *47*, 102745. [CrossRef]
49. Lee, B.Y. Removal of Antibiotics from Contaminated Waters Using Natural Zeolite. Master's Thesis, CUNY City College, New York, NY, USA, 2012.
50. Belviso, C.; Guerra, G.; Abdolrahimi, M.; Peddis, D.; Maraschi, F.; Cavalcante, F.; Ferretti, M.; Martucci, A.; Sturini, M. Efficiency in Ofloxacin Antibiotic Water Remediation by Magnetic Zeolites Formed Combining Pure Sources and Wastes. *Processes* **2021**, *9*, 2137. [CrossRef]
51. Tumrani, S.H.; Soomro, R.A.; Zhang, X.; Bhutto, D.A.; Buxa, N.; Ji, X. Coal fly ash driven zeolites for the adsorptive removal of the ceftazidime drug. *RSC Adv.* **2021**, *11*, 26110. [CrossRef]
52. Grela, A.; Kuc, J.; Bajda, T. A Review on the Application of Zeolites and Mesoporous Silica Materials in the Removal of Non-Steroidal Anti-Inflammatory Drugs and Antibiotics from Water. *Materials* **2021**, *14*, 4994. [CrossRef]
53. Ajo, P.; Preis, S.; Vornamo, T.; Mänttäri, M.; Kallioinen, M.; Louhi-Kultanen, M. Hospital wastewater treatment with pilot-scale pulsed corona discharge for removal of pharmaceutical residues. *J. Environ. Chem. Eng.* **2018**, *6*, 1569–1577. [CrossRef]
54. Shearer, L.; Pap, S.; Gibb, S.W. Removal of pharmaceuticals from wastewater: A review of adsorptive approaches, modelling and mechanisms for metformin and macrolides. *J. Environ. Chem. Eng.* **2022**, *10*, 108106. [CrossRef]
55. Zhao, J.; Yang, X.; Liang, G.; Wang, Z.; Li, S.; Wang, Z.; Xie, X. Effective removal of two fluoroquinolone antibiotics by PEG-4000 stabilized nanoscale zero-valent iron supported onto zeolite (PZ-NZVI). *Sci. Total Environ.* **2019**, *710*, 136289. [CrossRef]
56. Ayoub, H.; Roques-Carmes, T.; Potier, O.; Koubaissy, B.; Pontvianne, S.; Lenouvel, A.; Hamieh, T. Iron-impregnated zeolite catalyst for efficient removal of micropollutants at very low concentration from Meurthe river. *Environ. Sci. Pollut. Res.* **2018**, *25*, 34950–34967. [CrossRef]

57. Al-rimawi, F.; Daana, M.; Khamis, M.; Karaman, R.; Khoury, H.; Qurie, M. Removal of Selected Pharmaceuticals from Aqueous Solutions Using Natural Jordanian Zeolite. *Arab. J. Sci. Eng.* **2019**, *44*, 209–215. [CrossRef]
58. Smiljanić, D.; de Gennaro, B.; Daković, A.; Galzerano, B.; Germinario, C.; Izzo, F.; Rottinghaus, G.E.; Langella, A. Removal of non-steroidal anti-inflammatory drugs from water by zeolite-rich composites: The interference of inorganic anions on the ibuprofen and naproxen adsorption. *J. Environ. Manag.* **2021**, *286*, 112168. [CrossRef] [PubMed]
59. Pasti, L.; Sarti, E.; Cavazzini, A.; Marchetti, N.; Dondi, F.; Martucci, A. Factors affecting drug adsorption on beta zeolites. *J. Sep. Sci.* **2013**, *36*, 1604–1611. [CrossRef] [PubMed]
60. Farias, T.; Ruiz-Salvador, A.R.; Rivera, A. Interaction studies between drugs and a purified natural clinoptilolite. *Microporous Mesoporous Mater.* **2003**, *61*, 117–125. [CrossRef]
61. Lam, A.; Rivera, A. Theoretical study of the interaction of surfactants and drugs with natural zeolite. *Microporous Mesoporous Mater.* **2006**, *91*, 181–186. [CrossRef]
62. Wu, C.-H.; Xiong, Z.-H.; Lia, C.; Zhang, J. Zeolitic imidazolate metal organic framework ZIF-8 with ultra-high adsorption capacity bound tetracycline in aqueous solution. *RSC Adv.* **2015**, *5*, 82127–82137. [CrossRef]
63. Wanyonyi, F.S.; Pembere, A.; Mutua, G.K.; Orata, F.; Louis, H. Computational screening of zeolites for the adsorption of selected pharmaceutical pollutants. *SN Appl. Sci.* **2020**, *2*, 1901. [CrossRef]
64. Makkar, P.; Ghosh, N. A review on the use of DFT for the prediction of the properties of nanomaterials. *RSC Adv.* **2021**, *45*, 27897–27924. [CrossRef]
65. Braschi, I.; Gatti, G.; Paul, G.; Gessa, C.; Cossi, M.; Marchese, L. Sulfonamide Antibiotics Embedded in High Silica Zeolite Y: A Combined Experimental and Theoretical Study of Host−Guest and Guest−Guest Interactions. *Langmuir* **2010**, *26*, 9524–9532. [CrossRef]
66. Hessou, E.P.; Ponce-Vargas, M.; Mensah, J.-B.; Tielens, F.; Santos, J.C.; Badawi, M. Dibenzyl Disulfide Adsorption on Cationic Exchanged Faujasites: A DFT Study. *Nanomaterials* **2019**, *9*, 715. [CrossRef]
67. Fatouros, D.; Douroumis, D.; Nikolakis, V.; Ntais, S.; Moschovi, A.; Trivedi, V.; Khima, B.; Roldo, M.; Nazar, H.; Cox, P. In Vitro and In Silico investigations of drug delivery via zeolite BEA. *J. Mater. Chem.* **2011**, *21*, 7789–7794. [CrossRef]
68. Spanakis, M.; Bouropoulos, N.; Theodoropoulos, D.; Sygellou, I.; Ewart, S.; Moschovi, A.; Siokou, A.; Niopas, I.; Kachrimanis, K.; Nikolakis, V.; et al. Controlled release of 5-fluorouracil from microporous zeolites. *Nanomed. Nanotechnol. Biol. Med.* **2014**, *10*, 197–205. [CrossRef] [PubMed]
69. Fischer, M. Simulation-based evaluation of zeolite adsorbents for the removal of emerging contaminants. *Mater. Adv.* **2020**, *1*, 86–98. [CrossRef]
70. Purnomo; Setyarini, P.H.; Sulistyaningsih, D. Zeolite-based biomaterials for biomedical application: A review. *AIP Conf. Proc.* **2018**, *1977*, 030013. [CrossRef]
71. Mastinu, A.; Kumar, A.; Maccarinelli, G.; Bonini, S.A.; Premoli, M.; Aria, F.; Gianoncelli, A.; Memo, M. Zeolite Clinoptilolite: Therapeutic Virtues of an Ancient Mineral. *Molecules* **2019**, *24*, 1517. [CrossRef] [PubMed]
72. Pavelić, S.K.; Medica, J.S.; Gumbarević, D.; Filošević, A.; Pržulj, N.; Pavelić, K. Critical Review on Zeolite Clinoptilolite Safety and Medical Applications In Vivo. *Front. Pharmacol.* **2018**, *9*, 1350. [CrossRef] [PubMed]
73. Rininingsih, E.M.U. Correlation between Antioxidant Activity of Synthetic Zeolites Pillared Titanium Dioxide and Iron (III) Oxide with Adsorption DPPH. Master's Thesis, IPB University, Bogor, Indonesia, 2014. Available online: http://repository.ipb.ac.id/handle/123456789/68404 (accessed on 5 July 2022).
74. Janićijević, D.; Jevremović, A.; Režaić, A.J.; Vasiljević, B.N.; Uskoković-Marković, S.; Bajuk-Bogdanović, D.; Milojević-Rakić, M. Comparative assessment of pesticide adsorption capacity and antioxidant activity of Silver Dodecatungstophosphate/HBEA Zeolite composites. *J. Environ. Chem. Eng.* **2021**, *9*, 106341. [CrossRef]
75. Milojević-Rakić, M.; Popadić, D.; Ležaić, A.J.; Jevremović, A.; Vasiljević, B.N.; Uskoković-Marković, S.; Bajuk-Bogdanović, D. MFI, BEA and FAU zeolite scavenging role in neonicotinoids and radical species elimination. *Environ. Sci. Process. Impacts* **2022**, *24*, 265–276. [CrossRef]
76. Tomašević-Čanović, M.; Daković, A.; Rottinghaus, G.; Đuričić, M. Surfactant modified zeolites—New efficient adsorbents for mycotoxins. *Microporous Mesoporous Mater.* **2003**, *61*, 173–180. [CrossRef]
77. Ipek, H.; Avci, M.; Aydilek, N.; Yerturk, M. The effect of zeolite on oxidant/antioxidant status in healthy dairy cows. *Acta Vet. Brno.* **2012**, *81*, 43–47. [CrossRef]
78. Hao, J.; Lang, S.; Mante, F.; Pavelić, K.; Ozer, F. Antimicrobial and Mechanical Effects of Zeolite Use in Dental Materials: A Systematic Review. *Acta Stomatol. Croat.* **2021**, *55*, 76–89. [CrossRef] [PubMed]
79. Thom, D.C.; Davies, J.E.; Santerre, J.P.; Friedman, S. The hemolytic and cytotoxic properties of a zeolite-containing root filling material In Vitro. *Oral Surg. Oral Med. Oral Pathol. Oral Radiol. Endod.* **2003**, *95*, 101–108. [CrossRef] [PubMed]
80. Buchwald, Z. Calcium-Rich 13X Zeolite as a Filler with Remineralizing Potential for Dental Composites. *ACS Biomater. Sci. Eng.* **2020**, *6*, 3843–3854. [CrossRef] [PubMed]
81. Lee, J.H.; Lee, S.B.; Kim, K.N.; Kim, K.M.; Lee, Y.K. Antibacterial Effect of Silver-Zeolites in Glass-Ionomer Cements. *Key Eng. Mater.* **2007**, *330–332*, 831–834. [CrossRef]
82. Mabrouk, M.; Selim, M.; Beherei, H.H.; El-Gohary, M.I. Incorporation effect of silver and zinc-zeolites into commercial glass ionomer cement. *Int. Ceram. Rev.* **2013**, *62*, 50–54.

83. Hyun-Jin, K.; Sik, S.J.; Kyo-Han, K.; Tae-Yub, K. Antimicrobial Activity of Glass Ionomer Cement Incorporated with Chlorhexidine-Loaded Zeolite Nanoparticles. *J. Nanosci. Nanotechnol.* **2016**, *16*, 1450–1453. [CrossRef]
84. Patel, V.; Santerre, J.P.; Friedman, S. Suppression of Bacterial Adherence by Experimental Root Canal Sealers. *J. Endod.* **2000**, *26*, 20–24. [CrossRef]
85. Ghatole, K.; Gowdra, R.; Azher, S.S.; Sabharwal, S. Enhancing the antibacterial activity of the gold standard intracanal medicament with incorporation of silver zeolite: An In Vitro study. *J. Int. Soc. Prev. Community Dent.* **2016**, *6*, 75–79. [CrossRef]
86. Odabaş, M.E.; Çinar, Ç.; Akça, G.; Araz, I.; Ulusu, T.; Yücel, H. Short-term antimicrobial properties of mineral trioxide aggregate with incorporated silver-zeolite. *Dent. Traumatol.* **2011**, *27*, 189–194. [CrossRef]
87. Yokogawa, Y.; Sakanishi, M.; Morikawa, N.; Nakamura, A.; Kishida, I.; Varma, H.K. VCS Adsorptive Properties in Ion Exchanged Zeolite Materials in Gaseous and Aqueous Medium. *Procedia Eng.* **2012**, *36*, 168–172. [CrossRef]
88. Saravanan, M.; Kumar, V.; Padmanabhan, T.; Banu, F. Viscoelastic properties and antimicrobial effects of soft liners with silver zeolite in complete dental prosthesis wearers: An in vivo study. *Int. J. Prosthodont.* **2015**, *28*, 265–269. [CrossRef] [PubMed]
89. Naji, G.A.; Omar, R.; Yahya, R. Influence of sodalite zeolite infiltration on the coefficient of thermal expansion and bond strength of all-ceramic dental prostheses. *J. Mech. Behav. Biomed. Mater.* **2017**, *67*, 135–143. [CrossRef]
90. Aljafery, A.M.; Al-Jubouri, O.M.; Wally, Z.J.; Almusawi, R.M.; Abdulrudha, N.H.; Haider, J. The Effects of Incorporating Ag-Zn Zeolite on the Surface Roughness and Hardness of Heat and Cold Cure Acrylic Resins. *J. Compos. Sci.* **2022**, *6*, 85. [CrossRef]
91. Wang, J.; Wang, Z.; Guo, S.; Zhang, J.; Song, Y.; Dong, X.; Wang, X.; Yu, J. Antibacterial and anti-adhesive zeolite coatings on titanium alloy surface. *Microporous Mesoporous Mater.* **2011**, *146*, 216–222. [CrossRef]
92. Lopes, A.C.; Ribeiro, C.; Sencadas, V.; Botelho, G.; Lanceros-Mendez, S. Effect of filler content on morphology and physical–chemical characteristics of poly(vinylidene fluoride)/NaY zeolite-filled membranes. *J. Mater. Sci.* **2014**, *49*, 3361–3370. [CrossRef]
93. Dorozhkin, S.V. Calcium orthophosphates in dentistry. *J. Mater. Sci. Mater. Med.* **2013**, *24*, 1335–1363. [CrossRef] [PubMed]
94. Sandomierski, M.; Buchwald, Z.; Koczorowski, W.; Voelkel, A. Calcium forms of zeolites A and X as fillers in dental restorative materials with remineralizing potential. *Microporous Mesoporous Mater.* **2020**, *294*, 109899. [CrossRef]
95. Pratap, B.; Gupta, R.K.; Bhardwaj, B.; Nag, M. Resin based restorative dental materials: Characteristics and future perspectives. *Jpn. Dent. Sci. Rev.* **2019**, *55*, 126–138. [CrossRef] [PubMed]
96. Matinlinna, J.P.; Lung, C.Y.K.; Tsoi, J.K.H. Silane adhesion mechanism in dental applications and surface treatments: A review. *Dent. Mater.* **2018**, *34*, 13–28. [CrossRef]
97. Aydınoğlu, A.; Yoruç, A.B.H. Effects of silane-modified fillers on properties of dental composite resin. *Mater. Sci. Eng. C Mater. Biol. Appl.* **2017**, *79*, 382–389. [CrossRef] [PubMed]
98. Sandomierski, M.; Okulus, Z.; Voelkel, A. Active diazonium-modified zeolite fillers for methacrylate-based composites. *Compos. Interfaces* **2019**, *26*, 643–657. [CrossRef]
99. de Gennaro, B.; Catalanotti, L.; Cappelletti, P.; Langella, A.; Mercurio, M.; Serri, C.; Biondi, M.; Mayol, L. Surface modified natural zeolite as a carrier for sustained diclofenac release: A preliminary feasibility study. *Colloids Surf. B Biointerfaces* **2015**, *130*, 101–109. [CrossRef]
100. Pazarçeviren, E.; Erdemli, Ö.; Keskin, D.; Tezcaner, A. Clinoptilolite/PCL–PEG–PCL composite scaffolds for bone tissue engineering applications. *J. Biomater. Appl.* **2017**, *31*, 1148–1168. [CrossRef]
101. Al-Thawabeia, R.A.; Hodali, H.A. Use of Zeolite ZSM-5 for Loading and Release of 5-Fluorouracil. *J. Chem.* **2015**, *2015*, 403597. [CrossRef]
102. Abd-Elsatar, A.G.; Farag, M.M.; Youssef, H.F.; Salih, S.A.; Mounier, M.M.; El-Meliegy, E. Different zeolite systems for colon cancer therapy: Monitoring of ion release, cytotoxicity and drug release behavior. *Prog. Biomater.* **2019**, *8*, 101–113. [CrossRef]
103. Amorim, R.; Vilaça, N.; Martinho, O.; Reis, R.M.; Sardo, M.; Rocha, J.; Fonseca, A.M.; Baltazar, F.; Neves, I.C. Zeolite Structures Loading with an Anticancer Compound as Drug Delivery Systems. *J. Phys. Chem. C* **2012**, *116*, 25642–25650. [CrossRef]
104. Paradee, N.; Sirivat, A. Encapsulation of Folic Acid in Zeolite Y for Controlled Release via Electric Field. *Mol. Pharm.* **2015**, *13*, 155–162. [CrossRef]
105. Amani, S.; Garmarudi, A.B.; Rahmani, N.; Khanmohammadi, M. The β-cyclodextrin-modified nanosized ZSM-5 zeolite as a carrier for curcumin. *RSC Adv.* **2019**, *9*, 32348–32356. [CrossRef] [PubMed]
106. Wen, X.; Yang, F.; Ke, Q.-F.; Xie, X.-T.; Guo, Y.-P. Hollow mesoporous ZSM-5 zeolite/chitosan ellipsoids loaded with doxorubicin as pH-responsive drug delivery systems against osteosarcoma. *J. Mater. Chem. B* **2017**, *5*, 7866–7875. [CrossRef]
107. Potgieter, W.; Samuels, C.S.; Snyman, J.R. Potentiated clinoptilolite: Artificially enhanced aluminosilicate reduces symptoms associated with endoscopically negative gastroesophageal reflux disease and nonsteroidal anti–inflammatory drug induced gastritis. *Clin. Exp. Gastroenterol.* **2014**, *7*, 215–220.
108. Serri, C.; de Gennaro, B.; Quagliariello, V.; Iaffaioli, R.V.; de Rosa, G.; Catalanotti, L.; Biondi, M.; Mayol, L. Surface modified zeolite-based granulates for the sustained release of diclofenac sodium. *Eur. J. Pharm. Sci.* **2017**, *99*, 202–208. [CrossRef]
109. Pasquino, R.; Di Domenico, M.; Izzo, F.; Gaudino, D.; Vanzanella, V.; Grizzuti, N.; de Gennaro, B. Rheology-sensitive response of zeolite-supported anti-inflammatory drug systems. *Colloids Surf. B Biointerfaces* **2019**, *146*, 938–944. [CrossRef]
110. Krajisnik, D.; Dakovic, A.; Milojevic, M.; Malenovic, A.; Kragovic, M.; Bogdanovic, D.B.; Dondur, V.; Milic, J. Properties of diclofenac sodium sorption onto natural zeolite modified with cetylpyridinium chloride. *Colloids Surf. B Biointerfaces* **2011**, *83*, 165–172. [CrossRef] [PubMed]

111. Krajisnik, D.; Dakovic, A.; Malenovic, A.; Milojevic-Rakic, M.; Dondur, V.; Radulovic, Z.; Milic, J. Investigation of adsorption and release of diclofenac sodium by modified zeolites composites. *Appl. Clay Sci.* **2013**, *83–84*, 322–326. [CrossRef]
112. Serri, C.; de Gennaro, B.; Catalanotti, L.; Cappelletti, P.; Langella, A.; Mercurio, M.; Mayol, L.; Biondi, M. Surfactant-modified phillipsite and chabazite as novel excipients for pharmaceutical applications? *Microporous Mesoporous Mater.* **2016**, *224*, 143–148. [CrossRef]
113. Vargas, A.M.; Cipagauta-Ardila, C.C.; Molina-Velasco, D.R.; Ríos-Reyes, C.A. Surfactant-modified natural zeolites as carriers for diclofenac sodium release: A preliminary feasibility study for pharmaceutical applications. *Mater. Chem. Phys.* **2020**, *256*, 123644. [CrossRef]
114. Sağir, T.; Huysal, M.; Durmus, Z.; Kurt, B.Z.; Senel, M.; Isık, S. Preparation and in vitro evaluation of 5-fluorouracil loaded magnetite–zeolite nanocomposite (5-FU-MZNC) for cancer drug delivery applications. *Biomed. Pharmacother.* **2016**, *77*, 182–190. [CrossRef]
115. Sun, C.-Y.; Qin, C.; Wang, X.-L.; Yang, G.-S.; Shao, K.-Z.; Lan, Y.-Q.; Su, Z.-M.; Huang, P.; Wang, C.-G.; Wang, E.-B. Zeolitic imidazolate framework-8 as efficient pH-sensitive drug delivery vehicle. *Dalton Trans.* **2012**, *41*, 6906. [CrossRef] [PubMed]
116. Tiwari, A.; Singh, A.; Garg, N.; Randhawa, J.K. Curcumin encapsulated zeolitic imidazolate frameworks as stimuli responsive drug delivery system and their interaction with biomimetic environment. *Sci. Rep.* **2017**, *7*, 12598. [CrossRef]
117. Karimi, M.; Habibizadeh, M.; Rostamizadeh, K.; Khatamian, M.; Divband, B. Preparation and characterization of nanocomposites based on different zeolite frameworks as carriers for anticancer drug: Zeolite Y versus ZSM-5. *Polym. Bull.* **2018**, *76*, 2233–2252. [CrossRef]
118. Chen, J.; He, Z.-M.; Wang, F.-L.; Zhang, Z.-S.; Liu, X.; Zhai, D.-D.; Chen, W.-D. Curcumin and its promise as an anticancer drug: An analysis of its anticancer and antifungal effects in cancer and associated complications from invasive fungal infections. *Eur. J. Pharmacol.* **2016**, *772*, 33–42. [CrossRef] [PubMed]
119. Abadeh, Z.A.; Saviano, G.; Ballirano, P.; Santonicola, M.G. Curcumin-loaded zeolite as anticancer drug carrier: Effect of curcumin adsorption on zeolite structure. *Pure Appl. Chem.* **2019**, *92*, 461–471. [CrossRef]
120. Vilaça, N.; Bertão, A.R.; Prasetyanto, E.A.; Granja, S.; Costa, M.; Fernandes, R.; Figueiredo, F.; Fonseca, A.M.; De Cola, L.; Baltazar, F.; et al. Surface functionalization of zeolite-based drug delivery systems enhances their antitumoral activity In Vivo. *Mater. Sci. Eng. C* **2020**, *120*, 111721. [CrossRef]
121. Divband, B.; Rashidi, M.R.; Khatamian, M.; Eslamian, G.R.K.; Gharehaghaji, N.; Tabriz, F.D. Linde Type A and nano magnetite/NaA zeolites: Cytotoxicity and doxorubicin loading efficiency. *Open Chem.* **2018**, *16*, 21–28. [CrossRef]
122. Yang, F.; Wen, X.; Ke, Q.-F.; Xie, X.-T.; Guo, Y.-P. pH-responsive mesoporous ZSM-5 zeolites/chitosan core-shell nanodisks loaded with doxorubicin against osteosarcoma. *Mater. Sci. Eng. C* **2018**, *85*, 142–153. [CrossRef]
123. Abasian, P.; Radmansouri, M.; Habibi, M.; Ghasemi, M.V.; Mohammadi, A.; Irani, M.; Jazi, F.S. Incorporation of magnetic NaX zeolite/DOX into the PLA/chitosan nanofibers for sustained release of doxorubicin against carcinoma cells death in vitro. *Int. J. Biol. Macromol.* **2018**, *121*, 398–406. [CrossRef] [PubMed]
124. Zakeri, N.; Rezaie, H.R.; Javadpour, J.; Kharaziha, M. Effect of pH on cisplatin encapsulated zeolite nanoparticles: Release mechanism and cytotoxicity. *Mater. Chem. Phys.* **2021**, *273*, 124964. [CrossRef]
125. Zakeri, N.; Rezaie, H.R.; Javadpour, J.; Kharaziha, M. Cisplatin loaded polycaprolactone—Zeolite nanocomposite scaffolds for bone cancer treatment. *J. Sci. Adv. Mater. Devices* **2021**, *7*, 100377. [CrossRef]
126. Dasari, S.; Tchounwou, P.B. Cisplatin in cancer therapy: Molecular mechanisms of action. *Eur. J. Pharmacol.* **2014**, *740*, 364–378. [CrossRef]
127. Lázaro, I.A.; Lázaro, S.A.; Forgan, R.S. Enhancing anticancer cytotoxicity through bimodal drug delivery from ultrasmall Zr MOF nanoparticles. *Chem. Commun.* **2018**, *54*, 2792–2795. [CrossRef]
128. Yan, J.; Liu, C.; Wu, Q.; Zhou, J.; Xu, X.; Zhang, L.; Wang, D.; Yang, F.; Zhang, H. Mineralization of pH-Sensitive Doxorubicin Prodrug in ZIF-8 to Enable Targeted Delivery to Solid Tumors. *Anal. Chem.* **2020**, *92*, 11453–11461. [CrossRef]
129. Zheng, C.C.; Wang, Y.; Phua, S.Z.F.; Lim, W.O.; Zhao, Y.L. ZnO-DOX@ZIF-8 core-shell nanoparticles for pH-responsive drug delivery. *ACS Biomater. Sci. Eng.* **2017**, *3*, 2223–2229. [CrossRef]
130. Li, X.; Hou, S.; Chen, J.; He, C.E.; Gao, Y.E.; Lu, Y.; Jia, D.; Ma, X.; Xue, P.; Kang, Y.; et al. Engineering silk sericin decorated zeolitic imidazolate framework-8 nanoplatform to enhance chemotherapy. *Colloids Surf. B Biointerfaces* **2021**, *200*, 111594. [CrossRef]
131. Hao, J.; Milašin, I.S.; Eken, Z.B.; Mravak-Stipetic, M.; Pavelic, K.; Ozer, F. Effects of Zeolite as a Drug Delivery System on Cancer Therapy: A Systematic Review. *Molecules* **2021**, *26*, 6196. [CrossRef]
132. Proenza, Y.G.; Longo, R.L. Simulation of the Adsorption and Release of Large Drugs by ZIF-8. *J. Chem. Inf. Model.* **2020**, *60*, 644–652. [CrossRef]
133. Jevremović, A.; Stanojković, A.; Arsenijević, D.; Arsenijević, A.; Arzumanyan, G.; Mamatkulov, K.; Petrović, J.; Vasiljević, B.N.; Bajuk-Bogdanović, D.; Milojević-Rakić, M. Mitigating toxicity of acetamiprid removal techniques—Fe modified zeolites in focus. *J. Hazard. Mater.* **2022**, *436*, 129226. [CrossRef]
134. Popadić, D.; Gavrilov, N.; Ignjatović, L.; Krajišnik, D.; Mentus, S.; Milojević-Rakić, M.; Bajuk-Bogdanović, D. How to Obtain Maximum Environmental Applicability from Natural Silicates. *Catalysts* **2022**, *12*, 519. [CrossRef]
135. Gildernew, E.; Tareq, S.; Yang, S. Three-Dimensional Graphene with Preserved Channeling as a Binder Additive for Zeolite 13X for Enhanced Thermal Conductivity, Vapor Transport, and Vapor Adsorption Loading Kinetics. *Catalysts* **2022**, *12*, 292. [CrossRef]

MDPI
St. Alban-Anlage 66
4052 Basel
Switzerland
Tel. +41 61 683 77 34
Fax +41 61 302 89 18
www.mdpi.com

Catalysts Editorial Office
E-mail: catalysts@mdpi.com
www.mdpi.com/journal/catalysts

www.ingramcontent.com/pod-product-compliance
Lightning Source LLC
Chambersburg PA
CBHW040748250926
51850CB00007BB/1435
9783036581545